HANDBOOK OF PHYCOLOGICAL METHODS

DEVELOPMENTAL AND CYTOLOGICAL METHODS

Also sponsored by the Phycological Society of America

Handbook of phycological methods
Culture methods and growth measurements
Edited by Janet R. Stein
(published 1973)

Handbook of phycological methods
Physiological and biochemical methods
Edited by Johan A. Hellebust and J. S. Craigie
(published 1978)

HANDBOOK OF PHYCOLOGICAL METHODS

DEVELOPMENTAL AND CYTOLOGICAL METHODS

EDITED BY

ELISABETH GANTT

SENIOR RESEARCH BIOLOGIST
RADIATION BIOLOGY LABORATORY
SMITHSONIAN INSTITUTION, ROCKVILLE, MARYLAND

SPONSORED BY THE
PHYCOLOGICAL SOCIETY OF AMERICA, INC.

CAMBRIDGE UNIVERSITY PRESS

CAMBRIDGE

LONDON · NEW YORK · NEW ROCHELLE
MELBOURNE · SYDNEY

Published by the Press Syndicate of the University of Cambridge
The Pitt Building, Trumpington Street, Cambridge CB2 1RP
32 East 57th Street, New York, NY 10022, USA
296 Beaconsfield Parade, Middle Park, Melbourne 3206, Australia

First published 1980

Printed in the United States of America
Typeset by Progressive Typographers, Inc., Emigsville, Pa.
Printed and bound by Vail-Ballou Press, Inc., Binghamton, NY

Library of Congress Cataloging in Publication Data

Main entry under title:

Handbook of phycological methods.

Includes bibliographies.

CONTENTS: – [3] Developmental and cytological methods, edited by E. Gantt.

1. Algology – Technique.
I. Stein, Janet R. II. Hellebust, J. A. III. Craigie, J. S.
IV. Phycological Society of America.
QK565.2.S73 589'.3'028 73-79496
ISBN 0 521 22466 7

Contents

Contributors

Adamich, Marina, Department of Biological Sciences, University of California, Santa Barbara, California 93106 (Chapter 13)

Adshead, Patricia C., Department of Biology, University of Ottawa, Ottawa, Canada (Chapter 22)

Berger, Sigrid, Rosenhof, D6802 Ladenburg bei Heidelberg, Federal Republic of Germany (Chapter 5)

Duke, Eleanor L., Department of Biological Sciences, The University of Texas at El Paso, El Paso, Texas 79968 (Chapter 23)

Fagerberg, Wayne R., Department of Biology, The University of Texas at Arlington, Arlington, Texas 76019 (Chapter 32)

Floyd, Gary L., Department of Botany, Ohio State University, Columbus, Ohio 43210 (Chapter 23)

Gantt, Elisabeth, Radiation Biology Laboratory, Smithsonian Institution, Rockville, Maryland 20852 (Chapter 30)

Goff, Lynda J., Center for Coastal Marine Studies, University of California, Santa Cruz, California 95064 (Chapter 22)

Gordon-Mills, Elizabeth, Department of Botany, University of Adelaide, Adelaide, South Australia 5001, Australia (Chapter 27).

Green, Paul B., Department of Biological Sciences, Stanford University, Stanford, California 94305 (Chapter 21)

Haupt, Wolfgang, University of Erlangen-Nürnberg, D-8520 Erlangen, Federal Republic of Germany (Chapter 16)

Haury, John F., Biological Laboratories, Harvard University, Cambridge, Massachusetts 02137 (Chapter 18)

Hemmingsen, Barbara B., Department of Microbiology, San Diego State University, San Diego, California 92182 (Chapter 13)

Hill, Gerry J. C., Department of Biology, Carleton College, Northfield, Minnesota 55057 (Chapter 3)

Kataoka, Hironao, Institute for Agricultural Research, Tohoku University, Sendai 980, Japan (Chapter 17)

Kemp, C. Lindley, Department of Biological Sciences, Simon Fraser University, Burnaby, Canada V5A 1S6 (Chapter 25)

Kersey, Yolande, Department of Biological Sciences, Stanford University, Stanford, California 94305 (Chapter 14)

Kiermayer, Oswald, Botanical Institute, University of Salzburg, Salzburg, Austria (Chapter 1)

Lyman, Harvard, Biology Department, State University of New York at Stony Brook, Stony Brook, New York 11790 (Chapter 11)

McCandless, Esther L., Department of Biology, McMaster University, Hamilton, Ontario, Canada (Chapter 27)

McCully, Margaret E., Department of Biology, Carleton University, Ottawa, Canada (Chapter 22)

Moestrup, Øjvind, Institut for Sporeplanter, Ø. Farimagsgade 2D, 1353 Copenhagen K., Denmark (Chapter 31)

Nair, K. K., Department of Biological Sciences, Simon Fraser University, Burnaby, Canada V5A 1S6 (Chapter 25)

Neushul, Michael, Department of Biological Sciences and Marine Science Institute, University of California, Santa Barbara, California 93017 (Chapter 7)

Pickett-Heaps, Jeremy D., Department of Molecular, Cellular and Developmental Biology, University of Colorado, Boulder, Colorado 80309 (Chapter 29)

Polanshek, Alan R., Department of Biological Sciences, San Jose State University, San Jose, California 95192 (Chapter 8)

Quatrano, Ralph S., Department of Botany and Plant Pathology, Oregon State University, Corvallis, Oregon 97331 (Chapter 6)

Reimann, Bernhard E. F., Department of Biology, New Mexico State University, Las Cruces, New Mexico 88003 (Chapter 23)

Sanbonsuga, Yoshiaki, Department of Biological Sciences and Marine Science Institute, University of California, Santa Barbara, California 93017 (Chapter 7)

Schmitter, Ruth E., Biology Department, University of Massachusetts at Boston, Boston, Massachusetts 02125 (Chapter 26)

Schweiger, Hans-Georg, Rosenhof, D6802 Ladenburg bei Heidelberg, Federal Republic of Germany (Chapter 5)

Snell, William J., Department of Biology and the McCollum-Pratt Institute, Johns Hopkins University, Baltimore, Maryland 21218 (Chapter 4)

Staehelin, L. Andrew, Department of Molecular, Cellular and Developmental Biology, University of Colorado, Boulder, Colorado 80309 (Chapter 28)

Stevens, Catherine L. R., Department of Microbiology and Cell Biology, The Pennsylvania State University, University Park, Pennsylvania 16802 (Chapter 10)

Stevens, S. Edward Jr., Department of Microbiology and Cell Biology, The Pennsylvania State University, University Park, Pennsylvania 16802 (Chapter 10)

Sweeney, Beatrice M., Department of Biological Sciences, University of California, Santa Barbara, California 93106 (Chapter 19)

Tazawa, Masashi, Department of Botany, Faculty of Science, University of Tokyo, Hongo, Tokyo 113, Japan (Chapter 15)

Thomsen, Helge A., Institut for Sporeplanter, Ø. Farimagsgade 2D, 1353 Copenhagen K., Denmark (Chapter 31)

Trainor, Francis R., BSG, Botany Section, University of Connecticut, Storrs, Connecticut 06268 (Chapter 2)

Traverse, Karen, Biology Department, State University of New York at Stony Brook, Stony Brook, New York 11790 (Chapter 11)

Trelease, Richard N., Department of Botany and Microbiology, Arizona State University, Tempe, Arizona 85281 (Chapter 24)

Tuttle, Robert C., Department of Biology, University of California at San Diego, La Jolla, California 92093 (Chapter 12)

Vreeland, Valerie, Department of Botany, University of California, Berkeley, California 94720 (Chapter 27)

Waaland, J. Robert, Department of Botany, University of Washington, Seattle, Washington 98195 (Chapter 20)

Waaland, Susan D., Department of Botany, University of Washington, Seattle, Washington 98195 (Chapter 9)

West, John A., Department of Botany, University of California, Berkeley, California 94720 (Chapter 8)

Editor's preface

This handbook on *Developmental and Cytological Methods* is the third of a series sponsored by the Phycological Society of America. It follows the original intention of the first P.S.A. editorial committee constituted in 1967. The first volume, edited by Janet R. Stein, set the standards and general format for the subsequent volumes. It is hoped that this volume has not deviated too far from the high standard set by Professor Stein. Her advice and guidance in the preparation of this volume are greatly valued and appreciated. This volume, along with the first by Stein and the second by Hellebust and Craigie are complementary volumes, ranging from methods for culturing algae, techniques for studying their subcellular components, to new techniques for exploring their development and structure as whole cells.

Contributions were requested from a number of investigators of algal systems. Many others were considered, but it was not possible to ask for their contributions, because almost everyone who was asked agreed cheerfully and some limit had to be imposed on the length of the volume. The interest of the members of the Phycological Society of America and the offers of many to help have been particularly gratifying.

Production of a multi-author effort such as this, with its 32 chapters and 46 contributors, is subject to certain causes of delay, a major one being the editor's optimism and unfamiliarity with the process of publication. I am extremely grateful for the cooperation of the contributors and for the quality of the chapters that they have produced. It is after all their volume; my job as editor was chiefly to strive for as much uniformity and clarity as possible.

My experience with the editorial committee completely belies the old adage about the ineffectiveness of a committee. The advice of the committee members was invaluable. Their suggestions of topics and contributors broadened the coverage of the volume considerably, and they generously shared in the review of the submitted manuscripts.

Grateful acknowledgment is made of the help of the many phycologists who reviewed the various chapters, and my thanks go to Claudia A. Lipschultz, Karen Applestein, and Joan HajShafi of the Smithsonian Institution for their daily involvement and cheerful help in preparing the manuscripts for publication.

Elisabeth Gantt

Radiation Biology Laboratory
Smithsonian Institution
Rockville

Editorial Committee

Introduction

The general intent of this collection of chapters is to stimulate the investigation of algae and to recommend them as uniquely suitable experimental systems. It is hoped that the volume will aid students of phycology in applying new techniques and will entice experimentalists to explore algal material. It is an introduction to systems and methods from which the investigator can begin.

Chapters are grouped into two general categories. One deals with algae as developmental organisms. The second primarily covers microscopic tools for the study of living cells and the preparation and staining of fixed cells. Generally, the chapters reflect the present state of the art and thus are of necessity variable in content and format. Some will be adaptable for classroom exercises, whereas others require considerable sophistication in application. As introductory chapters they are not intended to be comprehensive. References to more advanced methods are given in most chapters, and a few additional references are included below.

In the chapters, a single species, or a group of species, are used as examples to which the methods have been successfully applied. The species, when appropriate, are identified with a source number of a culture collection listed in the Culture Collections appendix.

Sources of materials and equipment included in the chapters and appendix are for reference only and should not be construed as endorsements. Lists of suppliers are published annually in the United States in *Science* (American Association for the Advancement of Science, 1515 Massachusetts Ave., N.W., Washington, D.C. 20005); in Canada in *Research and Development* (MacLean Hunter, 418 University Ave., Toronto 101, Ontario) and *Laboratory Products News* (Southam Business Publications, Ltd., 1450 Don Mills Rd., Don Mills, Ontario).

General references

Bold, H. C. and M. J. Wynne. 1978. *Introduction to the Algae. Structure and Reproduction.* Prentice-Hall, Englewood Cliffs, N.J. 706 pp.

Hellebust, J. A. and Craigie, J. S. (eds.). 1978. *Handbook of Phycological Methods:*

Physiological and Biochemical Methods. Cambridge University Press, Cambridge. 512 pp.

Lewin, R. A. (ed.). 1976. *The Genetics of Algae.* University of California Press, Berkeley. 360 pp.

San Pietro, A. (ed.). 1971. *Methods in Enzymology,* vol. 23A, *Photosynthesis.* Academic Press, New York. 743 pp.

Stein, J. R. (ed.). 1973. *Handbook of Phycological Methods: Culture Methods and Growth Measurements.* Cambridge University Press, Cambridge. 448 pp.

Section I

Experimental algal systems and techniques

1: Control of morphogenesis in Micrasterias

OSWALD KIERMAYER

Botanical Institute,
University of Salzburg, Salzburg, Austria

CONTENTS

I. Objective

Micrasterias, a unicellular green alga, is an ideal choice for experiments in cell biology and development because of its relatively large size, distinctive shape, and the ease with which it can be cultured and experimentally manipulated. Apart from the opportunity it offers to study cytomorphogenesis, the differentiating cell may be used as a test object for detailed examination of such organelle systems as the Golgi complex and the diverse microtubule systems and their relation to cell function and development. Physiological and ultrastructural studies of the different stages in cell development (nuclear division, septum formation, cell wall growth, etc.) are possible because these developmental events occur consecutively in the cell cycle.

II. Test organism

A. Cytomorphology

The genus *Micrasterias* is classified in the family of Desmidiaceae. In the system of Fritsch (1961), Desmidiaceae are considered a suborder –Desmidioidae (placoderm Desmids)–of the Conjugales. Our test object, *Micrasterias denticulata* Bréb, a medium-sized species, will typically be 180–300 μm long, 165–300 μm wide, and 55–62 μm deep (Krieger 1937). The cell is organized in two semicells with a connecting region known as the isthmus. Both semicells are divided into one polar lobe and several lateral lobes. The nucleus, having a diameter of ca. 30 μm, is located in the isthmus.

B. Cytodifferentiation

1. Nuclear division. Algae ready for cell division are easily detected by their unusually dark, green chloroplast that tends to retract from the isthmus region, their expanded nucleus, and the changed appearance of the nucleoli. The changes seen in the nucleus upon division are described in detail by Waris (1950).

2. Septum growth. The septum grows centripetally so as to divide the two semicells. Cessation of free movement of crystals between the

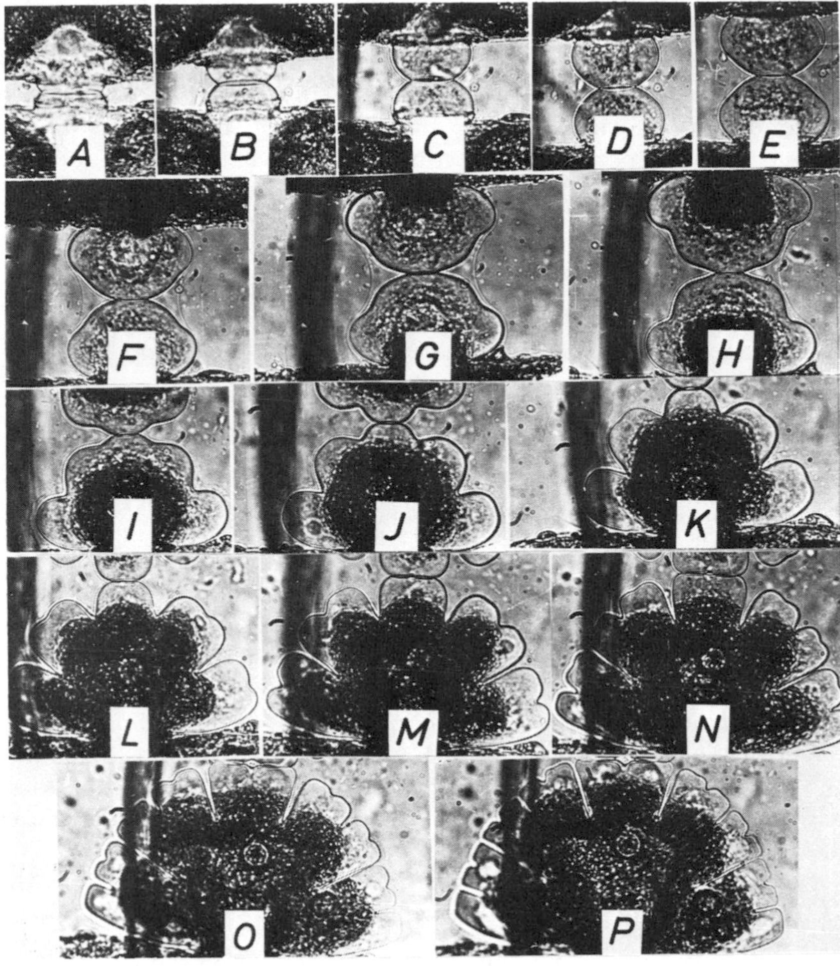

Fig. 1–1. Developmental sequence of the growing semicell of *Micrasterias denticulata,* time interval between pictures is 15 min (20° ± 1°C) made in a flow-through chamber (1 drop/min of medium) (Kiermayer 1964).

semicells indicates when the septum closes. During septum formation, a nucleus re-forms in both semicells, and a slight extension of the two semicells occurs in the isthmus region.

3. Cytomorphogenesis. Two small hemispherical bulges begin to form after the separation of the semicells by the septum (Fig. 1–1). The bulges are surrounded by a rather fine primary wall. It is in this region that further growth and differentiation of the new semicells is observed. As illustrated in Fig. 1–1, the hemispherical bulges pucker in the next stage of growth (three-lobe stage) into three lobes. The substructuring to a greater degree of lobing is illustrated in Fig. 1–1 A–P. After the primary wall is completed, a stronger and more rigid secondary wall forms. Swelling and subsequent shedding of the primary wall then occurs. The secondary wall, when formed, contains the pores and pore apparatus. The complete developmental process has been documented cinematographically (Kiermayer 1966).

During the development of hemispherical bulges, vigorous cytoplasmic streaming can be observed. Initially, the developing bulges contain only protoplasm. The nucleus does not enter until the three-lobe stage. At that time, the nucleus and an aggregation of dictyosomes surrounding it are especially conspicuous.

III. Culture methods

A. Growth conditions

1. Growth of natural isolates. The algae are taken to the laboratory from their natural habitat (peat bog). Samples may be kept in peat-bog water in unplugged flasks for several months in a relatively cool place (16–20°C). This can be in a window or on a cool sink, but not in direct sunlight. To obtain semisterile cultures of *Micrasterias,* a washing process is carried out under aseptic conditions. With the aid of a stereomicroscope (× 18 magnification) *Micrasterias* cells are transferred into small petri dishes (6 cm diameter) containing 10 ml of medium. After six such transfers, the cells are inoculated into Erlenmeyer flasks with sterile micropipettes (see below).

Cultures of several species of *Micrasterias* are available from the Culture Collection of Algae, University of Texas (UTEX).

2. Defined growth medium. The growth medium employed is essentially the one described by Waris (1950): 0.10 g KNO_3, 0.02 g $(NH_4)_2HPO_4$, 0.02 g $MgSO_4{\cdot}7H_2O$, 0.05 g $CaSO_4$, and 0.001 g $FeSO_4$ dissolved in 1000 ml doubly distilled water. Instead of $FeSO_4$, 5 ml of an iron stock complex may be added to the medium. It is prepared as follows: 0.4 g $FeSO_4{\cdot}7H_20$ plus 0.53 g Titriplex III (Merck) is dissolved in 160 ml of doubly distilled water, brought to a boil, and made up with water to 200 ml. The pH of the medium is 5.3.

Flasks (100 ml) are filled with 30 ml of medium, plugged with cotton, wrapped in foil, and steam-sterilized at 100°C for 1 h. After cooling, each flask is inoculated with 10–20 cells (bacteria are not fully excluded) and placed in a constant-temperature (20° ± 1°C) mirror incubator. The incubator is illuminated at 1,000 lux (95 ft-c) with 3 neon lamps (Osram-L. 40 W) which are vertically mounted in the middle. The flasks are placed at a distance of 25 cm from the lamps. Cultures should be transferred every 4–5 weeks, because cell defects increase and cell division rates decrease with the age of the cultures.

3. Light regime. The light-and-dark regime, which is controlled by an automatic clock, is adjusted so that formation of the new semicells occurs between 8 and 10 A.M. The light is on from 12 P.M. to 2 A.M. and off from 2 A.M. to 12 P.M. (Kiermayer 1964).

B. Induction of cell division

The time of the day for the onset of mitosis can be controlled by regulation of the light regime (see above). The normal daily rate of division is only 3% in 3- to 6-week-old cultures. This can be increased to ca. 8% by the light-shock method (Kiermayer 1970a) as follows: The day before use, 3- to 4-week-old *Micrasterias* cultures are shaken and transferred to petri dishes. These petri dishes are then exposed to continuous illumination of 1100–1400 lux (100–130 ft-c). Under a cyclic light regime, cell division occurs between 8 and 10 A.M. in the dark. Under continuous illumination, cell division also occurs between 8 and 10 A.M.

IV. Experimental procedures

The developmental stages, as identified in Fig. 1–1, can be obtained from stock cultures. Starting with septum formation, the time course of development can be predicted as outlined in Table 1–1. Once initiated the developmental stages are completed either in continuous light or in variable light–dark cycles, provided the temperature is maintained at about 20°C.

A. Microscopic observation of developmental stages

1. Cells from a stock culture emptied into a petri dish (6-cm diameter) are selected under a stereomicroscope.
2. With a micropipette 3–4 cells are then placed in a drop of peat-bog detritus (pH 6–7) on a microscope slide (previously washed in a chromate–sulfuric acid solution and thoroughly rinsed). The detritus is required for continued growth and development under a sealed coverslip (Kiermayer 1964, 1966).
3. A clean coverslip is carefully placed over the preparation, which is then sealed with petroleum jelly. Care should be taken to have a thin detritus layer.
4. Observation can be made under low light, sufficient for observation. The temperature around the specimens should not exceed 22°C.
5. Long-term observations can be made with such a setup. It is suit-

Table 1–1. *Time course of developmental processes in Micrasterias denticulata Bréb cells, starting with septum formation*

hours	(Nuclear division)	
0	Septum formation	
1		
2	Primary wall formation (cytomorphogenesis)	Nuclear migration
3		Chloroplast migration
4	Beginning of pore formation	
5		
6		
7		
8	Secondary wall formation	
9		
10		
11		
12		
13	Formation and differentiation of pores to complex pore apparatus	
14		
15		
16	Shedding of the primary wall	

able for time-lapse photography as in Fig. 1–1 where each frame represents a 15-min interval.

B. Hanging-drop method

With the hanging-drop method, the selection, microscopic observation, and micrography of developing cells can be carried out as above except that the medium with the cells is placed on a coverslip and sealed with petroleum jelly to a depression slide. This setup can be used for special applications, such as testing the effect of inhibitors and narcotic agents that are soluble in water. The modification is as follows:

1. Cells are transferred to the surface of a clean coverslip somewhat larger than the well of the depression slide.
2. The adhering medium is absorbed with a strip of filter paper placed as far away from the cells as possible.

Table 1–2. *Developmental studies in Micrasterias*

Cell structure or developmental event	Investigation on	References
Nuclear division	Chromosomes, microtubules	Waris (1950), Ueda (1972)
Primary wall	Growth; influence of turgor-pressure; "D-vesicles"	Kiermayer (1964, 1970a), Lacalli (1973)
Cytomorphogenesis	Formative events; effects of the nucleus and RNA on cell form	Waris (1950, 1951), Kallio (1951), Waris and Kallio (1964), Kiermayer (1964, 1966), Selman (1966)
Postmitotic nuclear migration	Microtubular systems	Kiermayer (1968, 1972)
Secondary wall	Formation of microfibrils; formation and incorporation of "F-vesicles"	Mix (1966), Dobberstein (1973), Kiermayer and Dobberstein (1973)
Slime secretion	Production and incorporation of "L-vesicles"	Kiermayer (1970a)
Golgi system	Structure of dictyosomes; vesicle production; dictyokinesis; ultracytochemistry	Drawert and Mix (1961), 1963), Kiermayer (1970a,b; 1977), Kiermayer and Dobberstein (1973), Menge (1976), Ueda and Noguchi (1976), Menge and Kiermayer (1977)

3. The cells are quickly rinsed with a drop of the test solution and are again quickly covered with a drop of the same solution.

4. The time, starting with the first application, is recorded with a stopwatch.

5. The coverslip is carefully turned over, placed on the depression slide, and sealed with petroleum jelly. Observations can be made over a period of hours or days.

Step 3 requires modification if a volatile solution is to be tested. The cells are suspended in a drop of the volatile solution, and the well of the depression slide is filled with the volatile test solution to within 2 mm of the upper surface. Coalescence with the drop should be avoided. Examples of substances to be tested are ethanol, ether, antimicrotubular substances, and herbicides.

C. Other variations

The influence of osmotically active substances can also be studied when cells adhere to the petri dish with their slime layer. Solution changes can easily be made under these conditions and then examined microscopically if desired.

Low-speed centrifugation experiments (100–5,000*g*) can also be carried out on *Micrasterias* cells.. By this technique the organelles, particularly the chloroplasts and different vesicles, can be shifted (Kiermayer and Dordel 1976). This technique may also prove suitable for studying cytoplasmic streaming in *Micrasterias* as detailed in Chap. 15.

D. Summary of studies

Differentiating cells of *Micrasterias* offer the possibility of studying a number of cell structures and developmental events under specific experimental conditions. Table 1–2 gives a general view of various physiological processes and submicroscopic structures that can be successfully studied using developing *Micrasterias* cells as test objects.

V. Acknowledgment

I am most grateful to Dr. Kathy Vammen for her help in translating this manuscript.

VI. References

Dobberstein, B. 1973. Einige Untersuchungen zur Sekundärwandbildung von *Micrasterias denticulata* Bréb. I. *Int. Desmidiaceensymp. Beih. Nova Hedwigia* 42, 83–90.

Drawert, H., and Mix, M. 1961. Licht- und elektronenmikroskopische Untersuchungen an Desmidiaceen. VII. Mitt.: Der Golgi-Apparat von *Micrasterias rotata* nach Fixierung mit Kaliumpermanganat und Osmiumtetroxyd. *Mikroskopie* 16, 207–12.

Drawert, H., and Mix, M. 1963. Licht- und elektronenmikroskopische Untersuchungen an Desmidiaceen. XI. Die Structur von Nucleolus und Golgi-Apparat bei *Micrasterias denticulata* Bréb. *Portug. Acta Biol.* 7, 17–28.

Fritsch, F. E. 1961. *The Structure and Reproduction of the Algae.* vol. I., Cambridge University Press, Cambridge, 939 pp.

Kallio, P. 1951. The significance of nuclear quantity in the genus *Micrasterias. Ann. Bot. Soc. Zool. Bot. Fenn. Vanamo* 24, 1–122.

Kiermayer, O. 1964. Untersuchungen über die Morphogenese und Zellwandbildung bei *Micrasterias denticulata* Bréb. *Protoplasma* 59, 76–132.

Kiermayer, O. 1966. *Micrasterias denticulata* (*Desmidiaceae*)-Morphogenese. *En-*

cyclopedia der Cinematographie E868, Institut für den Wissenschaftlichen Film, Göttingen, Germany.

Kiermayer, O. 1968. The distribution of microtubules in differentiating cells of *Micrasterias denticulata* Bréb. *Planta* 83, 223–36.

Kiermayer, O. 1970a. Elektronenmikroskopische Untersuchungen zum Problem der Cytomorphogenese von *Micrasterias denticulata* Bréb. I. Allgemeiner Überblick. *Protoplasma* 69, 97–132.

Kiermayer, O. 1970b. Causal aspects of cytomorphogenesis in *Micrasterias*. *Ann. N.Y. Acad. Sci.* 175, 686–701.

Kiermayer, O. 1972. Beeinflussung der postmitotischen Kernmigration von *Micrasterias denticulata* Bréb. durch das Herbizid Trifluralin. *Protoplasma* 75, 421–26.

Kiermayer, O. 1977. Biomembranen als Träger morphogenetischer Information. Untersuchungen bei der Grünalge *Micrasterias*. *Naturwiss. Rundschau* 30, 161–65.

Kiermayer, O., and Dobberstein, B. 1973. Membrankomplexe dictyosomaler Herkunft als "Matrizen" für die extraplasmatische Synthese und Orientierung von Mikrofibrillen. *Protoplasma* 77, 437–51.

Kiermayer, O., and Dordel, S. 1976. Elektronenmikroskopische Untersuchungen zum Problem der Cytomorphogenese von *Micrasterias denticulata* Bréb. II. Einfluss von Vitalzentrifugierung auf Formbildung und Feinstruktur. *Protoplasma* 87, 179–90.

Krieger, W. 1933–39. Die Desmidiaceen, *Rabenhorsts Kryptogamenflora* 13, Sect. 1, 1–712 and 1–117.

Lacalli, T. C. 1973. Cytokinesis in *Micrasterias rotata*. Problems of directed primary wall deposition. *Protoplasma* 78, 433–42.

Menge, U. 1976. Ultracytochemische Untersuchungen an *Micrasterias denticulata* Bréb. *Protoplasma* 88, 287–303.

Menge, U. and Kiermayer, O. 1977. Dictyosomen von *Micrasterias denticulata* Bréb – ihre Grössenveränderung während des Zellzyklus. *Protoplasma* 91, 115–23.

Mix, M. 1966. Licht- und elektronenmikroskopische Untersuchungen an Desmidiaceen. XII. Zur Feinstruktur der Zellwände und Mikrofibrillen einiger Desmidiaceen vom *Cosmarium*-Typ. *Arch. Mikrobiol.* 55, 116–33.

Selman, G. G. 1966. Experimental evidence for the nuclear control of differentiation in *Micrasterias*. *J. Embryol. Exp. Morph.* 16, 469–85.

Ueda, K. 1972. Electron microscopical observation on nuclear division in *Micrasterias americana*. *Bot. Mag.* (Tokyo) 85, 263–71.

Ueda, K., and Noguchi, T. 1976. Transformation of the Golgi apparatus in the cell cycle of a green alga, *Micrasterias americana*. *Protoplasma* 87, 145–62.

Waris, H. 1950. Cytophysiological studies on *Micrasterias*. I. Nuclear and cell division. *Physiol. Plant.* 3, 1–16.

Waris, H. 1951. Cytophysical studies on *Micrasterias*. III. Factors influencing the development of enucleate cells. *Physiol. Plant.* 4, 387–409.

Waris, H., and Kallio, P. 1964. Morphogenesis in *Micrasterias*. *Advan. Morphogen.* 4, 45–80.

2: Control of development in Scenedesmus

FRANCIS R. TRAINOR

BSG, Botany Section
University of Connecticut, Storrs, Connecticut 06268

CONTENTS

I. Introduction

In the past, the green alga *Scenedesmus* has usually been thought of as a colonial organism; in many descriptions of species it is taken for granted that this is the only possible pathway of development (Uherkovich 1966). However, we now know that there is no stability of form in many species of *Scenedesmus* (Trainor 1969). In fact, certain types of wall ornamentation and morphological forms develop only when specific nutrient conditions are realized.

Unicells resembling other genera can be easily produced by some strains of *Scenedesmus* (Trainor et al. 1976) and are now being found in some *Pediastrum* isolates (Millington, personal communication). Gametes may form in selected *Scenedesmus* species (Cain and Trainor 1976).

This chapter describes the conditions necessary for the development of both unicells and colonies as well as for the regulation of gamete production in *Scenedesmus.* The experiments described can be used as laboratory exercises.

II. Experimental organisms and culture methods

A. Sources

Many strains of algae useful as experimental material can be purchased from culture collections. *Scenedesmus* strains UTEX 1237, 1588, 1591, and 1592 are available from the Culture Collection of Algae, University of Texas. A strain of *Pediastrum boryanum* can be obtained from J. Davis, University of Florida, Gainesville, Florida 32601 or W. Millington, Marquette University, Milwaukee, Wisconsin 53233. New cultures are easily established and others are available from private sources. However, it would be necessary to run preliminary experiments to determine the capabilities of any new isolate.

B. Media

The formulations of revised medium C (Davis 1967; Millington and Gawlick 1975) for the culture of *Pediastrum,* as well as medium 3.07

Table 2–1. *Culture media for Scenedesmus and Pediastrum*

	Concentration (mg/liter)		
Compound added	Revised medium C	Bristol's medium	Medium 3.07
KNO_3	40.0	—	—
$NaNO_3$	—	250.0	2.0
$MgSO_4 \cdot 7H_2O$	50.0	75.0	1.0
K_2HPO_4	5.0	75.0	0.03
KH_2PO_4	—	175.0	—
K_2CO_3	50.0	—	—
$CaCl_2$	3.5	—	—
$CaCl_2 \cdot 2H_2O$	—	26.5	0.03
$Na_2SiO_3 \cdot 9H_2O$	10.0	—	—
$FeSO_4 \cdot 7H_2O$	1.75	—	—
$FeCl_3 \cdot 6H_2O$	—	5.0	5.0×10^{-3}
Na_2EDTA	2.35	—	—
$Na_2EDTA \cdot 2H_2O$	—	6.90	6.9×10^{-3}
$NaHCO_3$	20.0	—	—
TRIS	—	—	40.0
H_3BO_3	1.0	—	—
$ZnSO_4 \cdot 7H_2O$	1.0	0.04	4.0×10^{-5}
$MnSO_4 \cdot 4H_2O$	0.4	—	—
$CoCl_2 \cdot 6H_2O$	0.2	0.02	2.0×10^{-5}
$Na_2MoO_4 \cdot 2H_2O$	0.2	0.02	2.0×10^{-5}
$CuSO_4$	0.04	0.01	1.0×10^{-5}
$MnCl_2 \cdot 4H_2O$	—	0.3	3.0×10^{-4}
pH	7.5	7.0	8.0

and Bristol's medium for *Scenedesmus* are listed in Table 2–1. A culture in medium 3.07 should be transferred daily.

For preparation of soil extract, the soil selected should be free of applied chemicals (e.g., fertilizer, herbicides) and should have medium humus and low clay content. To 1 liter distilled water are added 100 g of the soil. The mixture is then autoclaved for 30 min, and allowed to settle and clear overnight. The sterile, clear supernatant is used as a medium.

A Morton filter apparatus (Corning 33990) with an ultrafine porosity sintered glass disk is used for preparing sterile medium 3.07 (Hamilton 1973; Shubert and Trainor 1973). The filter apparatus, soil extract, Bristol's medium, and medium C are sterilized by autoclaving.

C. Growth conditions

Organisms are grown in sterile test tubes or 50-ml Erlenmeyer flasks. It is convenient to work with inocula of 10^6 cells/ml in medium C and

Bristol's and 5 × 10^4 cells/ml in medium 3.07. With incubation at 25°C under fluorescent illumination (4000 lux; 370 ft-c) on a 15:9 h light–dark (LD) cycle, good growth can be achieved. All experiments can be performed in stationary culture. When attempting to synchronize cell division, one should gently shake cultures twice during the light period. However, a tissue-culture rotator (VWR Scientific or Lab-line Instrument Co.) may be used to subject the cultures to a gentle motion and ensure uniform conditions within culture tubes.

D. Synchronization of cultures

When *Scenedesmus* strains UTEX 1588 and 1591 are grown in Bristol's medium under the above conditions, two doublings per day will result in a complete turnover of four-celled colonies. To obtain a synchronous culture, the organism is transferred daily at the end of the dark period. If the daily inoculum is ca. 10^6 cells/ml and a 1:4 dilution is made, a synchronized culture is obtained within 3–4 days and may be maintained indefinitely.

To improve the synchrony to above 50%, the growth rate is adjusted so that exactly 2 doublings per day occur. Growth can be improved by increasing the light intensity or length of day, by raising the temperature, and by using a slightly lower cell inoculum. For isolation methods and further culture techniques with these species, the comprehensive summaries by Starr (1971) and Bishop and Senger (1971) should be consulted.

III. Control of development

A. Synchronized culture of Scenedesmus

Control of a particular phenomenon is not achieved by synchrony itself, but for classroom observations of cell division and release of new colonies or unicells, synchronized cultures are ideal (Trainor and Rowland 1968). The LD cycle is adjusted so that the dark period occurs during the laboratory period. An example for *Scenedesmus* strain UTEX 1588 is shown in Fig. 2–1. Increase in cell size occurs in the light period; this is followed by initiation of nuclear division, which is completed in the early part of the dark period. Cytoplasmic cleavage begins early in the dark period and is completed by the majority of colonies in 3–4 h in the dark. Release of newly formed colonies then takes place.

B. Induction of unicellular and colonial stages

Scenedesmus cultures UTEX 1588 and 1591 grown in medium 3.07 will form completely colonial populations if transferred daily with 1:2

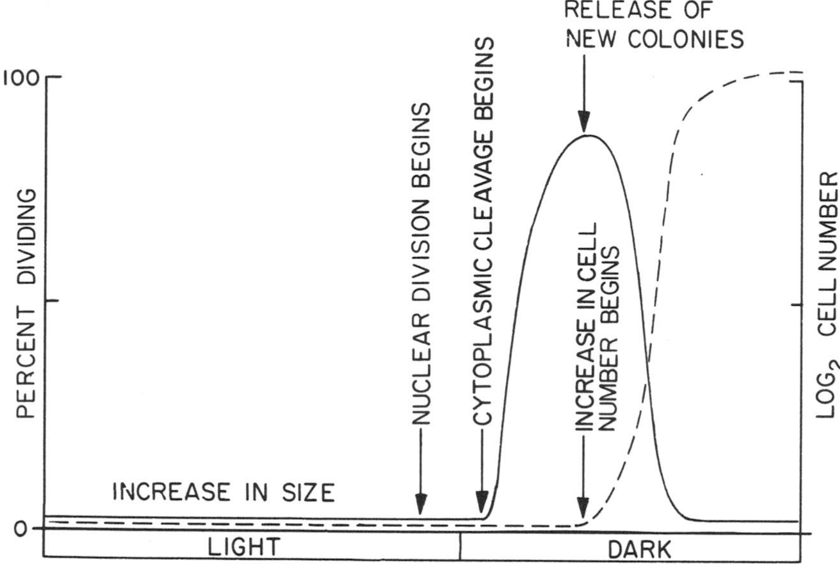

Fig. 2–1. Stages of growth and cell division in a synchronized culture of *Scenedesmus* strain UTEX 1588 grown in a 15:9 h LD cycle. Increase in cell size takes place in the light, followed by nuclear and cytoplasmic division, release of new colonies, and increase in cell number (----).

dilution (Figs. 2–2A and 2–3A). In Bristol's medium both unicells and colonies develop; the percentage of unicells is highly variable and depends on such factors as the history of the inoculum, cell density, and the rate of photosynthesis. However, a completely unicellular population is produced if either strain is grown in soil extract (Fig. 2–2C). During unicell formation the products of cell division fail to join before release from the parent cell (Fig. 2–2E). Any unicellular population transferred back to medium 3.07 becomes colonial (Fig. 2–2B), because the division products join laterally before release from the parent cell (Fig. 2–3E). Since medium is carried over in the transfer, 2–3 days should be allowed for the reversion.

C. Control of spine formation

Terminal cells of *Scenedesmus* UTEX 1591 characteristically bear at least two spines (Fig. 2–3A). Spine formation can be controlled by selective use of media. These appendages develop in cultures grown in Bristol's medium, but spine formation is turned off when such a culture is transferred to medium 3.07 (Fig. 2–3C). Transfer of spineless colonies back to Bristol's medium will result in spine formation in the next generation of colonies (Fig. 2–3B). Observations of synchronous

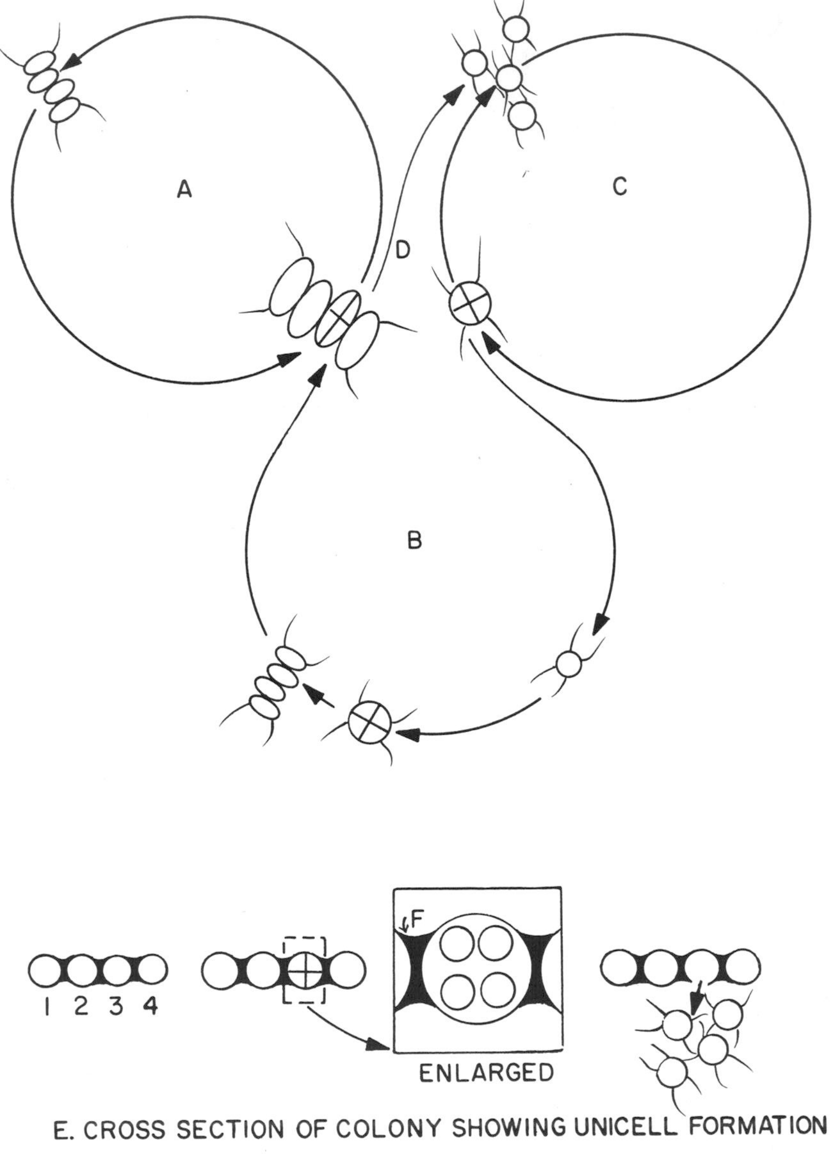

Fig. 2–2. Unicell production in *Scenedesmus* strain UTEX 1588. Individual cells divide twice and may form colonies (A) or unicells (C). Colonies develop in medium 3.07, and the conversion (D) to unicells takes place when the organism is grown in soil extract. A unicellular population (C) can be induced to form colonies (B) by growth in medium 3.07. A colony is shown in cross section (E), with one cell dividing and releasing four unicells. Additional wall material (F) was previously deposited between cells of the parent colony but is not produced during unicell formation (compare with Fig. 2–3E).

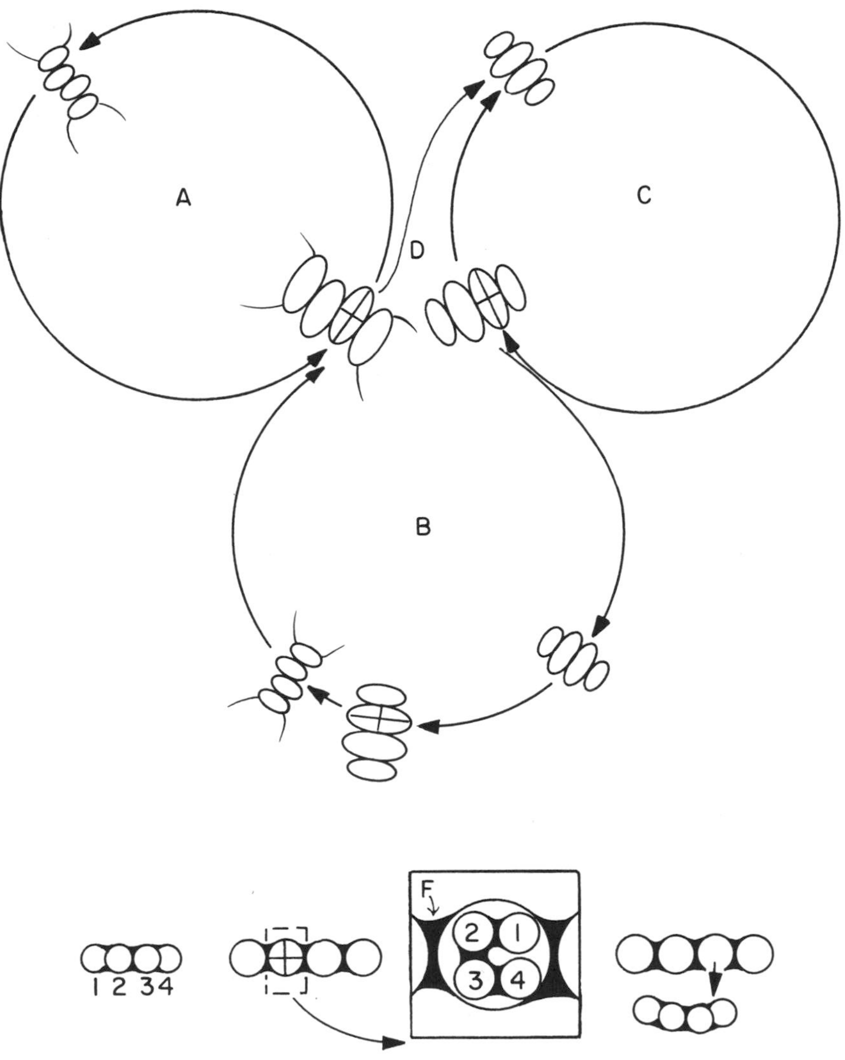

Fig. 2–3. Developmental stages in *Scenedesmus* strain UTEX 1591. Individual cells divide twice and form new colonies (A). Spiny colonies that form in Bristol's medium can be converted to spineless colonies (C) by transfer (D) to medium 3.07. The spineless condition can be again reversed by transfer to Bristol's medium (B), where spines are developed on the next generation of colonies. A colony is shown in cross section (E) with one cell dividing and releasing a new colony. Additional wall material (F) is deposited between cells of a colony (compare with Fig. 2–2E).

cultures show that spine formation takes place prior to colony release from parent cells.

D. Gamete induction

The first step is to grow *Scenedesmus* strain UTEX 1592 for 5 days in Bristol's medium, supplemented with 0.1% glucose for a denser culture. Cultures are then concentrated by centrifugation, washed, and resuspended in Bristol's medium minus $NaNO_3$ to a cell density of 2×10^7 cells/ml. Gametes form in this starvation medium within 36 h of incubation at 15°C. (A low temperature is essential for optimal results; fewer gametes develop at 22°C.)

E. Development in Pediastrum

Partial synchrony in cell division and colony formation can be obtained when *P. boryanum* is grown in medium C with frequent transfers. In stationary cultures in medium C, *P. boryanum* will form zoospores within 40 h. Within each parent cell, spores remain motile for about 15 min, but the process will be at different stages in various cells of the colony. Unicells may be produced by mechanical disruption of zoospores during colony formation (Pickett-Heaps 1975). Unicellular strains have also been found that arose by mutation.

IV. General comments

Whenever possible, axenic cultures should be used. Because there is great variability in strain behavior, the selection of strains for certain features is very important.

From past studies (Trainor et al. 1976) it is obvious that inorganic nutrients greatly influence the development of *Scenedesmus* and *Pediastrum*. Organic constitutents contained in yeast extract (1.5% w/v in Bristol's medium) can also greatly influence the morphology of *Scenedesmus* UTEX 1237, leading to forms resembling *Dactylococcus*. Further pursuit of nutritional studies appears promising and may lead to better control of specific developmental stages in several colonial green algae.

V. Acknowledgments

I thank J. S. Davis and W. Millington for their help in assembling these data.

VI. References

Bishop, N. I., and Senger, H. 1971. Preparation and photosynthetic properties of synchronous cultures of *Scenedesmus.* In San Pietro, A. (ed.), *Methods in Enzymology,* vol. 23A, pp. 53–66. Academic Press, New York.

Cain, J., and Trainor, F. 1976. Regulation of gametogenesis in *Scenedesmus obliquus* (Chlorophyceae). *J. Phycol.* 12, 383–90.

Davis, J. 1967. The life cycle of *Pediastrum simplex. J. Phycol.* 3, 95–103.

Hamilton, R. 1973. Sterilization. In Stein, J. (ed.), *Handbook of Phycological Methods: Culture Methods and Growth Measurements,* pp. 181–93. Cambridge University Press, Cambridge.

Millington, W. F., and Gawlick, S. R. 1975. Cell shape and wall pattern in relation to cytoplasmic organization in *Pediastrum simplex. Amer. J. Bot.* 62, 824–34.

Pickett-Heaps, J. 1975. *Green Algae: Structure, Reproduction and Evolution in Selected Genera.* Sinauer Associates, Sunderland, Mass. 606 pp.

Shubert, L., and Trainor, F. 1973. *Scenedesmus* morphogenesis. Control of the unicell stage with phosphorus. *Br. Phycol. J.* 8, 1–7.

Starr, R. 1971. Algal cultures – sources and methods of cultivation. In San Pietro, A. (ed.), *Methods in Enzymology,* vol. 23A, pp. 29–53. Academic Press, New York.

Trainor, F. 1969. *Scenedesmus* morphogenesis. Trace elements and spine formation. *J. Phycol.* 5, 185–90.

Trainor, F., and Rowland, H. 1968. Control of colony and unicell formation in a synchronized *Scenedesmus. J. Phycol.* 4, 310–17.

Trainor, F., Cain, J., and Shubert, L. 1976. Morphology and nutrition of the colonial green alga *Scenedesmus:* 80 years later. *Bot. Rev.* 42, 5–25.

Uherkovich, G. 1966. *Die Scenedesmus-Arten Ungarns.* Academiai Kiado, Budapest. 173 pp.

3: Mating induction in Oedogonium

GERRY J. C. HILL

Department of Biology,
Carleton College, Northfield, Minnesota 55057

CONTENTS

I. Objective

The objective of this chapter is to describe procedures for growing and inducing gamete formation in *Oedogonium* species and to facilitate analysis of chemical communication and quantitative determinations of chemotaxis and chemomorphogenesis in the sexual systems of this genus. This initial step can lead from control of mating induction in the genus to eventual genetic analysis. The system being described deals with both nannandrous and macrandrous species, which differ primarily in the type and process of their antheridia production. In the macrandrous species the antheridia are produced from vegetative filaments, whereas in the nannandrous species the antheridia are produced from dwarf male filaments. However, in both, large oogonia are formed, and during the development of the oogonial mother cell (Rawitscher-Kunkel and Machlis 1962) hormone-like sex factors are produced. Techniques for the isolation and partial purification of sperm and androspore attractants from several species are presented. These species have the advantage of being heterothallic (female and male gamete production occurs in separate filaments), allowing production of isolated female and male stages.

II. Experimental organisms and culture methods

A. Test organisms

Growth and gamete induction has been successful to date in the macrandrous heterothallic species, *Oedogonium cardiacum* Wittr (UTEX 39 and 40); and in two nannandrous heterothallic species, *Oedogonium borisianum* (le Cl.) Wittr and *Oedogonium donnellii* Wolle, both of which were isolated from soil collected from a vernal pool in Marin County of California in 1962. (These may be obtained from the author.) For isolation techniques, see Hoshaw and Rosowski 1973.

B. Culture technique

1. Culture media. The media used for growth and gamete induction are presented in Table 3–1. They are prepared in glass distilled water

Table 3–1. *Media for Oedogonium culture and gamete induction*

	Media dilutions (ml/liter)					
Nutrient stock[a]	A	ALN	A–N	ALNLP	DA	D/4A
KNO_3 (1.0 *M*)	20.0	0.1	0	0.1	5.0	0.02
$MgSO_4$ (0.1 *M*)	10.0	10.0	10.0	10.0	0.67	0.08
$CaCl_2$ (0.1 *M*)	5.0	5.0	5.0	5.0	0.33	0.08
KH_2PO_4 (0.1 *M*)	10.0	10.0	10.0	1.0	3.33	0.17
Vitamin B_{12} (2.5 μg/ml)	0.4	0.4	0.4	0.4	0.025	0.006
Micronutrient stock	20.0	20.0	20.0	20.0	1.35	0.33

Micronutrients[b]	Salt (g/100 ml)	Stock solution (ml/liter)
$N(CH_2COOH)_3$	1.910	20.0
$Na_2B_4O_7 \cdot 10H_2O$	4.770	10.0
$CuSO_4 \cdot 5H_2O$	0.250	10.0
$CoCl_2 \cdot 6H_2O$	0.238	10.0
$MnCl_2 \cdot 4H_2O$	0.198	10.0
$ZnSO_4 \cdot 7H_2O$	0.287	10.0
$(NH_4)_6Mo_7O_{24} \cdot 4H_2O$	0.177	10.0
$FeSO_4 \cdot 7H_2O$	0.278	50.0

[a] All media are adjusted to pH 6.8 with KOH before being autoclaved.
[b] The micronutrients are added sequentially to 750 ml distilled water; prior to addition of $FeSO_4 \cdot 7H_2O$, the pH is adjusted to 2.0 with H_2SO_4, and afterward the volume is brought to 1 liter. Storage is at 4°C.

or with soil water, in which case the soil provides a "soil factor" and the media are designated by the prefix SW. In the latter case the soil-water extract is prepared first by two consecutive 45-min extractions (with autoclaving) of 5.0 g soil in 1 liter distilled water (the solution being filtered through a 0.45 μm Millipore filter for clearing).

2. *Culture conditions.* The temperature is kept at 20–25°C in incubators with continuous illumination at 1.25×10^4 erg/cm²/sec (4300 lux; 400 ft-c) (Sylvania Grow-lux, 40 W lamps). Cultures are aerated with 2% CO_2 in air.

The macrandrous species, *Oe. cardiacum,* and the two nannandrous species, *Oe. borisianum* and *Oe. donnellii,* produce luxuriant vegetative growth in either 25 × 250-mm culture tubes (Fig. 3–1) containing 25 ml of medium or in 500-ml culture flasks containing 300 ml of medium. *Oe. cardiacum* and *Oe. donnellii* have a temperature optimum of 25°C, whereas *Oe. borisianum* has a temperature optimum of 20°C under the culture conditions used. *Oe. cardiacum* grows well in a fully defined medium (medium A), *Oe. donnellii* requires a soil-water supplement (medium SWA) for growth, and *Oe. borisianum* requires a di-

Fig. 3–1. Aerated test tube culture system. Drying tubes, when packed with nonabsorbent cotton and autoclaved, serve as sterile filters for the 2% CO_2 in air.

lute medium with soil-water supplement (medium SWDA) for optimum growth. Inocula may vary from small clumps of actively growing vegetative filaments to a blended sample from actively growing cultures or to a suspension of zoospores. The latter is preferred for nutritional studies, because the inoculum is more uniform and can be determined quantitatively. Time for maximum growth is dependent upon the size of the inoculum and the size of the culture vessel or quantity of medium. Seven days is sufficient for cultures in 25 × 250-mm tubes, and 2 weeks is sufficient for 500-ml culture flasks. Both unialgal and axenic cultures are successfully grown under these conditions; *Oe. cardiacum* (Machlis et al. 1974) is an exception.

III. Induction of sexual stages

Sexual stages are naturally produced after vegetative growth, or they can be induced by transfer of the vegetative plants to mineral-deficient media. Male sex attractants produced by developing oogonial mother cells are recoverable from the growth media of the female strains.

A. Oogonia from Oe. cardiacum

After reaching maximum vegetative growth in the culture medium, female cultures of *Oe. cardiacum* begin producing oogonia, which are easily recognizable (with a dissecting microscope) by their shape and size. Within 3 days (25 × 250-mm tubes) or one week (500-ml flasks) an optimum density of chemotactically and sexually active oogonia are present in the culture (Hill and Machlis 1970). With this technique, 1 oogonium per 10–15 vegetative cells on each filament can be obtained. A higher density of oogonia (1 oogonium per 5–8 vegetative cells) may be obtained by transferring the vegetative filaments to a culture with a reduced nitrate concentration (medium ALN) and allowing an additional 3 days to 1 week for optimum growth and development.

B. Sperm from Oe. cardiacum

The male filaments of *Oe. cardiacum* begin production of antheridia after reaching maximum vegetative growth. Within 3 days (tubes) or 1 week (flasks), an optimum density of antheridia is present, and sperm discharge begins. Sperm discharge is monitored with a dissecting microscope. At this time the male filaments are collected in a sterile stainless-steel strainer, washed with sterile distilled water, and placed in a flask or dish of culture medium totally lacking in nitrate (medium A–N). This treatment initiates the release of large numbers of sperm in a very short period of time (in excess of 50,000/ml in 4 h).

The released sperm are freed from the cellular debris by straining the solution through a sterile 400-mesh stainless-steel wire screen. The sperm pass through the mesh, but the antheridial remains and the vegetative cells are retained.

C. *Oogonial mother cells from Oe. donnellii*

The female filaments of *Oe. donnellii,* after reaching maximum vegetative growth, are collected, strained, washed, and transferred (as above) to a depletion medium (medium SWALN). Oogonial mother cells are produced within 9 days under continuous light at 25°C in a standing culture (no aeration or shaking).

D. *Androspores from Oe. donnellii*

After reaching maximum vegetative growth (in ca. 2 weeks), the male filaments are sampled and microscopically examined (with dissecting microscope) for the presence of androsporangia and some released androspores. If androsporangia are present in great numbers, the filaments are collected, strained, washed, and transferred to a depletion medium low in both nitrate and phosphate (medium SWALNLP). The newly inoculated standing cultures are incubated under continuous light at 25°C. Androspores are released within 2 h at a level in excess of 100,000/ml.

E. *Oogonial mother cells from Oe. borisianum*

After 2-weeks' growth in the culture medium, *Oe. borisianum* female cultures are collected, strained, washed, and transferred to 250 ml of depletion medium (medium SWD/4A). Under continuous light at 20°C in standing culture, chemotactically and sexually active oogonial mother cells are produced and reach maximum density in 2 weeks.

F. *Androspores from Oe. borisianum*

The male cultures, after 2-weeks' growth, are collected, washed, and transferred to fresh soil-water medium (SWDA). The cultures are placed under continuous light at 20°C without agitation or aeration. Each week for 7 consecutive weeks the male cultures are collected, washed, and transferred to fresh soil water. By the sixth transfer, occasional androspores and androsporangia can be observed in the cultures. The final transfer (week 8) is made into soil water buffered with 0.001 *M* TRIS to pH 6.8. Androspores are released over a period of 1–2 weeks, the density of the suspension seldom exceeding 5,000 androspores/ml. Higher densities may be obtained by passing the suspension through a glass-fiber filter by gravity flow; this raises the concentration to a level usable for the bioassay (Section V, below).

IV. Isolation of attractants

The sperm and androspore attractants of *Oedogonium* are isolated from the depletion medium in which the oogonia or oogonial mother cells are produced.

A. Sperm attractant

The *Oe. cardiacum* sperm attractant, which was found to be soluble in polar solvents (Machlis et al. 1974), is first concentrated by means of a Seitz filter (National Surgical Supply Co.) to remove the algae and cellular debris. The resulting filtrate is concentrated with a Diaflo membrane (UM05) in an Amicon model 402 ultrafiltration cell (Amicon Corp.). Depending upon the amount of original filtrate, a 10- to 25-fold concentration is obtained in 4–8 h at room temperature. This concentrate is partially purified by elution through a Sephadex G-25 column (2.5 × 45 cm, flow rate of 1.0 ml/min).

B. Androspore attractant

The androspore attractants for both *Oe. borisianum* and *Oe. donnellii* are preferentially soluble in nonpolar solvents and are concentrated by extraction with dichloromethane (×6 with volume equal to that of the original culture medium). Extraction can be achieved through 6 extractions in a separatory funnel or through the use of a continuous extractor [see Machlis et al. (1966) for the general design of a continuous extractor]. The resulting extract is concentrated (×200) by flash evaporation in a Büchi Rotovaporator-R (Buchler Instruments) at 37°C. This concentrate can be stored at 0°C in dichloromethane for several months without detectable loss of activity. The concentrate appears yellow in color and dries to an oily consistency.

V. Bioassay techniques for sperm and androspore attractants

A. Qualitative determination

The method following may be employed for *Oe. cardiacum* sperm attractant and *Oe. borisianum,* and *Oe. donnellii* androspore attractants. Equal volumes of an aqueous solution of the attractant and low-melting-point agar (4% w/v) are mixed (at 50°C) and poured into a flat depression slide 2 mm deep. After the solution solidifies, 2-mm plugs of agar are made with a flat-tipped hypodermic needle (No. 13). The agar plug is placed in a depression slide containing either sperm or androspores. A control plug, prepared from the appropriate depletion medium and agar, is introduced into the depression slide adjacent to the test plug. Presence of an active chemotactic agent results in clustering of the sperm or androspores on the surface and in close proximity to the test plug and can be observed with a dissecting microscope.

B. Quantitative determination

Specialized bioassay apparatuses have been designed to take into consideration the nature of the attractant and the response characteristics of the sperm or androspores.

1. In *Oe. borisianum* and *Oe. donnellii,* an androspore-settling response, elicited by the oogonial mother cell, is taken advantage of in this technique. The design of the apparatus is based on that used by Machlis (1969) for the water mold *Allomyces.* Each apparatus (Fig. 3–2) is loaded by pipetting an agar test solution, consisting of equal volumes of 4% (w/v) agar and androspore attractant (mixed into the agar at ca. 50°C) into the well of the assay apparatus while the lower surface rests on a cooled glass surface. Sufficient agar test solution is added to each well so that it overflows the upper end of the well, thus anchoring the agar into position. Then the apparatus is assembled into its tripod support and placed in a 35 × 10-mm tissue-culture dish containing 4.0 ml of androspore suspension in such a way that only the smooth surface of the agar block is in contact with the suspension. After 45–90 min of contact, the apparatus is removed from the suspension, dipped into water to remove any androspores contained in the attached drop on the surface of the agar, and inverted on a dissecting microscope for counting of attached androspores. An ocular reticule facilitates the counting of androspores over a reproducible area of the agar surface. This technique assures that the phenomenon under observation is chemotaxis and not a trapping or death response, which can occur with other bioassay systems (capillary tubes, agar blocks at bottom of dish, etc.).

2. Quantitative determination of the sperm attractant in *Oe. cardiacum* is less exact, because the sperm do not settle or attach to the surface of the attractant source but swim within the vicinity of it. It is not possible actually to count the number of sperm under such conditions, but it is possible to distinguish various densities of sperm at an interface. Fig. 3–3 shows the bioassay apparatus designed by Machlis et al. (1974). For loading this apparatus the sperm attractant is mixed with an equal volume of 4% melted agar at 50°C. This mixture is added to the numbered areas of the grooves, and water agar (control) is added to the spaces between the pairs of pegs in the assembled apparatus (Fig. 3–3C). After the agar solidifies, it is trimmed level with the surface of the lower component with a sharp scalpel. (The undercutting around each well serves to hold the agar in place during the trimming). The upper component is removed, a sperm suspension is

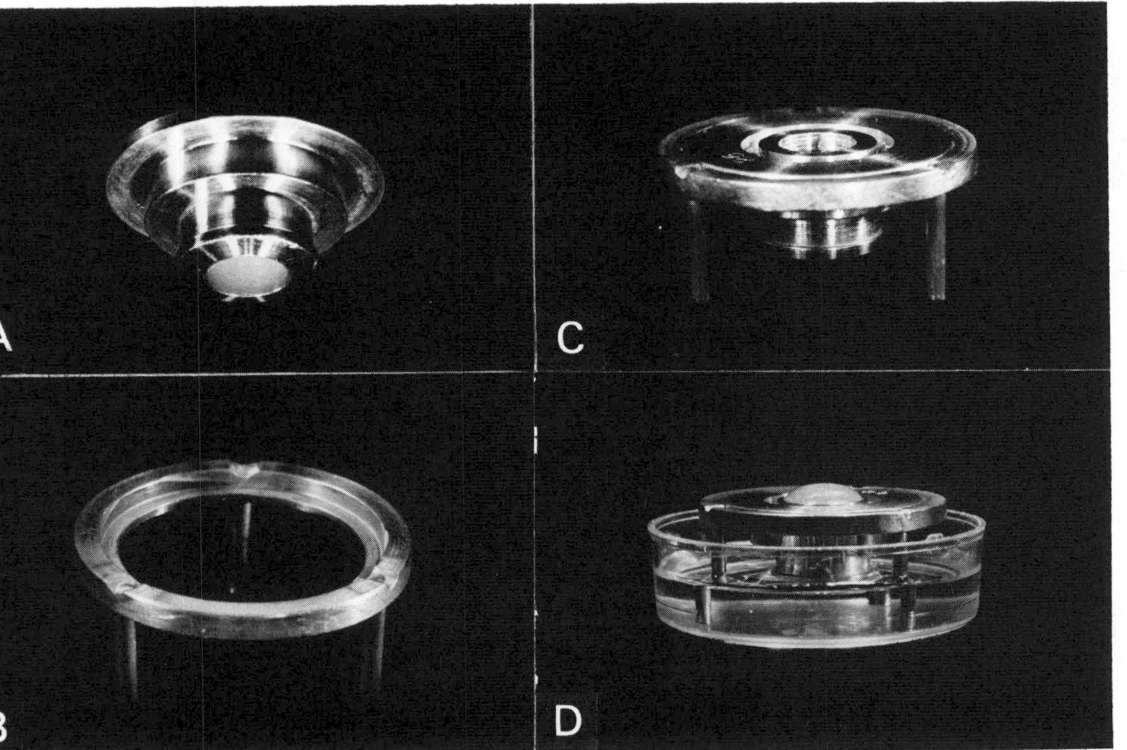

Fig. 3–2. Bioassay apparatus for androspores. A. Center unit containing a well 6.4 mm in diameter with the agar test block in position. Note that only a smooth agar face is exposed. B. Tripod support for the center unit. C. Assembled device. D. Fully assembled bioassay system in an androspore suspension.

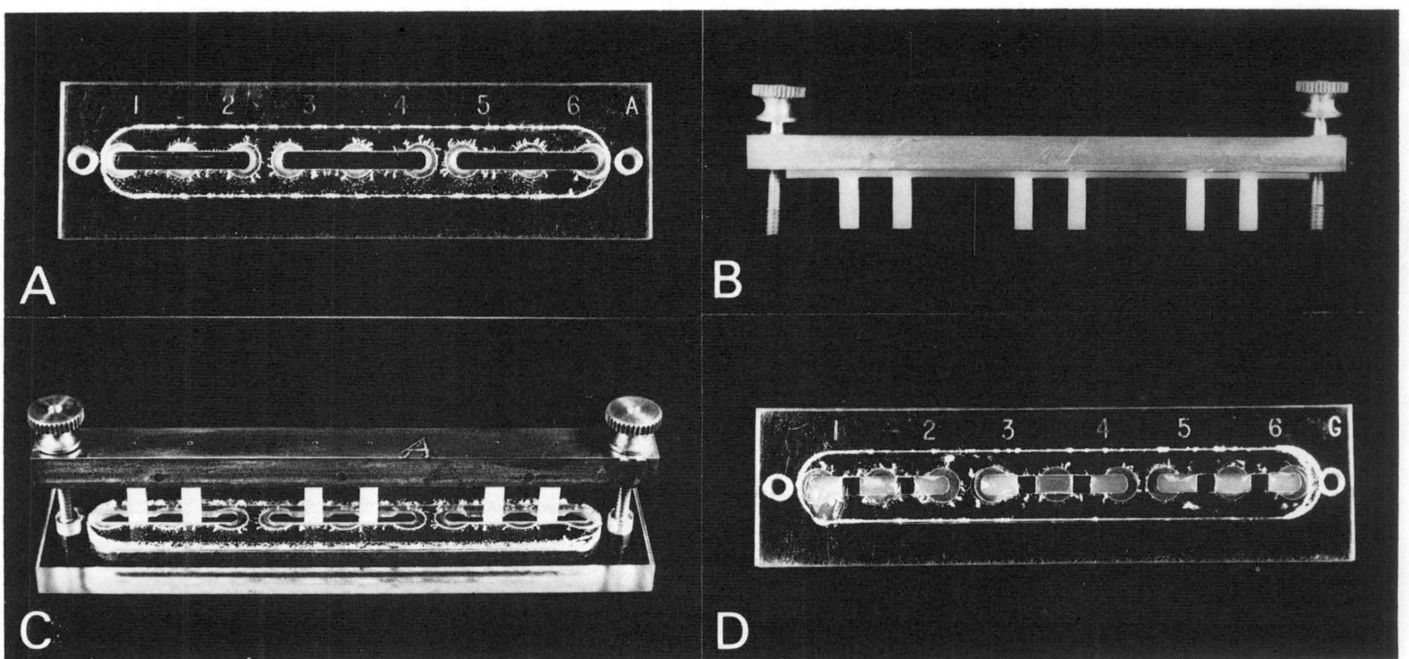

Fig. 3–3. Bioassay apparatus for sperm. A. The lower component of the assembly, made from Plexiglas, has three grooves 1.5 mm deep, 3 mm wide, and 2.54 cm long, with the partial circles being undercut another 1.5 mm deep into the 3-mm thick backing piece for the grooves. B. The upper component of the assembly consists of a series of 3-mm^2 pegs made of Teflon reinforced with stainless steel to prevent bending when assembled as shown in (C). C. A very small hole penetrates the steel and the Teflon pegs to prevent suction from disturbing the agar when the upper component is removed. D. The apparatus prepared for the bioassay with the agar test blocks in position and the empty wells for the sperm suspension.

added to the wells (Fig. 3–3D), and each groove is covered with a coverslip. The number of sperm attracted to the interface between the agar and the sperm suspension is scored on a relative scale (0, +, ++, +++) (Machlis et al. 1974).

VI. General Comments

The Ph.D. thesis of R. L. Hoffman (1961) contains extensive information on several species of *Oedogonium,* including methods for the induction of reproductive stages and zoosporogenesis and preliminary studies on chemotaxis. It proved to be a valuable aid in the early stages of our study.

There is every indication that bacterial "contaminants" are required for the growth and sexual induction process in *Oe. cardiacum* (Machlis 1974). Whether such will prove to be the case with the nannandrous species or other macrandrous species is as yet undetermined, because no rigorous attempts have thus far been made to insure or maintain axenic culture conditions in these species.

Induction of mating responses in the green algae (see also Chaps. 2 and 4) seems to be a result of decreased nitrogen and/or phosphorous sources in the medium (Ellis and Machlis 1968). The level of these factors required for induction of reproductive structures appears to be relative to the level required for the promotion of optimum vegetative growth of the species, if the differences between *Oe. borisianum* and *Oe. donnellii* can be taken as representative cases. Culture conditions can vary from those presented here, because these are designed for maximum growth rate per unit time (as measured by increase in dry weight).

The full character of the attractant remains unknown. However, the present isolation techniques allow for the use of the partially purified attractant in studies of the nature of the sperm chemotactic response and of the chemotactic determination of reproductive isolation in and between macrandrous and nannandrous species of the genus.

VII. Acknowledgments

First and foremost, I wish to acknowledge the fourteen years of support, guidance, encouragement, insight, and friendship provided by the late Prof. Leonard Machlis, Department of Botany, University of California, Berkeley, California. It was he who introduced me to *Oedogonium,* guided my graduate studies on its nutrition, hormonal integration, and ultrastructure, and displayed after my departure from Berkeley a continued and enthusiastic interest in the progress of this study. The work on the nutrition, mating induction, and initial stages of the assay system for *Oe. donnellii* was carried out while I was on sabbatical leave in his laboratory in the summer and fall of 1975.

Second, I would like to acknowledge the Research Corporation support in the form of a Cottrell College Science Grant, which provided for the establishment of a research program to continue the studies on *Oedogonium* here at Carleton College.

Mr. Laird D. Madison provided the photographs of the bioassay apparatus shown in Fig. 3–2 and carried out much of the work in quantifying the nannandrous bioassay system under a National Science Foundation Undergraduate Research Participation Program Grant (SMI76-03103) to the Biology Department during the summer of 1976.

VIII. References

Ellis, R. J., and Machlis, L. 1968. Control of sexuality in *Golenkinia. Amer. J. Bot.* 55, 600–10.

Hill, G. J. C., and Machlis, L. 1970. Defined media for growth and gamete production by green algae, *Oedogonium cardiacum. Plant Physiol.* 46, 224–6.

Hoffman, L. R. 1961. "Studies on the Morphology, Cytology, and Reproduction of *Oedogonium* and *Oedocladium.*" Ph.D. Thesis. University of Texas, Austin.

Hoshaw, R. W., and Rosowski, J. R. 1973. Methods for microscopic algae. In Stein, J. R. (ed.), *Handbook of Phycological Methods: Culture Methods and Growth Measurements,* pp. 53–68. Cambridge University Press, Cambridge.

Machlis, L. 1969. Zoospore chemotaxis in the water mold, *Allomyces. Physiol. Plant.* 22, 126–39.

Machlis, L. 1973. The effects of bacteria on the growth and reproduction of *Oedogonium cardiacum. J. Phycol.* 9, 342–4.

Machlis, L., Nutting, W. H., Williams, M. W., and Rapoport, H. 1966. Production, isolation, and characterization of sirenin. *Biochem.* 5, 2147–52.

Machlis, L., Hill, G. J. C., Steinback, E., and Reed, W. 1974. Some characteristics of the sperm attractant from *Oedogonium cardiacum. J. Phycol.* 10, 199–204.

Rawitscher-Kunkel, E., and Machlis, L. 1962. The hormonal integration of sexual reproduction in *Oedogonium. Amer. J. Bot.* 49, 177–83.

4: Gamete induction and flagellar adhesion in Chlamydomonas reinhardi

WILLIAM J. SNELL*

Department of Biology
and the McCollum-Pratt Institute,
Johns Hopkins University, Baltimore, Maryland 21218

CONTENTS

* Present address: Department of Cell Biology, Southwestern Medical School, The University of Texas, Health Science Center at Dallas, Dallas, Texas 75235

I. Introduction

Chlamydomonas reinhardi is an alga that is easily grown in defined media in large quantities and has been used as an experimental organism for a wide variety of genetic (see Sager 1972 for review), biochemical (Kates and Jones 1964), ultrastructural (Ringo 1967) and cell and developmental studies (Rosenbaum et al. 1969).

The life cycle of *C. reinhardi* is relatively simple. When zygotes germinate they release two haploid cells of the plus mating type (mt^+) and two of the minus mating type (mt^-). These cells are capable of vegetative (or asexual) growth, and it is in this phase of their life cycle that they are grown and maintained in the laboratory. If vegetatively growing cells are put into medium lacking a nitrogen source (Sager and Granick 1954), they undergo a final division (Jones 1966) and enter the gametic (or sexual) phase of their life cycle. If mt^+ gametes are mixed with mt^- gametes there is a rapid agglutination of the cells caused by cross linking of their flagella. Large clumps of cells are formed, and within these clumps, pairs composed of cells of opposite mating types eventually fuse to form zygotes, thus completing the life cycle.

There are several interesting aspects of this complex mating reaction:

1. Only gametes are adhesive; vegetative cells are not. Moreover, gametes of the same mating type will not adhere to each other but only to gametes of the opposite mating type.
2. The flagella are the only parts of the cells involved in the initial recognition and adhesion.
3. Interaction between specialized membrane areas of gametic cell bodies initiates cell fusion (Friedman et al. 1968; Weiss et al. 1977).
4. After the gametes fuse, which takes only a few minutes, the flagella of the resulting zygote are no longer adhesive to each other or to other unmated gametes in the mixture.

Table 4–1. *Growth medium for C. reinhardi*

Stock solution	Concentration % (w/v)	Stock (ml/liter)
Trace metal solution (see below)	—	1.0
Na citrate·2H_2O	10	5.0
$FeCl_3$·6H_2O	1	1.0
$CaCl_2$	4	1.0
$MgSO_4$·7H_2O	10	3.0
NH_4NO_3	10	3.0
KH_2PO_4	10	1.0
K_2HPO_4	10	1.0
Trace metal stock	**g/liter**	
H_3BO_3	1.00	
$ZnSO_4$·7H_2O	1.00	
$MnSO_4$·4H_2O	0.40	
$CoCl_2$·6H_2O	0.20	
$NaMoO_4$·2H_2O	0.20	
$CuSO_4$	0.04	

II. Organism and culture methods

A. Test organisms

Strains 21 gr (mating type +) and 6145 c (mating type −) of *Chlamydomonas reinhardi* originally obtained from Dr. Ruth Sager (Hunter College, New York) have been used by the author for the kinds of experiment described. Several strains of *C. reinhardi* are also available from the University of Texas Culture Collection.

B. Growth media

1. Medium I. The constituents are prepared as stock solutions which are then added to distilled water (final pH 6.8) in the sequence shown in Table 4–1 (Sager and Granick 1954).

2. Medium II. This is of the same composition as medium I except that for each liter of medium the phosphate stock solutions (KH_2PO_4 and K_2HPO_4) are each increased to 3.0 ml, and 10 ml of 2.2 *M* Na acetate is added.

3. M–N medium. This is a nitrogen-deficient medium. It contains the components of medium I, but NH_4NO_3 is omitted, and only enough K_2HPO_4 is added to bring the pH to 7.6.

C. Culturing of cells

All cultures are grown and maintained axenically.

1. Stock cultures. Stock cultures are maintained on 1.5% agar slants or petri dishes in medium I of Sager and Granick (1954) and transferred every 2 months to fresh slants with a transfer loop. For faster growth, the agar can be supplemented with 2 g/liter of sodium acetate and 4 g/liter of yeast extract. The stocks are kept at 10–15°C in an incubator equipped with fluorescent lamps set on a LD cycle of 12:12 h. For details on general growth facilities suitable for *Chlamydomonas*, Starr (1973) should be consulted.

2. Liquid cultures. Liquid cultures are maintained at 24°C on a LD cycle of 12:12 h with illumination by fluorescent lamps (1,000 ft-c, 11,000 lux) from the side. Cells can be grown asynchronously in continuous light, if necessary, but synchronous cultures produced by the LD cycle are more satisfactory because the cells are more homogenous in their morphology and physiology (Surzycki 1970). The cultures are bubbled with air from laboratory air lines, but 5% CO_2 can also be used (Kates 1966). A combination pressure regulator and dust trap is inserted in the air line, and subsequent to that a Millipore catridge-type Lifeguard filter (No. CP1501003, Millipore Filter Corp.) is used to reduce contamination. The air supply for each culture container is regulated with brass needle valves obtainable from tropical fish supply stores or from scientific supply companies.

For small batches, cells are grown in 250-ml Erlenmeyer flasks in 125 ml of medium I. The flasks are stoppered with disposable plugs (Identi-plugs, Gaymar Industries, Inc.) through which a cotton-plugged, 12.5-cm long Pasteur pipette is inserted. The aeration rate is adjusted so that the cells are just barely kept in suspension as they grow. Under these conditions, flasks inoculated with 0.5 ml of cells from a fully grown suspension begin to enter stationary phase in 2–4 days at approximately $2–4 \times 10^6$ cells/ml. These cells are used as inocula for larger cultures grown in 4-liter quantities in 5-liter diptheria toxin bottles stoppered with an Identi-plug through which is inserted a coarse gas-dispersion tube. A piece of Tygon tubing is attached to the end of the gas-dispersion tube, and a drying tube filled with cotton is attached to the other end of the Tygon tubing to keep the contents of the bottle sterile after autoclaving. Medium II is used for these large cultures because the acetate permits more rapid growth; this leads to a higher saturation density and a decrease in cell synchrony. It is important not to continue to culture the cells after they have reached saturation (ca. 8×10^6 cells/ml) because the pH of the medium will rise as high as 8.5 and the cells will lose their flagella.

III. Gamete production

A. Induction of gametes

Withdrawal of nitrogen from the medium triggers gamete production (Sager and Granick 1954; Kates and Jones 1964), and this is taken advantage of to induce separate cultures of the two mating types to form gametes. Vegetatively growing cultures that have nearly reached stationary phase (ca. 4 days for cells in medium I, 3 days for cells in medium II) are transferred to M–N medium after the cells have been in the light part of their cycle for 6 h. Cells are harvested from 125-ml cultures by centrifugation at 1000*g* for 3–5 min in a table-top centrifuge (50-ml conical polycarbonate centrifuge tubes) at room temperature. The sedimented cells are resuspended in 125 ml of M–N and centrifuged again. These washed cells are again resuspended in 125 ml of M–N, transferred to 250-ml Erlenmeyer flasks and aerated overnight in continuous light.

Cells of different mating types are separately harvested from larger cultures in a continuous-flow preparative centrifuge or in a cream separator (Hemerick 1973). The cells are washed out of the unit with a squirt bottle filled with M–N and further concentrated by centrifugation for 5 min, in 250-ml polycarbonate centrifuge bottles at 1000*g*. They are then resuspended in a volume of fresh M–N equivalent to the original culture volume and placed in continuous light (as above) with aeration for 15–18 h.

B. Measuring gamete formation

The essential characteristic of a gamete that distinguishes it from a vegetative cell is that it is able to agglutinate and fuse with a gamete of the opposite mating type to form a zygote. Thus, the ability to form zygotes can be used as an assay for gametes. Five ml of mt^+ gametes are mixed with 5 ml of mt^- gametes; the density of each suspension should be equivalent to ca. 2×10^6 cells/ml. After 1 h of incubation with gentle aeration, a portion of the suspension is fixed in Lugol's iodine (1% iodine, 2% potassium iodide) or 1% glutaraldehyde in M–N, and the proportion of cells with 4 flagella (zygotes) is determined with a phase contrast microscope. If the gametes have been successfully induced, 90–100% of the single cells should have fused to form zygotes.

An alternate assay can also be used in which the cell number is determined before mixing the cells and again 1 h after mixing. Upon complete fusion, the cell density is reduced to half its original value. The formula for percent mating is

$$\text{Percent mating} = 2 \times \frac{\text{final cell concentration}}{\text{initial cell concentration}} \times 100$$

IV. Isolation of flagella and agglutination

A. Flagellar isolation procedure

Flagella can be isolated from gametic cells (Snell 1976a) using the sucrose–pH method originally described by Witman et al. (1972) for vegetative cells. Gametes grown in 4- to 16-liter batches are quickly harvested from their medium as described above (III). They are then resuspended in 10 m*M* TRIS buffer, pH 7.8, at 25°C and concentrated by centrifugation at 1100*g* for 6–8 min at 25°C in 250-ml polycarbonate centrifuge tubes. The cells are immediately rinsed again by resuspension in TRIS buffer and centrifugation. The sedimented cells are resuspended in 100–200 ml of 7% sucrose–TRIS buffer at 2–4°C. All subsequent steps are carried out at 2–4°C. In all of the steps up to this point it is essential to work quickly. Sedimented cells left in a pellet rapidly lyse, releasing their internal constituents, which will contaminate the final preparation of flagella.

The suspension of cells in 7% sucrose–TRIS buffer is vigorously stirred with a magnetic stirrer while the pH is rapidly lowered to 4.5 with 0.5 *N* acetic acid. After 30–90 sec, 90–100% of the cells should be deflagellated, as determined by phase-contrast microscopy. The pH of the suspension is then raised to 7.8 by the addition of 0.5 *N* KOH.

For separation of the flagella, 15-ml aliquots are transferred to chilled 50-ml conical polycarbonate centrifuge tubes and underlayed with 20 ml of 25% sucrose–TRIS buffer. A 50-ml volumetric pipette is used to underlay the sucrose by inserting the tip of the filled pipette through the suspension of cells to the very bottom of the centrifuge tube. If the 25% sucrose–TRIS buffer is allowed to drain slowly from the pipette, the interface between the cells and the sucrose will be sharp; this allows good separation of cells from flagella in the subsequent centrifugation step. The tubes are centrifuged at 2000*g* for 11 min; this forces the cell bodies to sediment at the bottom of the tubes and leaves the flagella in the 7% sucrose–TRIS buffer layer. This layer is withdrawn by aspiration down to and including the interface. A 25-ml pipette fitted with a propipette can be used for aspiration, but vacuum aspiration into a chilled vacuum flask is more convenient and quicker for large volumes. If the latter method is used, it is important to use only a slight vacuum to minimize foaming of the flagellar suspension.

To remove any remaining cell bodies, 20-ml portions of the flagellar suspensions are underlayed with 4 ml of 25% sucrose–TRIS buffer, as above, and centrifuged at 1800*g* for 9 min. The supernatant, containing the detached flagella, is collected as above and further centrifuged in 50-ml round-bottomed polycarbonate centrifuge tubes at 27,000*g* for 20 min.

The sedimented flagella, resuspended in a small volume of 7% sucrose–TRIS buffer to a protein concentration determined by the method of Lowry et al. (1951) of approximately 1 mg/ml, can be stored frozen (−20°C) for several months without losing their ability to agglutinate live cells of the opposite mating type (see B. below). If the flagella are kept in the 7% sucrose–TRIS buffer, which is isotonic for them, they will maintain their extended morphology. However, media of lower osmolarity will cause the flagella to assume a spherical shape if their membranes are still intact (Witman et al. 1972).

B. Assessment of adhesiveness of isolated flagella

1. The adhesiveness of the flagella can be qualitatively determined by mixing isolated flagella with whole cells. One ml of cells (2×10^6 cells/ml) in M–N medium are mixed with 25 μl of flagella (in 7% sucrose–TRIS buffer). Flagella mixed with the cells of the opposite mating type should reveal on examination with a phase microscope that 90–100% of the cells are agglutinated. If the flagella are mixed with gametes of the same mating type or vegetative cells of either mating type, there is no agglutination.

Wiese (1974) has used such an assay to titrate the agglutinating activity of mating factors (called *gamone*) that are found in the medium of *Chlamydomonas* gametes (Bergman et al. 1975; McLean et al. 1974; Snell 1976a). A similar procedure can be used to titrate the agglutinating ability of flagellar suspensions.

2. A more quantitative flagellar binding assay has been developed to determine both the adhesiveness of isolated flagella and the adhesiveness of the flagella of intact cells. This assay measures the binding of radioactive flagella of one mating type to unlabeled cells of the opposite mating type. To do this, one mixes radioactive mt^+ flagella with mt^- gametes and with mt^+ gametes, and after various incubation times, the suspensions are centrifuged. The amount of radioactivity sedimenting with the mt^- gametes less the amount sedimenting with the mt^+ gametes is defined as *specific binding* and can be shown to be dependent on cell number as well as flagellar concentration. This assay is described in detail in Snell 1976b.

V. General comments

Kates and Jones (1964) have extensively studied growth of *Chlamydomonas* in liquid culture with respect to the extent of synchrony, growth media concentrations, and intensity of illumination. Kates' dissertation (1966) is an excellent source of information about growth and gametogenesis.

Whenever live cells are being examined by phase microscopy, it is necessary to place a spot of petroleum jelly under each corner of the coverslip to minimize damage to the cells. Excessive pressure from the coverslip can sometimes cause the flagella to change in morphology or even to detach.

Both vegetative and gametic cells have a tendency to stick to almost any surface that they touch for any length of time. This causes severe problems, for example, in determining cell density. To avoid these problems, whenever cells are being pipetted the entire contents of the pipette should be delivered and all operations should be carried out as quickly as possible.

VI. Acknowledgment

I would like to thank Dr. Joel Rosenbaum, Yale University, for helping me to learn about culturing and working with *Chlamydomonas*.

VIII. References

Bergman, K., Goodenough, U. W, Goodenough, D. A., Jawitz, J., and Martin, H. 1975. Gametic differentiation in *Chlamydomonas reinhardtii.* II. Flagellar membranes and the agglutination reaction. *J. Cell Biol.* 67, 606–22.

Friedman, I., Colwin, A. L., and Colwin, L. H. 1968. Fine-structural aspects of fertilization in *Chlamydomonas reinhardi. J. Cell Sci.* 3, 115–28.

Hemerick, G. 1973. Mass Culture. In Stein, J. R (ed.), *Handbook of Phycological Methods: Culture Methods and Growth Measurements,* pp. 255–66, Cambridge University Press, Cambridge.

Jones, R. F. 1966. Physiological and biochemical aspects of growth and gametogenesis in *Chlaymdomonas reinhardi. Ann. N.Y. Acad. Sci.* 175, 648–59.

Kates, J. 1966. "Biochemical Aspects of Synchronized Growth and Differentiation in *Chlamydomonas reinhardtii.*" Ph.D. Thesis. Princeton University, Princeton, New Jersey.

Kates, J. R., and Jones, R. F. 1964. The control of gametic differentiation in liquid cultures of *Chlamydomonas. J. Cell Comp. Physiol.* 63, 157–64.

Lowry, O., Rosenbrough, N. J.. Farr, A. L., and Randall, R. J. 1951. Protein measurement with the Folin phenol reagent. *J. Biol. Chem.* 193, 265–75.

McLean, R. J., Laurendi, C. J., and Brown, R. M., Jr. 1974. The relationship of gamone to the mating reaction in *Chlamydomonas moewusii. Proc. Nat. Acad. Sci. U.S.* 71, 2610–13.

Ringo, D. L. 1967. Flagellar motion and fine structure of the flagellar apparatus in *Chlamydomonas. J. Cell Biol.* 33, 543–71.

Rosenbaum, J. L., Moulder, J. E., and Ringo, D. L. 1969. Flagellar elongation and shortening in *Chlamydomonas.* The use of cycloheximide and colchicine to study the synthesis and assembly of flagellar proteins. *J. Cell Biol.* 41, 600–19.

Sager, R. 1972. *Cytoplasmic Genes and Organelles.* Academic Press, New York, 405 pp.

Sager, R., and Granick, S. 1954. Nutritional control of sexuality in *Chlamydomonas reinhardi*. *J. Gen. Physiol.* 37, 729–42.

Snell, W. J. 1976a. Mating in *Chlamydomonas:* a system for the study of specific cell adhesion. I. Ultrastructural and electrophoretic analyses of flagellar surface components involved in adhesion. *J. Cell. Biol.* 68, 48–9.

Snell, W. J. 1976b. Mating in *Chlamydomonas*: a system for the study of specific cell adhesion. II. A radioactive flagella-binding assay for quantitation of adhesion. *J. Cell Biol.* 68, 70–9.

Starr, R. C. 1973. Apparatus and Maintenance. In Stein, J. R. (ed.), *Handbook of Phycological Methods: Culture Methods and Growth Measurements,* pp. 171–80, Cambridge University Press, Cambridge.

Surzycki, S. J. 1970. Synchronously grown cultures of *Chlamydomonas reinhardi.* In San Pietro, A. (ed.), *Methods in Enzymology,* 23A, pp. 67–73, Academic Press, New York.

Wiese, L. 1974. Nature of sex specific glycoprotein agglutinins in *Chlamydomonas. Ann. N.Y. Acad. Sci.* 234, 383–95.

Weiss, R. L., Goodenough, D. A., and Goodenough, U. W. 1977. Membrane differentiations at sites specialized for cell fusion. *J. Cell Biol.* 72, 144–60.

Witman, G. B., Carlson, K., Berliner, J. and Rosenbaum, J. L. 1972. *Chlamydomonas* flagella. I. Isolation and electrophoretic analysis of microtubules, matrix, membranes and mastigonemes. *J. Cell Biol.* 54, 507–39.

5: *Acetabularia:* techniques for study of nucleo-cytoplasmic interrelationships

SIGRID BERGER AND HANS-GEORG SCHWEIGER

Rosenhof, D6802 Ladenburg bei Heidelberg
Federal Republic of Germany

CONTENTS

I. Introduction

Acetabularia, because of its large size and distinctive life cycle, through the pioneering work of Hämmerling (1932, 1963) has become a favorite experimental organism for cell biologists. The distinct morphogenesis and among other factors the availability of techniques to enucleate cells and to exchange nuclei are the basis for using *Acetabularia* cells for detailed study of the mechanism of morphogenesis. A combination of methods including biochemistry, cell biology and biophysics, used on a single cell can provide more insight into the role of nucleo-cytoplasmic interactions in the mechanism of differentiation.

The various species of *Acetabularia* are, like the other members of the family Dasycladaceae, found in tropical and subtropical oceans. They grow in shallow lagoons or old coral reefs, usually on small pieces of wood, or on shells of marine organisms (Schweiger et al. 1974). Sometimes, they form dense carpetlike lawns. The best known places for collecting these algae are the Mediterranean, the Caribbean and the western and southern Pacific.

The zygote (Fig. 5–1), which initially carries four flagella, is formed by the fusion of a (+) and a (−) isogamete. The zygote elongates, forming a rhizoid on one side and a stalk on the other. The rhizoid contains a single primary nucleus throughout the entire vegetative phase. Depending on the species, the stalk may attain a length from a few millimeters to more than two hundred millimeters. When the cap attains its maximum size, the generative phase begins. The primary nucleus starts to form secondary nuclei. The site of meiosis has not been convincingly demonstrated, but several observations indicate that it is in the primary nucleus (Green 1973; Berger et al. 1975a; Koop 1975b). After several mitotic divisions, the secondary nuclei are transported through the stalk into the cap. A cell wall is formed around each secondary nucleus to produce cysts. Within the cysts, the secondary nuclei continue to divide and eventually form gametes. Finally, the cysts are released, and they open, releasing the (+) and (−) gametes.

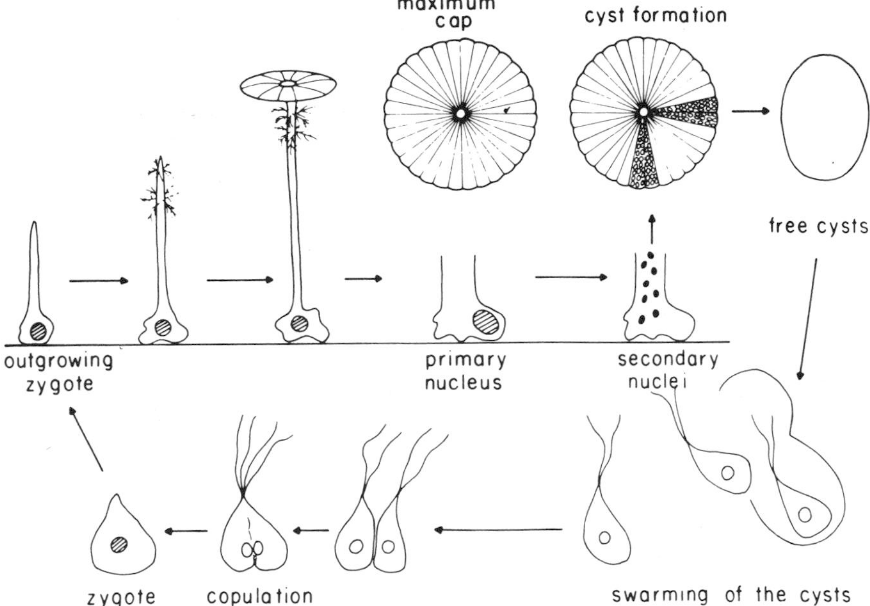

Fig. 5–1. The life cycle of *Acetabularia,* showing development from a zygote to a large cell with a single basal nucleus in the rhizoid, long stalk, and segmented cap. Cysts are formed in cap segments, which then release flagellated gametes.

II. Culture methods

Various conditions for growth and selection of appropriate experimental stages are described by Chapman (1973) and Page (1973) in the first volume of this series. Those primarily used in our laboratory will be described here.

A. Culture medium

Many efforts have been made to culture *Acetabularia* in synthetic media, because it can provide defined and repeatable conditions. One of the most successful was developed by Shephard (1970). However, the disadvantages of this medium are a slower rate of growth and loss of capability to form caps in anucleate cells. Recently, very good results were obtained with the synthetic medium (Schweiger et al. 1977) developed by Provasoli et al. (1957) and modified by Müller (1962).

A chemically undefined Erd-Schreiber medium is usually found to give the best growth in most laboratories and is routinely used by us, as modified from Føyn (1934):

1. *Prepartion of seawater.* Seawater is sterilized by filtration through

a 0.22 μm pore size filter or by autoclaving after addition of 1/10 v distilled water (to prevent precipitation).

2. *Preparation of soil extract stock.* A stock solution of soil extract is prepared by autoclaving humus-rich soil for 1 h at 2 atm. To 100 g of this soil are added 500 ml seawater and the mixture is boiled for 2 h, keeping the water level the same. The suspension is then filtered through filter paper (Selectra No. 595½, Schleicher and Schüll).

3. *Final medium.* The following solutions are added together sequentially: 1500 ml seawater, 40 ml of 47 m*M* $NaNO_3$, 40 ml of 2.2 m*M* Na_2HPO_4, and 40 ml soil extract stock. The final mixture is then filtered through a set of filters with pore sizes of 0.45 μm (upper) and 0.2 μm (lower filter).

B. Growth conditions

All species of *Acetabularia* can be grown at temperatures between 20°C and 28°C (Schweiger et al. 1977). The optimum growth rate is at ca. 20°C for Mediterranean species, like *A. mediterranea* and *A. wettsteinii*, and 25°C for most other species. The algae are maintained in a 12:12 h LD cycle at 2,000 lux (190 ft-c). They are grown in large petri dishes (10 cm in diameter and 5.5 cm high) containing 150 ml medium each.

It is desirable to start cultures from cysts or gametes because field-collected cells are covered with other algae and bacteria which usually grow more rapidly than *Acetabularia*. It is difficult to keep the slowly growing *Acetabularia* axenic. Although for many experiments it is not necessary to use axenic cultures, every effort should be made to work under aseptic conditions.

C. Selection of cysts

1. Nonaxenic. Cyst formation can be detected by the dark green color and speckled appearance of the cap. If the mature cells are retained for several weeks, the caps disintegrate and the cysts are released. The cyst-bearing caps can also be opened by cutting off their outer margin with a pair of iridectomy scissors or with a razor blade and squeezing out their contents with bent forceps (Dumont No. 7 available from Balzers, Ladd, and others) into Erd-Schreiber medium.

2. Axenic. Methods for preparing axenic cultures have been described by Gibor and Izawa (1963), Berger (1967), and Shephard (1970). The modified method of Gibor for obtaining axenic cultures that has proven useful in our laboratory is the following. Isolated cysts are treated for 2 h in silver-protein (Argyrol) prepared as a 10% solution w/v in seawater. Following the treatment, the cysts are washed several times with sterile seawater. They are then incubated in the dark in an antibiotic solution (200 mg streptomycin sulfate, 100 mg penicillin, 20 mg chloramphenical, 20 mg neomycin, 20,000 units mycostatin in 100 ml seawater). Prior to use, the antibiotic solution is passed through a Millipore filter (0.45 μm). After 5 days in the antibiotic solution, the cysts are washed several times with sterilized seawater and are ready for germination.

D. Techniques for collecting gametes

1. To prepare cysts for swarming, freshly collected cysts are washed several times with seawater. Each rinsing should be accompanied by a short, gentle stirring, after which the cysts will settle within a few minutes. The last rinse should be performed in distilled water, because the resulting osmotic shock accelerates gametogenesis (Hämmerling 1934). This method has proven successful with all *Acetabularia* species presently in culture. The cysts are maintained in Schreiber medium (this is Erd-Schreiber medium without soil extract) in the normal LD cycle. If release of gametes does not occur within three weeks, a dark period of a few days and another osmotic shock may induce release.

2. Another method to encourage release of gametes from *A. mediterranea,* by exposure of the cysts to the atmosphere, was described by Koop (1975a). This species usually releases gametes in large quantities very easily, but the method might be usefully applied to *Acetabularia* species in which swarming from the cysts cannot be easily provoked. The method is as follows. Cysts of mature caps are harvested and maintained under standard culture conditions for two weeks. They are then placed in the dark (ca. 17 weeks) and kept in the dark even while the medium is being changed. For release of gametes, the cysts are brought into light and are layered onto filter paper (3-cm diameter) on top of 0.5% agar (Erd-Schreiber medium) in glass dishes (4-cm diameter). The agar plates with cysts are then maintained under standard culture conditions until release of gametes, which must be observed under a binocular microscope.

3. If one uses the method of Hämmerling (1934), swarming of gametes can be readily detected in Boveri dishes, which are useful in most operations. The phototactically positive gametes will migrate to the light and will gather at the lighted side of the dish. To collect the healthiest gametes and to eliminate contaminating algae, the gametes are transferred 3 or 4 times with a fine pipette to the unlighted side of a Boveri dish containing sterile Schreiber medium. They will swim through the dish and gather at the lighted side within a few minutes. This step is particularly important if material from natural sources must be prepared. Gametes usually conjugate within 30 min to 4 h after release from the cysts, forming zygotes, which gather at the un-

lighted side of the dish. To prevent zygotes from clumping after conjugation, they should be kept in the dark for a few hours or overnight. During this time the zygotes will lose their flagella and settle to the bottom of the dish.

E. Starting cultures

Elongating zygotes can be detected under a binocular microscope after 2–3 weeks. To prevent them from firmly attaching to the bottom of the container, they are brushed off with a sterile fine brush every week. As soon as the germlings attain a length of ca. 1 mm, they can be transferred to seawater and stored in the dark as stock cultures. Such stock cultures require a short light exposure of 3 days and a transfer to fresh medium every 3 months. In this way the germlings can be stored for years.

To start a working culture, a small sample (ca. 400 cells) of the stock culture is transferred to a petri dish or Erlenmeyer flask and illuminated. After ca. 2 weeks, when the cells have reached a length of 4 mm, 50 individual cells are transferred to a petri dish filled with 150 ml medium. The medium is renewed every 2 weeks. The time required for a zygote to develop into a mature cyst-bearing cell varies from species to species and depends on culture conditions. The zygote of *A. mediterranea,* the most often used species, may form fully developed cells in ca. 14 weeks. Another few weeks are required for cyst formation.

III. Experimental techniques

A. Fragmentation

Cells in the stage just prior to cap formation are selected. They are cut into fragments of the desired length with scissors or a razor blade in sterile growth medium. Normally, a loss of cytoplasm results from fragmentation. This loss can be prevented by ligation of the stalk at the intended place of the cut prior to fragmentation. The fragments can survive for many weeks.

Three different types of fragments are usually distinguished: the rhizoid, the anucleate basal part, and the anucleate apical part. Rhizoids have the capability of regenerating and are excellent subjects for studying nuclear-dependent regeneration. The apical fragment has the capability of continuing growth and finally forming a species-specific cap. The third fragment, the basal anucleate fragment, still has the capability of surviving, but it has lost its capability of growth and cap formation.

B. *Transplantation of fragments*

Transplantation (in the same medium as that used for culturing) of cell fragments can be conducted under a binocular microscope. When the stalk of *Acetabularia* cell is cut, a small part of the cytoplasm streams out of the cell before the cytoplasm closes the opening by forming a new cell surface. For a certain time period this membrane retains the capacity to fuse when it contacts a similarly newly formed cytoplasmic surface. One may take advantage of this feature by combining fragments of the same or of different species. Not only complementary (e.g., basal and apical parts), but also noncomplementary fragments (e.g., two rhizoids or two apical fragments) can be combined.

The transplantation procedure is carried out in sterile Erd-Schreiber medium in petri dishes. By means of fine forceps, the end having the smaller diameter is pushed into the larger one in a telescopelike manner until the stalks overlap by a few millimeters. The grafts are then exposed to continuous light for at least 12 h and should not be disturbed during this time. Only ca. 10 grafts should be placed in each petri dish.

Grafts with discontinuity in their cytoplasm are discarded 12 h after transplantation because fusion of cytoplasm will not occur after this time. Even at a later time, some grafts may separate spontaneously; this indicates that fusion of the two cytoplasmic surfaces was only simulated.

C. *Isolation of nuclei*

Depending on the species, the single primary nucleus of an *Acetabularia* cell attains a diameter of up to 250 μm. The location and size simplify isolation of the cell nucleus.

To isolate the nucleus, the rhizoid is first cut off from the stalk. Under a dissecting microscope, the end of the rhizoid is then grasped with fine forceps, and, by means of a rather thick needle (ca. 0.8 mm diameter), the cytoplasm is squeezed out of the rhizoid. Since the cytoplasm tends to stream back into the rhizoid, the forceps should follow the needle down the stalk. After the cytoplasm has been squeezed into the medium, two very fine needles (ca. 0.2 mm diameter) are used to locate the bright, shiny nucleus. The nuclei can be picked up by suction and collected. Different media are employed, depending on the intended experimental use of the isolated nuclei.

1. For biochemical or electron microscopic investigations, nuclei are isolated in 0.1 *M* KCl, pH 7.0. Indeed, isolation in KCl is more efficient because the cytoplasm does not clump or adhere as much to the nucleus as it does in implantation buffer.

2. For implantation experiments, nuclei are isolated in a buffered medium containing: 0.67 *M* phosphate buffer (pH 7.4), 0.3 *M* sucrose, 0.005 *M* Mg acetate, and 0.1% bovine serum albumin.

D. Implantation of isolated nuclei

In the implantation buffer (above), the nucleus may survive for up to 24 h (Berger et al. 1975b). Although only some nuclei will survive that long, slow and careful handling of the nuclei is more important than speed.

If the isolated nucleus is to be implanted into a cell fragment, it is best to prepare the recipient fragment several days in advance. This will eliminate the waste of time and effort that would result from introducing a nucleus into a fragment that will not survive surgery. For the implantation, a thin cotton thread is loosely tied around the cell or cell fragment. The cell or fragment is opened at one end with fine scissors or a sharp razor blade while being held with forceps. With a fine pipette, the nucleus is placed near the open end of the cell. By means of a round-tipped glass needle, the nucleus is slowly and carefully pushed into the cell. Immediately after this procedure, the thread is tightened. The cytoplasm of the recipient will surround the nucleus, thus preventing extrusion of the nucleus by the cytoplasm. As with grafts, implants should be exposed to continuous light for at least 12 h and should not be transferred into fresh culture medium for a least 2 days.

IV. Examples of cell biological studies on Acetabularia

The experimental techniques described have been developed during the study of nucleo-cytoplasmic interrelationships in *Acetabularia,* an organism particularly well suited for such investigations.

A. Influence of the nucleus on the cytoplasm

Early experiments had shown that differentiation in this organism depends upon morphogenetic substances (Hämmerling 1932). Although their chemical nature is still unknown, evidence is accumulating that these substances are mRNA. Isolation of a mRNA fraction and translation in a cell-free wheat germ system into *Acetabularia*-specific proteins has been recently performed (Kloppstech and Schweiger 1975a; 1976).

The genes for rRNA, the transcription and processing of rRNA, as well as its transport throughout the cell and its turnover have been studied in detail (Berger and Schweiger 1975b; Kloppstech and Schweiger 1975b; Kloppstech et al. 1977).

The role of the nucleus in enzyme regulation has been investigated during differentiation in nucleate and anucleate cells (Schweiger et al. 1967; Bannwarth and Schweiger 1975; Yamakawa et al. 1977). By means of nuclear exchange experiments, the coding sites of several organelle proteins, including the proteins of 70 S chloroplast ribosomes, have been shown to be in the nuclear genome (Apel and Schweiger 1973; Kloppstech and Schweiger 1973).

B. *Influence of the cytoplasm on the nucleus*

The ultrastructure of nuclei in young and old cells exhibit substantial differences in *Acetabularia*. Nuclear-exchange experiments have revealed the role of the cytoplasm in determining the functional age of the nucleus (Berger and Schweiger 1975a).

C. *Autonomy of the cytoplasm and nucleus*

The ease of isolating chloroplasts free of nuclei has facilitated studies on the genetic apparatus of these organelles (Green 1973; 1976).

The nucleus without cytoplasm can be kept alive in a buffered solution for 24 h (Berger et al. 1975b). During this time, it is capable of synthesizing RNA (Papaphilis, unpublished results).

D. *Circadian rhythm*

Circadian oscillations of photosynthesis can be followed in individual cells and in nucleate as well as in anucleate cell fragments of *Acetabularia* (Mergenhagen and Schweiger 1975; see also Chap. 19). Recently, it has been demonstrated that this rhythm is closely related to gene expression (Schweiger and Schweiger 1977).

V. Acknowledgments

The authors are most grateful for the help of Drs. Dan Zellmer and Bill Cairns in preparing the manuscript.

VI. References

Apel, K., and Schweiger, H.-G. 1973. Sites of synthesis of chloroplast-membrane proteins. Evidence for three types of ribosomes engaged in chloroplast-protein synthesis. *Eur. J. Biochem.* 38, 373–83.

Bannwarth, H., and Schweiger, H.-G. 1975. Regulation of thymidine phosphorylation in nucleate and anucleate cells of *Acetabularia. Proc. Roy. Soc., Ser. B,* 188, 203–19.

Berger, S. 1967. "RNA-Synthese in isolierten Chloroplasten von *Acetabularia.*" Ph.D. Thesis. University of Cologne. 71 pp.

Berger, S., and Schweiger, H.-G. 1975a. Cytoplasmic induction of changes in the ultrastructure of the *Acetabularia* nucleus and perinuclear cytoplasm. *J. Cell Sci.* 17, 517–29.

Berger, S., and Schweiger, H.-G. 1975b. Ribosomal DNA in different members of a family of green algae (Chlorophyta, Dasycladaceae): an electron microscopical study. *Planta* 127, 49–62.

Berger, S., Herth, W., Franke, W. W., Falk, H., Spring, H., and Schweiger, H.-G. 1975a. Morphology of the nucleocytoplasmic interactions during the development of *Acetabularia* cells. II. The generative phase. *Protoplasma* 84, 223–56.

Berger, S., Niemann, R., and Schweiger, H.-G. 1975b. Viability of *Acetabularia* nucleus after 24 hours in an artificial medium. *Protoplasma* 85, 115–18.

Chapman, A. R. O. 1973. Methods for macroscopic algae. In Stein, J. R. (ed.), *Handbook of Phycological Methods: Culture Methods and Growth Measurements,* pp. 87–104. Cambridge University Press, Cambridge.

Føyn, B. 1934. Lebenszyklus, Cytologie and Sexualität der *Chlorophycee Cladophora suhriana Kützing. Arch. Protistenk.* 83, 1–56.

Gibor, A., and Izawa, M. 1963. The DNA content of the chloroplasts of *Acetabularia. Proc. Nat. Acad. Sci. U.S.* 50, 1164–69.

Green, B. R. 1973. Evidence for the occurrence of meiosis before cyst formation in *Acetabularia mediterranea* (Chlorophyceae, Siphonales). *Phycologia* 12, 233–5.

Green, B. R. 1976. Covalently closed minicircular DNA associated with *Acetabularia* chloroplasts. *Biochim. Biophys. Acta* 447, 156–66.

Hämmerling, J. 1932. Entwicklung und Formbildungsvermögen von *Acetabularia mediterranea.* II. Das Formbildungsvermögen kernhaltiger und kernloser Teilstücke. *Biol. Zentralbl.* 52, 42–61.

Hämmerling, J. 1934. Uber die Geschlechtsverhältnisse von *Acetabularia mediterranea* und *Acetabularia wettsteinii. Arch. Protistenk.* 83, 57–97.

Hämmerling, J. 1963. Nucleo-cytoplasmic interactions in *Acetabularia* and other cells. *Annu. Rev. Plant Physiol.* 14, 65–92.

Kloppstech, K., and Schweiger, H.-G. 1973. Nuclear genome codes for chloroplast ribosomal proteins in *Acetabularia.* II. Nuclear transplantation experiments. *Exp. Cell Res.* 80, 69–78.

Kloppstech, K., and Schweiger, H.-G. 1975a. Polyadenylated RNA from *Acetabularia. Differentiation* 4, 115–23.

Kloppstech, K., and Schweiger, H.-G. 1975b. 80 S ribosomes in *Acetabularia major.* Distribution and transportation within the cell. *Protoplasma* 83, 27–40.

Kloppstech, K., and Schweiger, H.-G. 1976. *In vitro* translation of poly (A) RNA from *Acetabularia. Cytobiologie* 13, 394–400.

Kloppstech, K., Richter, G., and Schweiger, H.-G. 1977. Maturation of ribosomal RNA in *Acetabularia. Cytobiologie* 14, 301–9.

Koop, H.-U. 1975a. Germination of cysts in *Acetabularia mediterranea. Protoplasma* 84, 137–46.

Koop, H.-U. 1975b. Uber den Ort der Meiose bei *Acetabularia mediterranea. Protoplasma* 85, 109–4.

Mergenhagen, D., and Schweiger, H.-G. 1975. Circadian rhythm of oxygen evolution in cell fragments of *Acetabularia mediterranea. Exp. Cell Res.* 92, 127–30.

Müller, D. 1962. Uber jahres- und lunarperiodische Erscheinungen bei einigen Braunalgen. *Bot. Mar.* 4, 140–55.

Page, J. Z. 1973. Methods for coenocytic algae. In Stein, J. R. (ed.), *Handbook of Phycological Methods: Culture Methods and Growth Measurements,* pp. 105–26. Cambridge University Press, Cambridge.

Provasoli, L., McLaughlin, J. J. A., and Droop, M. R. 1957. The development of artificial media for marine algae. *Arch. Mikrobiol.* 25, 392–428.

Schweiger, H.-G., and Schweiger, M. 1977. Circadian rhythms in unicellular organisms: an endeavour to explain the molecular mechanism. *Int. Rev. Cytol.* 315–42.

Schweiger, H.-G., Master, R. W. P., and Werz, G. 1967. Nuclear control of a cytoplasmic enzyme in *Acetabularia. Nature* (London) 216, 554–57.

Schweiger, H.-G., Berger, S., Kloppstech, K., Apel, K., and Schweiger, M. 1974. Some fine structural and biochemical features of *Acetabularia major* (Chlorophyta, Dasycladaceae) grown in the laboratory. *Phycologia 13,* 11–20.

Schweiger, H.-G., Dehm, P., and Berger, S. 1977. Culture conditions for *Acetabularia.* In Woodcock, C. L. F. (ed.), *Progress in Acetabularia Research,* pp. 319–30. Academic Press, New York.

Shephard, D. C. 1970. Axenic culture of *Acetabularia* in a synthetic medium. In Prescott, D. (ed.), *Methods in Cell Physiology,* Vol. 4, pp. 49–69. Academic Press, New York.

Yamakawa, M., Ikehara, N., and Schweiger, H.-G. 1977. The occurrence of a 5-methylthioadenosine nucleosidase in *Acetabularia mediterranea.* In Woodcock, C. L. F. (ed.), *Progress in Acetabularia Research,* pp. 33–38. Academic Press, New York.

6: Gamete release, fertilization, and embryogenesis in the Fucales

RALPH S. QUATRANO

Department of Botany and Plant Pathology, Oregon State University, Corvallis, Oregon 97331

CONTENTS

I. Introduction

Zygotes of the Fucales have been used for over 75 years to study the experimental control of polarity (Whitaker 1940) and more recently to probe the biochemical, cytological, and biophysical basis of polar axis determination (Jaffe 1968; 1970; Jaffe et al. 1974; Quatrano 1974). In addition, *Fucus* is being used to study the mechanism of cell-wall assembly around the wall-less egg in response to fertilization (Quatrano and Stevens 1976; Peng and Jaffe 1976). Because the pattern of polar development and embryogenesis in *Fucus* is similar to that of many different plant embryos, including angiosperms (Wardlaw 1968), and because some events triggered by fertilization (e.g., cell wall formation) are common to both *Fucus* and certain angiosperms, this system can be used to study mechanisms in embryogenesis that may be similar in both algae and higher plants. The major advantages of *Fucus* are that zygotes can be obtained aseptically and that the embryos develop synchronously in defined media. This chapter discusses methods and techniques for obtaining gametes and synchronously developing zygotes and embryos of *Fucus* for investigative and instructional purposes.

II. Test organism

The brown alga *Fucus* (Phaeophyta, Fucales, Fucaceae) grows attached to intertidal rocks in northern coastal areas. It possesses a strikingly diplontic life cycle with only a few haploid cells comprising the transient gametophyte generation. Haploid gametes are produced from the meiotic products (megaspores or microspores) by 1, 2, 4, or 8 mitotic divisions. Sexual reproduction is oogamous. The large wall-less, nonmotile egg (ca. 75 μm) is fertilized in the open sea by a small (ca. 5 μm), motile, laterally biflagellate sperm. The diploid zygote of *Fucus* deposits a cell wall, adheres to the substratum, and begins an embryogenic sequence that is morphologically similar to patterns in other plant phyla including angiosperms (Wardlaw 1968). With the proper photoperiod and temperature (Bird and McLachlan

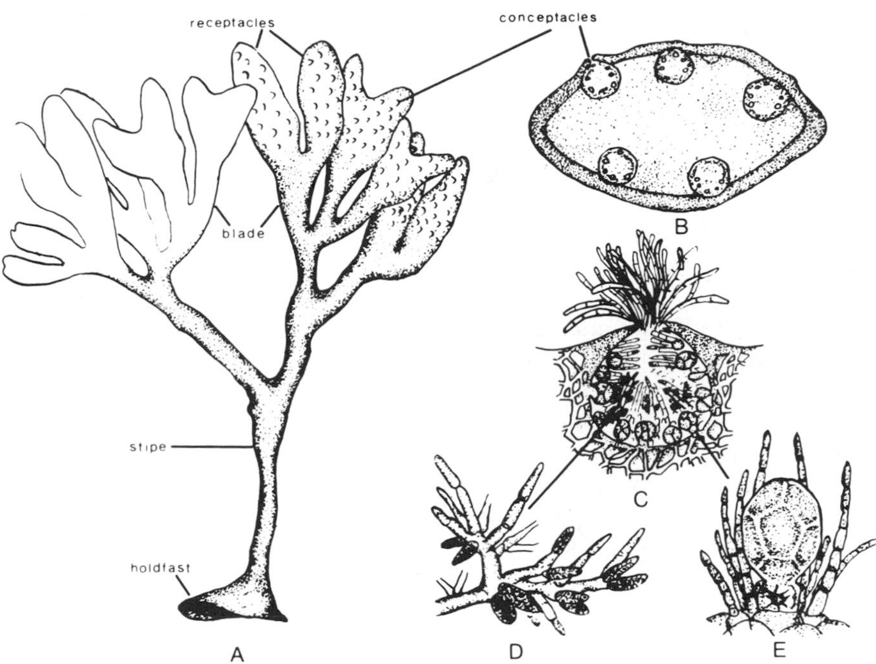

Fig. 6–1. Reproductive structures found in the mature *Fucus* plant. The swollen tips of the blades are receptacles (A). When viewed in cross section (B), the receptacles are seen to contain numerous cavities called conceptacles. Each conceptacle (C) has an opening in which sterile filaments are located. The filaments are attached to the conceptacle wall as are the male reproductive structures (antheridia), which contain many sperm (D), and the female reproductive structures (oogonia), which contain several eggs (E) (from Quatrano 1974).

1976), the tips of mature vegetative fronds swell and initiate reproductive development by the formation of receptacles within which is located the sporogenous tissue that eventually gives rise to mature haploid gametes.

Upon wounding, rhizoidal filaments of sporelings and holdfast tissue of mature plants can be induced to regenerate a whole plant. Callus is not formed, but new meristematic regions in the wounded epidermal layers form adventive embryos (Fulcher and McCully 1969; McLachlan and Chen 1972). Additional studies have elucidated other environmental factors that influence vegetative and reproductive development as well as geographical distribution (McLachlan 1974; Bird and McLachlan 1976).

Receptacles from reproductive fronds possess small openings on the surface, each of which leads to a cavity (conceptacle) containing either eggs or sperm in dioecious species and both eggs and sperm in monoecious species (Fig. 6–1). The common species along the eastern North American coast, *Fucus vesiculosus* L. is dioecious, whereas the common west coast species, *Fucus distichus* subsp. *edentatus* (De la Phylaie) Powell, is monoecious. In nature, the incoming tide washes the slightly desiccated receptacles, causing them to expand and extrude the gametes into the open sea, where fertilization occurs.

III. Methods

A. Collection of receptacles

Material can be obtained throughout the year if receptacles, attached to fronds, are collected from the middle or low intertidal region. Receptacles that are relatively large, flat, mucilagenous, and have conspicuous conceptacles (a bumpy surface) are ideal for gamete release in the laboratory. In summer months, exposure of receptacles during low tides on warm, sunny days will dramatically decrease the number of viable gametes. One should collect specimens when the temperature between layered thalli in the intertidal region is below 17°C, because at higher temperatures it is difficult to obtain material for recovery of gametes (Pollock 1970). Receptacles collected soon after being exposed by the outgoing tide are placed in plastic bags set in crushed ice. To prevent premature shedding of gametes, excess seawater should be removed from the receptacles as soon as possible after collecting. Receptacles are best stored in a dark, well-aerated refrigerator (4°C) for use in the subsequent 2–3 weeks. Care must be taken during this time to avoid desiccation. With the above precautions, receptacles can be shipped successfully by air from coastal regions to distant inland locations. Various marine stations can be contacted for such a service (e.g., Marine Biology Laboratory, Woods Hole, MA).

B. Obtaining gametes and culturing zygotes

One can elicit gamete discharge at desired times during the 2–3 weeks following collection by cutting the receptacles away from the attached fronds, washing them in cold (4°C) tap water (removing prematurely discharged eggs, abnormal embryos and epiphytes), and drying them by blotting. The washing–drying cycle (2–5 min each) should be repeated at least twice before the receptacles are placed in a seawater medium in a lighted incubator (500–1500 ft-c; 5,400–16,000 lux) at 12–15°C. Optimal discharge of gametes occurs when the surface of the receptacles are slightly exposed to air and not completely submerged.

For gamete release, fertilization and embryo development to the 3- to 4-cell stage several media are suitable. Some of these media are filtered natural seawater; commercially available seawater mixtures (e.g., Instant Ocean, Aquarium Systems); artificial seawater media of Weisenseel and Jaffe (1972), Müller (1962), McLachlan (1973); or a simple salt solution (450 m*M* NaCl, 350 m*M* $MgCl_2$, 16 m*M* $MgSO_4$, 10 m*M* KCl, 10 m*M* $CaCl_2$, 5 m*M* $NaHCO_3$ and 5 m*M* TRIS, buffered at pH 7.5–8.0). For further growth of the embryo and development into whole plants, a fortified seawater mixture must be used. The SWM-3 medium of McLachlan contains vitamins and liver extracts added to natural seawater and is sufficient for completion of the life cycle (McLachlan et al. 1971).

1. Dioecious species (*Fucus vesiculosus*). After being washed in seawater, each receptacle is placed, without being submerged, in a separate small petri dish (6-cm diameter) in 5–10 ml of medium and exposed to light (500–1500 ft-c; 5,400–16,000 lux) at 12–15°C. In several hours, each dish will contain either a sperm suspension or from several hundred to several thousand eggs. The egg solution is passed through a nylon mesh (ca. 80–100 μm Nitex; Tobler, Ernst and Traber, Inc.) to remove large debris and oogonia. Eggs are then collected by centrifugation (100*g*, 2 min), and resuspended in a minimal amount of seawater. For fertilization, the eggs are added with a Pasteur pipette to a sperm suspension being held in petri dishes. The dishes are placed in a lighted incubator (12–15°C) on a black cloth or paper. This reduces the light intensity from the bottom, and the sperm, being negatively phototactic, migrate toward the bottom of the dish, where the eggs have settled and where fertilization then occurs. If a dense sperm solution (milky appearance) is used, intense spinning of the egg during sperm attachment can be observed within 5 min. Once a sperm enters the egg, most of the unsuccessful ones fall off, and the egg stops spinning.

Thirty minutes after the eggs stop spinning, sperm can be removed by passing the sperm–egg suspension through another nylon mesh (35–50 μm) that permits sperm to pass but retains zygotes. The zygotes are washed several times with seawater containing either penicillin–streptomycin (25–200 μg/ml of streptomycin and 25–800 μg/ml of penicillin at 1625 units/mg) or chloramphenicol (40 μg/ml); they are then pipetted into sterile petri dishes containing the same seawater–antibiotic mixture and covered. Neither antibiotic mixture alters subsequent development, and both are effective in eliminating bacterial contamination (Peterson and Torrey 1968; Quatrano 1968; McLachlan et al. 1971). The time of fertilization ($\pm$15 min) is recorded 15 min after the sperm–egg suspension is mixed. Four to 6 h after fertilization, unfertilized eggs that do not adhere to the dish are removed from the population by a gentle washing of the culture with fresh seawater. As a result of this procedure, the population within each dish will consist entirely of zygotes that will develop synchronously into embryos.

2. *Monoecious species* (*Fucus distichus*). In the monecious species, fertilization occurs during or shortly after shedding. To produce synchronously developing zygotes, shedding must be rapid and as synchronous as possible. This can be achieved by washing the receptacles with cold, running tap water for 2–3 min and then drying them between layers of toweling. The receptacles are placed in seawater in small petri dishes in light as above. Gamete shedding will occur within 30–60 min after the receptacles are placed in seawater. Receptacles are removed 30 min after the first signs of shedding, and the released zygotes are treated as described above for *F. vesiculosus* (B.1.). The time of fertilization (±15 min) is recorded 15 min after the first signs of shedding.

Large numbers of eggs of *F. distichus* can also be obtained from receptacles if sperm activation is inhibited with agents such as EDTA (1 m*M*), *m*-chlorocarbonylcyanide phenylhydrazone (1 m*M*), 1% (w/v) chloral hydrate, or acidified seawater (pH 5.5, using HCl). Moreover, if gamete shedding is initiated at 2–4°C instead of at 15°C, egg release is prevented. However, discharge of intact oogonia results but without release of the eggs. The oogonia thus released can be easily washed and separated from sperm and zygotes by filtration through appropriate nylon meshes. Fertilization will not occur until the eggs are discharged from the oogonia (Pollock 1970); one achieves this by raising the temperature to 15°C.

In addition to the general methods and techniques described above, several others have been successfully used by a number of investigators to obtain synchronously developing zygotes and embryos (e.g., Peterson and Torrey 1968; Pollock 1970); for fine-structural studies (Quatrano 1972); and for localization of ion fluxes across the rhizoid and thallus membranes (Robinson and Jaffe 1975).

IV. Sample data

The above methods have been successfully used to obtain synchronously developing populations of eggs, zygotes, and embryos (Fig. 6–2). In Fig. 6–3, the time course is given for cell wall assembly (0–4 h), rhizoid appearance (14–18 h) and the formation of two-celled embryos (20–24 h) following fertilization in *F. vesiculosus.* The initiation of these events will vary as much as ±5 h depending on the species, the time of the year receptacles are collected, the length of storage, and the culture conditions of the zygotes. However, the degree of synchrony at each event in a given population is comparable. For example, a population of zygotes can be observed to initiate rhizoid for-

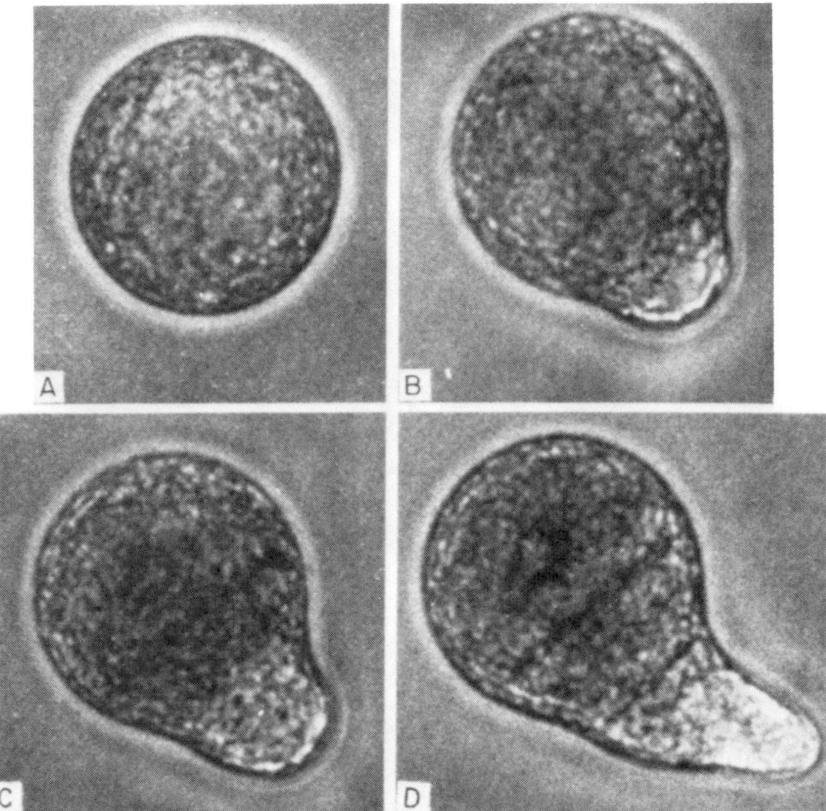

Fig. 6–2. Photomicrographs of zygote development in *Fucus*. The apolar zygote (A) forms a localized protuberance (rhizoid) 14 h after fertilization (B), which is subsequently partitioned from the rest of the cell by the first division at about 20 h (C) and the second division at about 28 h (D) (from Quatrano 1974). (Photomicrographs courtesy of Dr. G. Benjamin Bouck.)

mation over a 4-hour period, but the species and environmental conditions will determine when that will occur after fertilization. For any species in the Fucales, postfertilization development is very reproducible once these conditions are standardized.

In addition to the easily observable morphological changes, events in cellular differentiation can be delineated. By the use of nontoxic inhibitors, such as sucrose, the following events can be separated from each other in a synchronous population and followed independently: establishment of a polar axis; the molecular and organellar localizations occurring in a polar cell; events of cell division; and the differences in composition and structure between the rhizoid cell and the vegetative cell (Quatrano 1973). To ascertain the cytological and biochemical basis of these developmental events associated with polar cell

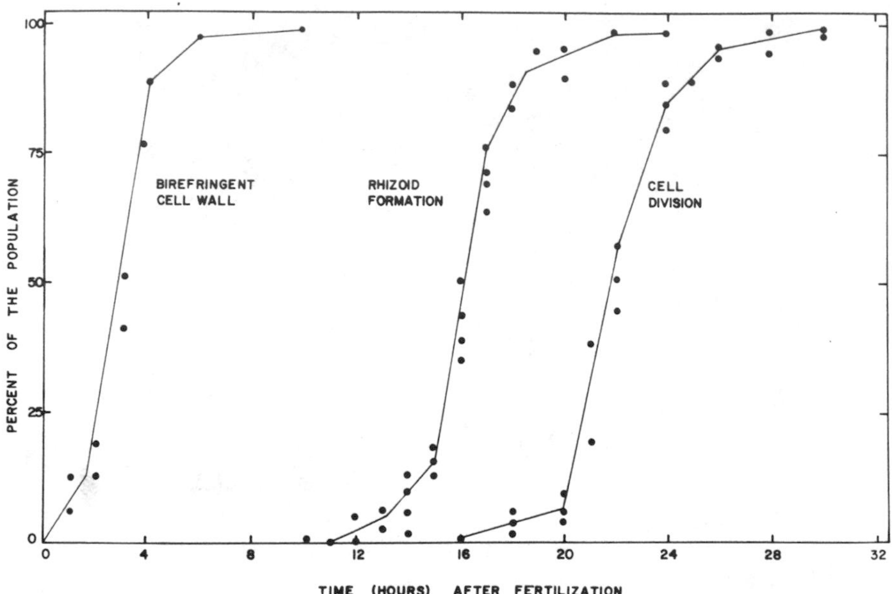

Fig. 6–3. The synchrony and reproducibility of postfertilization events in developing zygote populations of *F. vesiculosus.* Each point represents a percent of the population taken from counts of 100–350 zygotes undergoing oriented cell wall deposition (indicated by birefringence), rhizoid formation, and cell division. The results of four experiments carried out at different times during the year are shown for rhizoid formation and cell division, and two experiments are plotted for birefringent properties.

formation and cellular differentiation as well as cell wall assembly, a number of recent studies have employed these culturing techniques to describe:

1. the fine structure of eggs, the fertilization process, and early zygote and embryo development (Pollock 1970; Quatrano 1972; Brawley et al. 1976a, b; 1977; Peng and Jaffe 1976).

2. changes in ion fluxes, membrane potentials, and the ionic composition of developing zygotes (Jaffe and Nuccitelli 1977; Jaffe et al. 1974; Weisenseel and Jaffe 1972).

3. patterns of RNA and protein synthesis and the role of these macromolecules in later development (Peterson and Torrey 1968; Quatrano 1968).

4. isolation of the cell wall and characterization of its components at different times during assembly (Quatrano and Stevens 1976).

5. changes in the amount and the postsynthetic modification of various polysaccharides found in the eggs and the cell walls of the rhizoid and thallus cells of two-celled embryos (Quatrano and Crayton 1973; Quatrano and Stevens 1976; Hogsett and Quatrano 1975).

These data, considered with the results of earlier investigations describing the effect of various physical and chemical gradients on orienting polar development (Whitaker 1940; Jaffe 1968), may lead to the development of models to explain how specific microenvironments can influence the expression of genes involved with polar cell formation and cellular differentiation in plants.

V. Acknowledgments

Investigations carried out in the author's laboratory that are reported in this article were supported by grants from the National Science Foundation (GB 37149) and the Public Health Service (GM 19247).

VI. References

Bird, N. L., and McLachlan, J. 1976. Control of formation of receptacles in *Fucus distichus* L. subspecies *distichus* (Phaeophyceae, Fucales). *Phycologia* 15, 79–84.

Brawley, S. H., Wetherbee, R., and Quatrano, R. S. 1976a. Fine structural studies of the gametes and embryo of *Fucus vesiculosus* L. (Phaeophyta)I. Fertilization and pronuclear fusion. *J. Cell Sci.* 20, 233–54.

Brawley, S. H., Wetherbee, R., and Quatrano, R. S. 1976b. II. The cytoplasm of the egg and young zygote. *J. Cell Sci.* 20, 255–71.

Brawley, S. H., Quatrano, R. S., and Wetherbee, R. 1977. III. Cytokinesis and the multicellular embryo. *J. Cell Sci.* 24, 275–94.

Fulcher, R. G., and McCully, M. E. 1969. Laboratory culture of the intertidal brown alga *Fucus vesiculosus. Can. J. Bot.* 47, 219–22.

Hogsett, W., and Quatrano, R. S. 1975. Isolation of polysaccharides sulfated during early embryogenesis in *Fucus. Plant Physiol.* 55, 25–9.

Jaffe, L. F. 1968. Localization in the developing *Fucus* egg and the general role of localizing currents. *Advan. Morphogen.* 7, 295–328.

Jaffe, L. F. 1970. On the centripetal course of development, the *Fucus* egg, and self-electrophoresis. *Develop. Biol. Suppl.* 3, 83–111.

Jaffe, L. F., and Nuccitelli, R. 1977. Electrical controls of development. *Ann. Rev. Biophys. Bioeng.* 6, 445–76.

Jaffe, L. F., Robinson, K. R., and Nuccitelli, R. 1974. Local cation entry and self-electrophoresis as an intracellular localization mechanism. *Ann. N.Y. Acad. Sci.* 238, 372–389.

McLachlan, J. 1973. Growth media–marine. In Stein, J. R. (ed.), *Handbook of Phycological Methods: Culture Methods and Growth Measurements,* pp. 25–51. Cambridge University Press, Cambridge.

McLachlan, J. 1974. Effects of temperature and light on growth and development of embryos of *Fucus endentatus* and *F. distichus* subsp. *distichus. Can. J. Bot.* 52, 943–51.

McLachlan, J., and Chen, L. C.-M. 1972. Formation of adventive embryos from rhizoidal filaments in sporelings of four species of *Fucus* (Phaeophyceae). *Can. J. Bot.* 50, 1841–44.

McLachlan, J., Chen, L. C.-M., and Edelstein, T. 1971. The culture of four species of *Fucus* under laboratory conditions. *Can. J. Bot.* 49, 1463–9.

Müller, D. 1962. Über jahres- und lunarperiodische Erscheinungen bei einigen Braunalgen. *Bot. Mar.* 4, 140–55.

Peng, H. B., and Jaffe, L. F. 1976. Cell wall formation in *Pelvetia* embryos. A freeze-fracture study. *Planta* 133, 57–71.

Peterson, D. M., and Torrey, J. G. 1968. Amino acid incorporation in developing *Fucus* embryos. *Plant Physiol.* 43, 941–7.

Pollock, E. G. 1970. Fertilization in *Fucus*. *Planta* 92, 85–99.

Quatrano, R. S. 1968. Rhizoid formation in *Fucus* zygotes: dependence on protein and ribonucleic acid syntheses. *Science* 162, 468–70.

Quatrano, R. S. 1972. An ultrastructural study of the determined site of rhizoid formation in *Fucus* zygotes. *Exp. Cell Res.* 70, 1–12.

Quatrano, R. S. 1973. Separation of processes associated with differentiation of two-celled *Fucus* embryos. *Develop. Biol.* 30, 209–13.

Quatrano, R. S. 1974. Developmental biology: development in marine organisms. In Mariscal, R. E. (ed.), *Experimental Marine Biology*, pp. 303–64. Academic Press, New York.

Quatrano, R. S., and Crayton, M. A. 1973. Sulfation of fuciodan in *Fucus* embryos. I. Possible role in localization. *Develop. Biol.* 30, 29–41.

Quatrano, R. S., and Stevens, P. T. 1976. Cell wall assembly in *Fucus* zygotes. I. Characterization of the polysaccharide components. *Plant Physiol.* 58, 224–321.

Robinson, K. R., and Jaffe, L. F. 1975. Polarizing fucoid eggs drive a calcium current through themselves. *Science* 187, 70–2.

Wardlaw, C. W. 1968. *Morphogenesis In Plants.* Methuen, London. 451 pp.

Weisenseel, M. H., and Jaffe, L. F. 1972. Membrane potential and impedance of developing fucoid eggs. *Develop. Biol.* 27, 555–74.

Whitaker, D. M. 1940. Physical factors of growth. *Growth Suppl.* 75–90.

7: Hybridization and genetics of brown algae

YOSHIAKI SANBONSUGA* AND MICHAEL NEUSHUL

*Department of Biological Sciences
and Marine Science Institute,
University of California, Santa Barbara, California 93017*

CONTENTS

* Present address: Hokkaido Regional Fisheries Research Laboratory, Kushiro, Hokkaido 085, Japan.

I. Introduction

Progress in genetic studies in brown algae has been limited because of several technical difficulties (Lewin 1976); but, because of their diverse life histories and growth forms, they present a fascinating challenge to the geneticist. The morphological similarity of the gametophytic phases of members of the Laminarriales led Schreiber (1930) to attempt crossing of *Laminaria* species without success, but recently several controlled interspecies crosses have been made (Tokida and Yabu 1962; Yabu 1964). The feasibility of these interspecies crosses and those with other brown algal genera such as *Undaria* and *Agarum* (Migita 1967; Saito 1972) have raised questions about the validity of species-specific differences.

This chapter focuses on the commercially important California kelp. It describes the methods for making interspecies crossings and for selection of inbred lines. Such techniques will be useful in assessing the validity of species differences, and more important, their application may prove helpful in improving commercially useful lines. The techniques have worked well on plants from the Pacific Ocean but have as yet not been used elsewhere.

II. Test organisms

Strains of *Macrocystis, Pelagophycus,* and *Nereocystis* were collected from the southern California coast (from Goleta to San Diego). The methods described below are applicable to genera grown in the laboratory. They were grown in enriched seawater medium (ES) (McLachlan 1973) that was adjusted to $35^0/_{00}$ salinity, unless otherwise noted. The reader should also consult the techniques described by Chapman (1973) for specifics on culturing and surface sterilization of macroscopic algae.

III. Method

A. Hybrid crossing procedure

1. Collection of plants and induction of zoospore release. Fertile material is collected, washed in filter-sterilized seawater, and wiped with soft paper to remove attached organisms. The fertile, sorus-bearing blade is then rolled in paper toweling and stored overnight at room temperature (overdrying should be avoided). Pieces of sporophyll (ca. 100 cm^2) are cut, and bent down so as to touch the sterile seawater medium (500 ml) in a deep petri dish. To avoid mucilage drainage, the cut surfaces should not touch the seawater. After incubation in 10–15°C zoospores are released into the dish in 10–30 min. Spore release can be observed with a compound microscope, the sporophyll is removed, and the uppermost portion (ca. 100 ml) of the water in the dish is decanted into a volumetric cylinder (100 ml) to insure that only swimming spores are collected. One can achieve further spore selection by allowing the cylinder to stand in a cold room (5°C) in the dark for 5 h (Kain 1964). During this time potentially contaminating diatoms sink away from the actively swimming kelp zoospores. The final zoospore sample is then collected with a sterile Pasteur pipette from about 5 cm below the surface of the cylinder. At room temperature, drops of the zoospore samples are placed on small (1 × 1.5 cm) pieces of glass cut from microscope slides. Within 5 min the squares, held with tweezers, should be washed with sterile seawater and transferred into small (ca. 5-cm diameter) pyrex petri dishes containing 10 ml of sterilized enriched seawater. Samples are then incubated at 15°C on a 12 : 12 h LD photoperiod at 2,500 lux (230 ft-c).

2. Isolation of gametophytes. About 5 days after inoculation, one-celled gametophytes are formed. These are isolated with a Pasteur pipette drawn out to a diameter of ca. 30 μm. Each gametophyte is gently disloged with the pipette tip and transferred into liquid culture (Figs. 7–1 and 7–3) or onto agar plates (1% w/v with enriched seawater) (Fig. 7–2). To prevent desiccation, the agar plates are sealed with tape.

3. Cultivation and storage of gametophytes. Gametophytes are grown under cold-room conditions (15°C) on a 12 : 12 LD photoperiod. Cool white fluorescent tubes provide light intensities of 2,500 lux (230 ft-c). Under these conditions, most of the gametophytes remain sterile and produce a tufted ball of branches reaching several millimeters in diameter in 2–4 months. Identification of the male plants is made by their denser branching and darker color. Under the microscope the female gametophytes are seen to be 50% thicker and darker than those of the male.

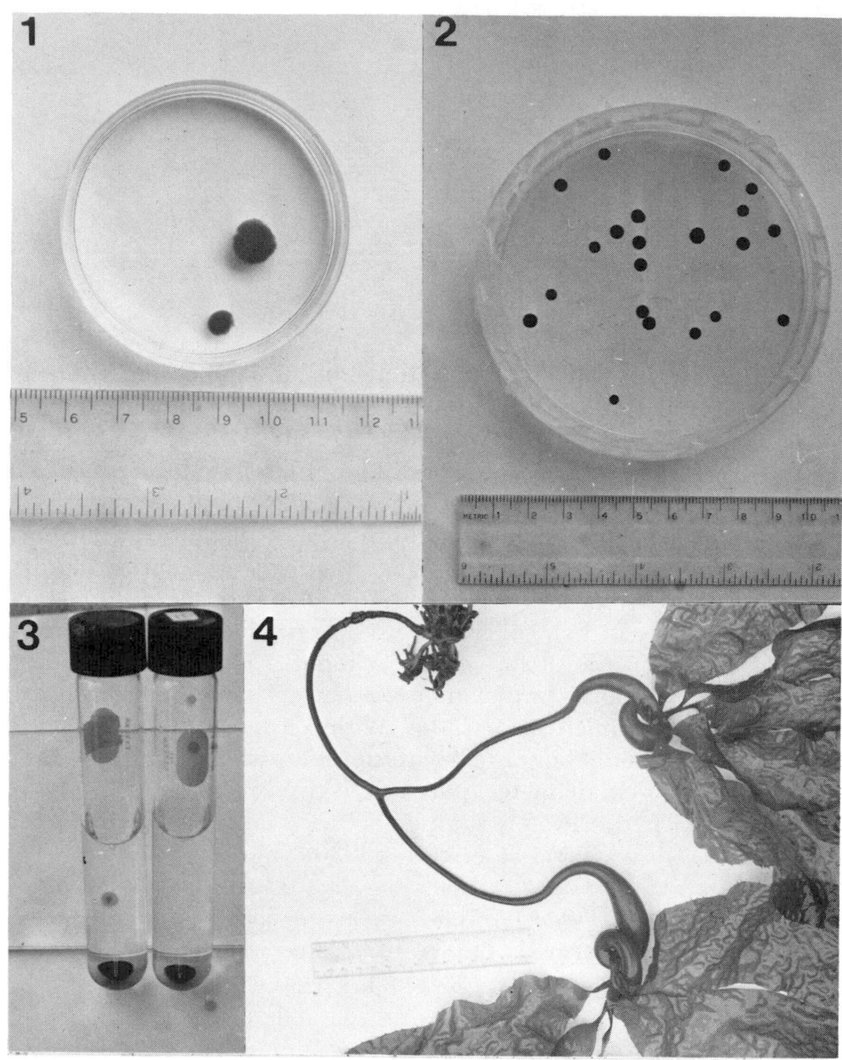

Figs. 7–1 to 7–4. 7–1: Gametophytes of *Macrocystis angustifolia* in liquid culture. The clumps shown are from single isolated gametophytes. The larger clump is female, the smaller male. 7–2: Gametophytes of *Macrocystis angustifolia* on agar. The plants shown are all derived from individual single-cell isolations. Microscopic examination is required to determine the sex. 7–3: *Macrocystis angustifolia* gametophytes in liquid culture for long-term storage. 7–4: A morphologically distinctive intergeneric hybrid between *Macrocystis* and *Pelagophycus,* produced using the crossing method described in this chapter (Sanbonsuga and Neushul 1978).

4. Induction of gametogenesis and gamete release. Gametogenesis is induced within a few days if one places gametophytes into media with salinity of around 30‰. Salinities below 25‰ should be avoided because they produce growth abnormalities in the sporophytes.

5. Crossing. Crossing is accomplished by first dividing the unisexual gametophytic clumps under a dissecting microscope into 4 or 5 nearly equal parts. These are distributed as follows: (a) one portion is placed in a culture tube for further growth, (b) another is placed in a small (ca. 5-cm diamater) petri dish to test for parthenogenesis, and (c) the remaining pieces are placed into small petri dishes with similarly derived pieces of the opposite sex. The gametophytic filaments placed in the dishes are further fragmented to increase the number of gametes formed (Sundene 1958; Vadas 1972). All dishes used for crossing, selfing, and tests for parthenogenesis are incubated at 12°C, under a 12:12 LD photoperiod (5000 lux; 45 ft-c) for 1 to 3 weeks.

6. Verification. Crossing, or parthenogenesis, has taken place when a sporophyte developes. The sporophyte is identified by its darker color and characteristic shape. Parthenogenetic sporophytes can be morphologically distinguished from normal ones within a week as explained below. One can obtain further verification of crossing by growing the sporophytes to a larger size or by determining that they are diploid by counting chromosomes (Evans 1966). Continued growth of enlarging sporophytes is carried out in surge tanks (Sanbonsuga and Neushul 1978) or in still water tanks (Sanbonsuga and Hasegawa 1967). Further cultivation of sporophytes is usually undertaken in the sea.

B. Hybrid crossing results

In Table 7–1 are shown the results of self- and intergeneric crosses of *Pelagophycus* and *Macrocystis*. The self-crossing of a series of female gametophytic strains of *Pelagophycus* with different male strains resulted in a large number of normal sporophytes. In hybrid crosses, the total number of normal sporophytes, although reduced, was very substantial (over 50%). Parthenogenically produced sporophytes were always abnormal.

A normal hybrid sporophyte (Fig. 7–4) from a cross between *Pelagophycus* (female gamete) and *Macrocystis* (male gamete) possesses some morphological characteristics of both genera and exhibits a single primary dichotomy, two vesicles, and corrugated blades with marginal spines.

Table 7–1. *Intergeneric hybridization and crossing of Pelagophycus porra and Macrocystis angustifolia*

1. Crosses of *Pelagophycus*[a]				2. Intergeneric crosses[b]				3. Test for parthenogenesis			
	nS	abS	$\frac{nS}{nS + abS}$		nS	abS	$\frac{nS}{nS + abS}$		nS	abS	$\frac{nS}{nS + abS}$
P♀ × P♂	47	13	0.78	P♀ × M♂	3	6	0.33	P♀	0	3	0
P♀ × P♂	23	7	0.77	P♀ × M♂	2	6	0.25	P♀	0	3	0
P♀ × P♂	39	9	0.81	P♀ × M♂	4	1	0.80	P♀	0	8	0
P♀ × P♂	46	10	0.82	P♀ × M♂	6	21	0.22	P♀	0	10	0
P♀ × P♂	103	18	0.85	P♀ × M♂	40	22	0.61	P♀	0	8	0
P♀ × P♂	57	11	0.84	P♀ × M♂	17	10	0.63	P♀	0	0	0
P♀ × P♂	119	7	0.94	P♀ × M♂	22	7	0.76	P♀	0	4	0
P♀ × P♂	126	6	0.95	P♀ × M♂	5	3	0.63	P♀	0	4	0
P♀ × P♂	67	11	0.86	P♀ × M♂	29	32	0.48	P♀	0	4	0
Totals	627	92	0.87		128	108	0.54		0	44	0

[a] *Pelagophycus* P♀ female and P♂ male strains.
[b] *Pelagophycus* P♀ female, and *Macrocystis* M♂ male strains showing the number of normal sporophytes (nS), abnormal sporophytes (abS).
Source: Sanbonsugo and Neushul (1978).

IV. Alternative methods

The Chinese, by strain selection of *Laminaria japonica,* have produced lines with high temperature tolerance, high iodine content, and other desirable characteristics. A review of this work (in English) can be found in *Scientia Sinica* (Anon. 1976). In their technique, Fang et al. (1962) begin with highly selected strains and then allow crossing within the gametophytic populations to produce inbred lines.

Other techniques for the selection of better lines involve using the half-frond method, where the investigator uses one half of a frond for chemical or physiological analysis and retains the other half for spore production and further propagation if warranted. The culturing of the gametophyte generation, as described in Section III.A., suggests that mutagenic agents could be employed. This coupled with self-crossing might also lead to the production of desirable lines.

V. Problem areas

One problem is to define conditions under which it would be possible to produce many genetically uniform diploid plants from the haploid gametophytic strains. Controlled release of gametes in sufficient quantities, as in *Fucus* (see Chap. 6), would greatly facilitate the genetic crossing. It would also be advantageous if cytological or biochemical methods were devised that could be used to rapidly assess desirable strain characteristics in a very young sporophyte. Above all, it is important for one to be able to determine whether a sporophyte arose from a genetic cross or apogamously (Nakahara and Nakamura 1973).

VI. References

Anon. 1976. The breeding of new varieties of Haidai (*Laminaria japonica* Aresch.) with high production and high iodine content. *Scientia Sinica.* 19, 243–52.

Chapman, A. R. O. 1973. Method for macroscopic algae. In Stein, J. R. (ed.), *Handbook of Phycological Methods: Culture Methods and Growth Measurements,* pp. 87–104. Cambridge University Press, Cambridge.

Evans, L. V. 1966. The Phaeophyceae. In Godward, M. B. E. (ed.), *The Chromosomes of the Algae.* pp. 122–48. Edward Arnold London.

Fang, T. C., Wu, C. Y., Jiang, B. Y., Li, J. J., and Ren, K. Z. 1962. The breeding of a new breed of Haidai (*Laminaria japonica,* Aresch.) and its preliminary genetic analysis. *Acta Bot. Sinica.* 3, 197–209.

Kain, J. M. 1964. Aspects of the biology of *Laminaria hyperborea.* III. Survival and growth of gametophytes. *J. Mar. Biol. Assoc. U.K.* 44, 415–33.

Lewin, R. A. (ed.). 1976. *The Genetics of Algae.* University of California Press, Berkely. 360 pp.

McLachlan, J. 1973. Growth media–marine. In Stein, J. R. (ed.), *Handbook of Phycological Methods: Culture Methods and Growth Measurements,* pp. 25–51. Cambridge Univeristy Press, Cambridge.

Migita, S. 1967. Studies on artificial hybrids between *Undaria pinnatifida* (Kjellm.) Okam. and *Undaria peterseniana* (Harv.). *Sur. Bull. Fac. Fish., Nagasaki Univ.* 24, 9–20 (in Japanese, with English abstract).

Nakahara, H. and Nakamura, Y. 1973. Parthenogenesis, Apogamy and Apospory in *Alaria crassifolia* (Laminariales). *Mar. Biol.* 18, 327–32.

Saito, Y. 1972. On the effects of environmental factors on morphological characteristics of *Undaria pinnatifida* and the breeding of hybrids in the genus *Undaria.* In Abbott, I. A. and Kurogi, M. (eds.), *Contributions to the Systematics of Benthic Marine Algae of the North Pacific,* pp. 117–30. Japan Soc. Phycol. (Toyko).

Sanbonsuga, Y., and Hasegawa, Y. 1967. Studies on Laminariales in culture. I. On the formation of zoosporangia on the thalli of *Undaria pinnatifida* and *Costaria costata* in culture. *Bull. Hokkaido. Reg. Fish. Res. Lab.* 32, 41–8.

Sanbonsuga, Y., and Neushul, M., 1978. Hybridization of *Macrocystis* with other float-bearing kelps. *J. Phycol.* 14, 214–224.

Schreiber, E. 1930. Untersuchungen über Parthenogenesis, Geschlechtsbestimmung und Bastardierungsvermögen bei Laminarien. *Planta* 12, 331–53.

Sundene, O. 1958. Interfertility between forms of *Laminaria digitata. Nytt Mag. Bot* 6, 121–8.

Tokida, J., and Yabu, H. 1962. Some observations on *Laminaria* gametophytes and sporophytes. *Proc. Ninth Pacific Sci. Congr. Bot.* 4, 222–8.

Vadas, R. L. 1972. Ecological implications of culture studies on *Nereocystis leutkeana. J. Phycol.* 8, 196–203.

Yabu, H. 1964. Early development of several species of Laminariales in Hokkaido. *Mem. Fac. Fish. Hokkaido Univ.* 12, 1–72.

8: Hybridization of marine red algae

ALAN R. POLANSHEK

Department of Biological Sciences, San Jose State University, San Jose, California 95192

JOHN A. WEST

Department of Botany, University of California, Berkeley, California 94720

CONTENTS

I. Introduction

Hybridization is a basic technique used in genetic and systematic studies of sexually reproducing organisms. With regard to species relationship, studies of higher organisms, especially animals, resulted in the concept of a biological species, that is, the concept that individuals of the same species are capable of genetic exchange, whereas individuals of different species are not. This generalization does not always apply, because studies of higher plants indicate that some species cannot be defined experimentally by interfertility or reproductive isolation. Too few red algae have been studied to know whether the concept of a biological species applies to this group.

Testing the interfertility of red algal species depends ideally upon the ability to culture each species through its entire life history; therefore, a working knowledge of general culture techniques is essential. The methods described in Section II below were used to test the interfertility of two species of *Petrocelis* (Polanshek and West 1975). Because the field-collected crustose phase is the tetrasporophyte, gametophytes were obtained by culturing tetraspores.

II. Hybridization method for Petrocelis franciscana and P. middendorffii

A. Gametophyte cultures

Tetrasporangiate plants of *Petrocelis franciscana* were collected from the California coast and those of *P. middendorffii* from Amchitka Island in Alaska. In the following procedures, although aseptic techniques were used, axenic cultures were not attained.

1. Each tetrasporangiate specimen is placed in a crystallizing dish (9-cm diameter) containing 100 ml of 30‰ sterile seawater.

2. The specimens are placed in a culture chamber at 10°C on an 8:16 LD photoperiod under cool white fluorescent light (500 lux, 45 ft-c).

3. Tetraspores are released on the bottom of the dish, usually after 1–24 h.

4. Within 1–2 h following release the tetraspores are transferred with a pipette to another crystallizing dish containing 4 No. 2 cover slips and 150 ml of enriched seawater (ES medium, McLachlan 1973).

5. Germanium dioxide (5 mg/liter) is added to each culture to eliminate diatoms. Penicillin G (50 mg/liter) is added to control bacterial and blue-green algal growth.

6. Tetraspore cultures are maintained in a chamber at 15°C on a 16:8 LD photoperiod in cool white fluorescent light (200–300 lux; 20–30 ft-c).

7. The coverslips, to which the tetraspores adhere, are transferred to new medium every 2 weeks.

8. During the first 4 weeks of growth, algal contaminants and tetraspore germlings in excess of 4–6 per coverslip are removed by scraping with a microdissection scalpel. Tetraspore germlings with radial growth (indicating origin from single tetraspores) are selected for continued culture.

9. Two weeks later, individual germlings are separated by cutting the cover glasses with a diamond marking pencil.

10. Each germling is placed in a separate dish and cultured to sexual maturity (6–8 months) on a rotary shaker under conditions as in step 6 above.

11. Vigorously growing male and female plants are selected for use in hybridization experiments as described below.

B. Hybridization experiments

Fertile blades of female gametophytes are characterized by multiple papillate outgrowths. The outer cortex of each papilla contains many female gametangia that are produced continuously as the papillae grow. Under a dissecting microscope the gametangia stand out as white spots against the otherwise red pigmented papilla.

Fertile blades of male gametophytes are covered with a white film of released spermatia. The presence of elongate colorless spermatia ($2 \times 6–8\ \mu m$) can be verified by examination of wet mounts using phase contrast optics.

1. Fertile blades are cut from male and female plants.

2. For each experiment, one fertile male and one fertile female blade are placed in a crystallizing dish (70-cm diameter) containing 75 ml ES medium and then covered with plastic wrap held in place with a rubber band. An isolated female blade in a similar dish serves as the control.

3. The dishes are maintained for 15–30 days on a rotary shaker (100 rpm) at 15°C on a 16:8 LD photoperiod under cool white fluorescent light (500 lux; 45 ft-c).

4. Female blades are reisolated, cultured, and examined after 15–30 days. Swollen cystocarps in papillae of crossed female blades, but not in the control, indicate that fertilization has occurred.

5. Carpospores from cystocarps are cultured to test their viability and development into tetrasporophytes.

C. Sample results

Petrocelis middendorffii and *P. franciscana* are interfertile to the extent that cystocarps, viable carpospores, and vigorously growing second-generation sporophytes are produced. The results, in addition to similarities of anatomy, morphology, and life history, indicate that these plants represent a single species, that is, *P. middendorffii.* Validation of this interpretation must await further study because tetrasporangia formation cannot yet be induced consistently in tetrasporophytes cultured from carpospores. It is unknown whether hybrid tetrasporophytes produce viable tetraspores and fertile second-generation gametophytes. It is possible that these species are reproductively isolated by a failure of meiosis or reduced viability of tetraspores.

Hybridization experiments similar to the above have been used to evaluate species relationships in several other genera of red algae. Edwards (1970) and Rueness (1973) reported different degrees of reproductive isolation between *Polysiphonia* species. Sundene (1975) demonstrated that varieties of *Antithamnion plumula* from the North Atlantic were interfertile but were reproductively isolated from Mediterranean varieties. McLachlan et al. (1977) attempted unsuccessfully to hybridize *Gracilaria foliifera* and *Gracilaria* sp. Spermatial filtrates and heat-killed spermatial controls were used in the latter study.

III. Hybridization experiments and life-history relationships

Current life-history studies of red algae center upon species for which naturally occurring alternate phases were unknown. Often, such species have morphologically different alternate phases. In some instances, the alternate phase obtained in culture is represented in nature by a single species; in other cases by more than one species. Hybridization experiments can be used to determine which naturally occurring species represents the alternate phase in a particular heteromorphic life history. The following techniques have been used to determine which of two species, *Gigartina agardhii* or *G. papillata,* represents the naturally occurring gametophyte of *Petrocelis middendorffii* in California (Polanshek and West 1977; West et al. 1978).

A. Petrocelis gametophyte cultures

Male and female gametophytes are cultured from tetraspores of *P. middendorffii* collected at California localities as described in Section II above.

B. Culture of field-collected Gigartina blades

Female plants of *G. agardhii* and *G. papillata* are collected at the same California localities as *Petrocelis*.

1. The 2–3 cm apical portion of growing blades are cut off, surface sterilized 30 sec in a 2% (v/v) solution of 6% sodium hypochlorite–seawater, and washed 10 sec in distilled water.
2. All papillae containing gametangia are excised.
3. Blade apices are cultured in ES medium with germanium dioxide (5 mg/liter) at 15°C, 16:8 LD photoperiod under cool white fluorescent light (500–1000 lux; 45–90 ft-c).
4. New papillate outgrowths bearing gametangia develop in 6–8 weeks. Male plants of both species, especially *G. agardhii,* collected and cultured as above, frequently fail to produce gametangia. In such cases fertile, field-collected male blades are substituted in crosses.

C. Hybridization experiments

All crosses are made using the methods described in Section II. In crosses using female *Gigartina* blades, a single blade is split longitudinally; half is used in the cross, the other half serves as the control. Positive crosses are those in which crossed female blades develop cystocarps but control blades do not. Carpospores of positive crosses are cultured to test their viability and development.

D. Sample results

Petrocelis and *Gigartina* strains collected both at the same locality and at different localities were tested in the following combinations:

Gigartina male × *Petrocelis* female
Petrocelis male × *Gigartina* female

Approximately 30% of the crosses between *G. papillata* and *Petrocelis* gametophytes were positive (Polanshek and West 1977). Crosses between *G. agardhii* and *Petrocelis* gametophytes were always negative (West et al. 1978). The results support the hypothesis that *G. papillata* represents the naturally occurring gametophyte of *P. middendorffii* in central California. *G. agardhii* is reproductively isolated from *G. papillata–P. middendorffii* and represents a distinct species.

E. Possible problems

Some female plants of both *Gigartina* species reproduce apomictically. Although apomictically produced cystocarps are usually detected before field-collected female blades are used in crosses, the use of cloned female blades as controls is essential.

IV. Additional uses of hybridization experiments

Genetics studies of red algae are just beginning. Rueness and Rueness (1975) demonstrated genetic control of branching patterns in two varieties of *Antithamnion plumula.* Studies on *Gracilaria* mutants, carried out by Van der Meer and coworkers (Van der Meer and Bird 1977; Van der Meer and Todd 1977; Van der Meer 1978), have shown that pigmentation is inherited in two ways: only through nuclear genes (Mendelian inheritance); and through cytoplasmic genes (maternal; non-Mendelian). In the future, hybridization experiments using select mutants may be used to study cellular processes of red algae. They may also provide a means of studying problems related to spermatial viability, fertilization, carposporophyte development and mechanisms of reproductive isolation.

V. References

Edwards, P. 1970. Attempted hybridization in the red algal genus *Polysiphonia. Nature* 226, 467–8.

McLachlan, J. 1973. Growth media–marine. In Stein, J. R. (ed.), *Handbook of Phycological Methods: Culture Methods and Growth Measurements,* pp. 25–51. Cambridge University Press, Cambridge.

McLachlan, J., Van der Meer, J. P., and Bird, N. L. 1977. Chromosome numbers of *Gracilaria foliifera* and *Gracilaria* sp. (Rhodophyta) and attempted hybridizations. *J. Mar. Biol. Assn. U.K.* 57, 1137–41.

Polanshek, A. R., and West, J. A. 1975. Culture and hybridization studies on *Petrocelis* (Rhodophyta) from Alaska and California. *J. Phycol.* 11, 434–9.

Polanshek, A. R., and West, J. A. 1977. Culture and hybridization studies on *Gigartina papillata* (Rhodophyta). *J. Phycol.* 13, 141–9.

Rueness, J. 1973. Speciation in *Polysiphonia* (Rhodophyceae, Ceramiales) in view of hybridization experiments: *P. hemisphaerica* and *P. boldii. Phycologia* 12, 107–9.

Rueness, J., and Rueness, M. 1975. Genetic control of morphogenesis in two varieties of *Antithamnion plumula* (Rhodophyceae, Ceramiales). *Phycologia* 14, 81–5.

Sundene, O. 1975. Experimental studies on form variation in *Antithamnion plumula* (Rhodophyceae). *Norweg. J. Bot.* 22, 35–42.

West, J. A., Polanshek, A. R., and Shevlin, D. E. 1978. Field and culture studies on *Gigartina agardhii* (Rhodophyta). *J. Phycol.* 14, 416–26.

Van der Meer, J. 1978. Genetics of *Gracilaria* sp. (Rhodophyceae, Gigartinales) III. Non-Mendelian gene transmission. *Phycologia* 17, 311–15.

Van der Meer, J. P., and Bird, N. L. 1977. Genetics of *Gracilaria* sp. (Rhodophyceae, Gigartinales) I. Mendelian inheritance of two spontaneous green variants. *Phycologia* 16, 159–61.

Van der Meer, J. P., and Todd, E. R. 1977. Genetics of *Gracilaria* sp. (Rhodophyceae, Gigartinales) IV. Mitotic recombination and its relationship to mixed phases in the life history. *Can. J. Bot.* 55, 2810–17.

9: Development in red algae: elongation and cell fusion

SUSAN D. WAALAND

Department of Botany,
University of Washington, Seattle, Washington 98195

CONTENTS

I. Introduction

Until recently, red algae have often been overlooked as subjects for the study of vegetative morphogenesis. However, many filamentous, uncorticated members of the Ceramiaceae, for example, *Griffithsia, Callithamnion,* and *Antithamnion,* can be easily cultured in the laboratory. Because they have relatively simple morphologies and relatively large cells, these red algae are well suited for experimental studies of cell elongation and cell differentiation.

To study development, it is useful to start with a uniform, unicellular inoculum. Many workers have used spores as an inoculum for single cells because they are all at the same stage of development (Boney and Corner 1962; Chemin 1937; Dixon 1973; Edwards 1970). With species in the genus *Griffithsia,* it is possible to isolate single-shoot (thallus) cells that will rapidly regenerate entire plants (Duffield et al. 1972).

Techniques described in this chapter for obtaining regeneration from isolated shoot (thallus) cells have been successfully used with several species of *Griffithsia.* These techniques, and those used for studying cell elongation and somatic cell fusion in several filamentous, uncorticated genera in the Ceramiaceae, will probably also be applicable to other genera and families.

II. Methods

A. Culture conditions

Several tropical and temperate isolates of species of *Griffithsia* are grown in enriched seawater medium ES or f/2 (McLachlan 1973) in stationary deep culture dishes at ca. 20°C. Cultures are irradiated with cool white fluorescent illumination at 500–3300 lux (45–300 ft-c) on a 16:8 LD photoregime; the medium is changed at $1\frac{1}{2}$–2 week intervals. In *Griffithsia pacifica,* plants grown at low intensities (ca. 440 lux; 40 ft-c) develop as unbranched filaments; but as the light intensity for growth is increased, branch formation increases and pigmentation decreases (Waaland and Cleland 1972; Waaland et al. 1974). A maximum number of branches per plant are formed at ca. 300–400 ft-c. At higher intensities, plants grow for 1–2 months but eventually die. Cultures of *Griffithsia pacifica* can be obtained from the University of Texas Culture Collection.

B. Cell isolation

The general procedure for isolating intercalary shoot cells may be accomplished under a dissecting microscope. Aseptic conditions are not required but are desirable. Several filaments are removed from a rapidly growing stock culture; these are placed in a drop of culture medium on a microscope slide. An intercalary cell 5–8 cells below a filament apex is selected for isolation. The chosen cell is removed from the filament by severing adjacent cells with a sharp razor blade. The isolated shoot cell may be picked up by grasping the cell-wall remnants of adjacent cells with fine watchmakers' forceps. Each cell is placed in ca. 10 ml enriched seawater medium in an individual plastic petri dish (6-cm diameter). Culture conditions described above (II.A.) are used for the study of regeneration and subsequent development. Small plants produced in this way can be used to study environmental control of development, cell elongation (III.A.), cell fusion (III.B.), and phototropism (Waaland et al. 1977).

III. Studies on development

A. Cell elongation

In the higher red algae, intercalary cell divisions do not occur. There is a gradient in cell age along a filament, with the youngest cells at the apex and the oldest at the base. Elongation of growing cells can be quantitatively studied by staining the cell wall with fluorescent dyes or by attaching particles to the surface as reference markers.

1. Vital staining with fluorescent dyes. This technique involves staining walls of living cells with a fluorescent dye that binds to the cell wall. When subsequent cell elongation occurs in the absence of the dye, new cell-wall material is distinguinished by the unstained regions (Waaland and Waaland 1975). Several fluorescent brighteners will serve as vital stains for cell walls; these include Calcofluor White ST, Calcofluor White M2R, and other Calcofluor compounds (American Cyanamid Co.).

This technique allows the localization of cell elongation in cells with nongrowing sheaths; it also permits the analysis of elongation patterns in small cells and in many cells at the same time. It does not yield information about relative rates of elongation within a growth zone.

a. A 0.01% solution (w/v) of Calcofluor is prepared in culture medium.

b. A small plant, or the apical portion of a larger plant, is selected either from single cell regenerates (II.B.) or from rapidly growing stock cultures. The algal filament is immersed in dye solution for 15–30 min.

c. The filament is then rinsed with agitation in three successive changes (5 min each) of dye-free medium.

d. The filament is immediately photographed under a compound microscope both in visible and in ultraviolet light. For UV light observations, a UV light source with a 360 nm exciter filter is used. Any light source which emits 360 nm light can be used from a 4 W fluorescent mineral light (Model UVL-21, Ultraviolet Products) to a 200 W Zeiss UV microscope lamp. A UV absorbing barrier filter which transmits wavelengths above 410 nm (e.g., pale yellow Zeiss 41 filter) is inserted between the specimen and the observer's eye and/or the camera. If a low intensity UV light source is used, it will be necessary to use a sensitive film and long exposures. Kodak Tri-X film has an ASA rating of 1200 when developed with Acufine and has been successfully used by us.

e. The filament is returned to fresh medium and allowed to grow under standard culture conditions (II.A.).

f. The filament should be rephotographed at intervals during growth with both visible and ultraviolet light.

g. Analysis of cell elongation can be made using prints or projected images. Photographs made with visible light are used to determine how much elongation has occurred; photographs made with ultraviolet light are used to localize addition of new (unstained) wall material.

h. If cell wall synthesis appears to be localized in one region of the cell, it is important to compare the width of the band of new wall material with the incremental increase in cell length.

2. *Surface markers*

With markers applied to the surface of a cell, one can determine where along a cell elongation is occurring by observing the change in distance between marks. From this, relative rates of elongation can be determined within a growing zone. This technique has been described in detail by Green (1973). For marine algae, charcoal grains are a more suitable marker than ion-exchange resin beads. Powdered charcoal may be sieved through lens paper or some other sieve and applied to the cell surface with a hair loop. For data analysis, sequential bright-field photographs or time-lapse photography are usually used. This technique is not suitable for small cells and will give information only a few cells at a time. A more serious problem can arise if an alga is encased in a sheath that does not expand as the cells below

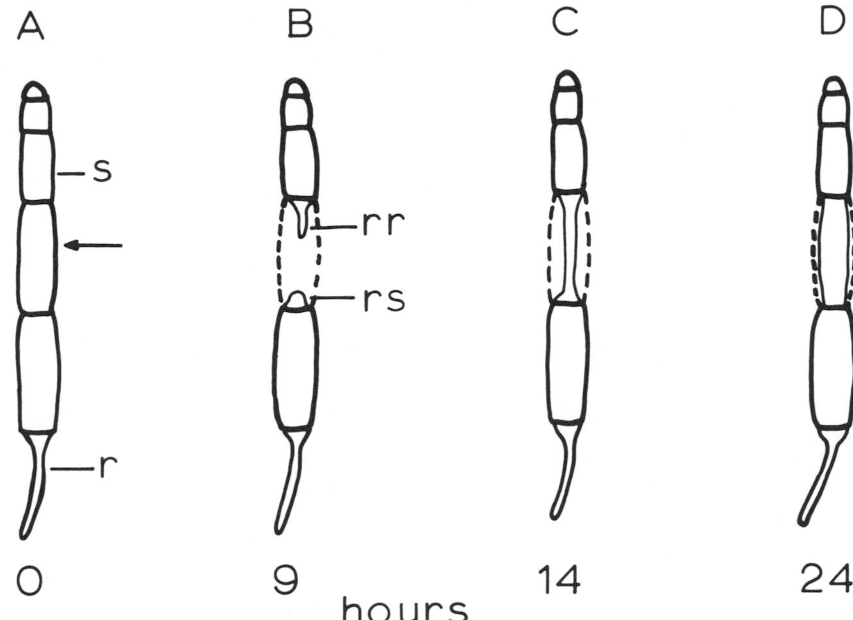

Fig. 9–1. In vivo cell repair in *Griffithsia.* At time = 0: Arrow indicates cell to be punctured. Time = 9 h: Cell above the dead cell has divided to produce a rhizoid; cell below the dead cell has divided to produce a repair shoot cell. Time = 14 h: Fusion between the repair rhizoid and repair shoot is complete. Time = 24 h: Lateral expansion of the fused cell is complete. r = rhizoid, s = shoot cell, rr = repair rhizoid, rs = repair shoot cell.

it. Such a problem may be suspected if charcoal grains that appear to be attached to one cell at the beginning of an experiment are found on a more basal cell after some cell elongation has occured.

B. Somatic cell fusion: in vivo and in vitro

In a number of red algae, dead intercalary shoot cells may be replaced by the process of cell repair. This process has been observed in four species of *Griffithsia* and in two species of *Antithamnion;* it may also occur in species of *Callithamnion* (Höfler 1934; Lewis 1909; Waaland and Cleland 1974; Waaland 1975; Whittick and South 1972). Cell repair is induced when an intercalary shoot (thallus) cell is killed without severing of the filament. The cell above the dead cell divides to produce a "repair" rhizoid; the cell below the dead cell divides to produce a highly modified "repair shoot" cell. These two new cells grow toward each other through the lumen of the dead cell. When they meet, they press together; their adjacent walls are dissolved and cytoplasmic fusion occurs. The cell resulting from the fusion expands laterally until it fills the wall of the killed cell (Fig. 9–1) (Waaland and

Cleland 1974). In *Griffithsia,* the process of cell repair appears to be controlled by a species-specific cell-fusion hormone, rhodomorphin. This hormone is produced by the repair rhizoid; it appears to accelerate cell division in the cell below the dead cell, to regulate the morphology and growth of the repair shoot cell, and to act as a chemotropic hormone attracting the repair shoot to the repair rhizoid for fusion (Waaland 1975; Waaland 1978b; Waaland and Cleland 1974).

1. In vivo cell repair

a. Best results will be obtained if rapidly growing plants are used. Therefore, either filaments from healthy stock cultures or plants grown from isolated cells for 5–7 days are used for these experiments.

b. An intercalary shoot cell 4–7 cells below the apical cell (in *Griffithsia*) is punctured by means of a sharp piece of glass or a dull razor blade. One should be careful not to sever the cell.

c. After puncture, plants are again incubated under standard culture conditions.

d. If cell repair has been induced successfully, the cell superjacent to the dead cell should divide in 4–6 h to produce a repair rhizoid; and the cell subjacent to the dead cell should divide in 4–9 h to produce a repair shoot. Fusion of these cells will begin in 10–14 h, depending on the length of the dead cell.

2. Artificial cell-fusion system

It is possible to examine the process of cell repair in an artificial cell-fusion system. In this way, the interaction between repair rhizoids and repair shoots can be studied. With this system, it is also possible to obtain cell fusion between repair shoot cells from different plants. In *Griffithsia,* such artificially produced hybrid cells can be isolated, and these will regenerate whole new plants. Although interspecific fusion has not yet been obtained, hybrids have been formed between plants of the same species isolated from widely separated localities and between male, female, and tetrasporic plants (Waaland 1975; Waaland 1977; Waaland 1978a).

a. For these experiments, plants of *Griffithsia* grown 5–7 days under standard conditions are used. Two types of filaments are required: short apical filaments with small rhizoids and freshly decapitated filaments.

b. To obtain apical filaments with short rhizoids, filaments 3–4 cells long are excised from *Griffithsia* plants 18–24 h before the beginning of the experiment; these will regenerate rhizoids at their bases (Fig. 9–2).

c. To obtain decapitated filaments, one cuts off the 3–5 apical cells

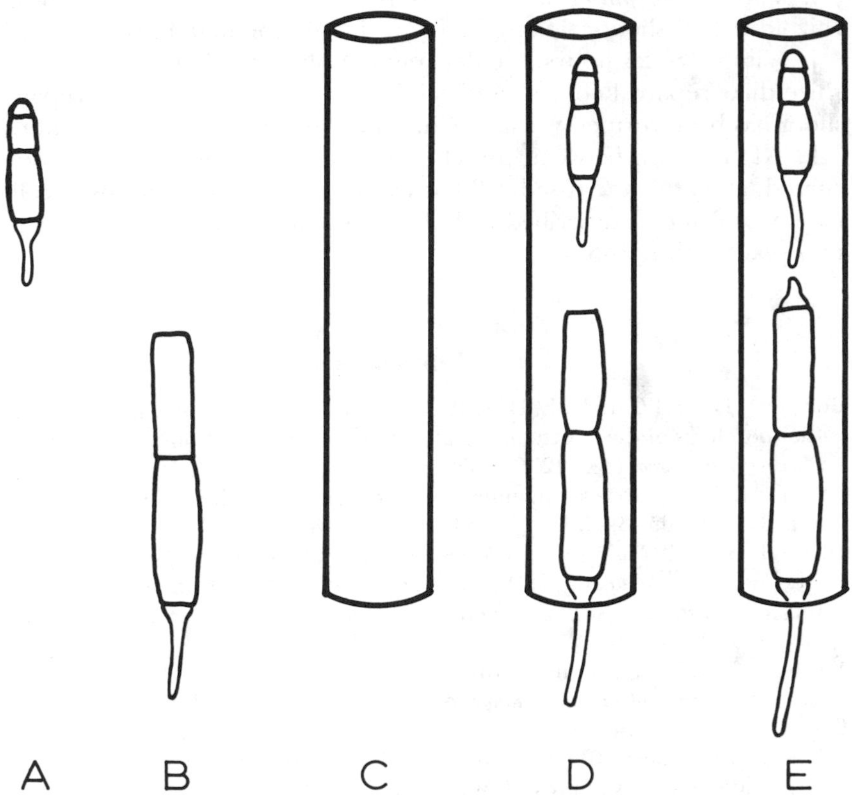

Fig. 9–2. Artificial cell fusion system. A. Apical filament that has regenerated a rhizoid; B. newly decapitated plant; D. (A) and (B) inserted into a cleaned cylinder (C) of *Nitella* cell wall; E. Cell repair in progress within the artificial cell-fusion system.

of *Griffithsia* plants, leaving a two-celled, decapitated filament. This must be done just prior to the cell fusion experiment.

d. An internodal cell of *Nitella* sp. is laid on a glass slide and is cut into several pieces with a razor blade. The pieces of cell are transferred to a dish of distilled water, and their cytoplasmic contents are removed by squeezing with a glass rod; this leaves a cleaned cylinder of cell wall.

e. The cleaned cylindrical cell walls are transferred to petri dishes containing enriched seawater and again squeezed with a glass rod.

f. The *Nitella* wall is used to hold filaments of *Griffithsia* close together so that cell repair will occur.

g. Into one end of the wall cylinder, an apical filament of *Griffithsia* bearing a short rhizoid is inserted, rhizoid end first.

h. Into the opposite end of the cylinder a *freshly* decapitated shoot filament is inserted, apical end first.

i. The two *Griffithsia* filaments are positioned so that the tip of the rhizoid is 1–2 shoot-cell-lengths from the decapitated apex.

j. Within 18–24 hours, the decapitated filament should produce a rhizoidlike repair shoot cell, and cell fusion should occur. If no repair shoot has been formed within 24 h, none is likely to be formed later.

k. After fusion between the rhizoid and the repair shoot has occurred, and after the fused cell has expanded laterally, the fused cell may be isolated as described in II.B.; it should begin to regenerate a new plant within 2–3 days.

IV. References

Boney, A. D., and Corner, E. D. S. 1962. The effect of light on the growth of the sporelings of the intertidal red alga, *Plumaria elegans* (Bonnem.) Schm. *J. Marine Biol. Assoc. U.K.* 42, 65–92.

Chemin, E. 1937. Le développement des spores chez les Rhodophycées. *Revue gen. Bot.* 49, 205–34, 300–27, 353–74, 424–48, 478–536.

Dixon, P. S. 1973. *Biology of the Rhodophyta.* Macmillan (Hafner), N.Y. 285 pp.

Duffield, E. C. S., Waaland, S. D., and Cleland, R. 1972. Morphogenesis in the red alga, *Griffithsia pacifica:* regeneration from single cells. *Planta* (Berlin) 105, 185–95.

Edwards, P. 1970. Field and cultural observations on the growth and reproduction of *Polysiphonia denudata* from Texas. *Brit. Phycol. J.* 5, 145–53.

Green, P. 1973. Intracellular growth rates. In Stein, J. R. (ed.), *Handbook of Phycological Methods: Culture Methods and Growth Measurements,* pp. 369–74. Cambridge University Press, Cambridge.

Höfler, K. 1934. Regenerationsvorgänge bei *Griffithsia Schousboei. Flora* 27, 331–44.

Lewis, I. F. 1909. The life history of *Griffithsia Bornetiana. Ann. Bot.* 23, 639–90.

McLachlan, J. 1973. Growth media – marine. In Stein, J. R. (ed.), *Handbook of Phycological Methods: Culture Methods and Growth Measurements,* pp. 25–51. Cambridge University Press, Cambridge.

Waaland, J. R., Waaland, S. D., and Bates, G. 1974. Chloroplast structure and pigment composition in the red alga *Griffithsia pacifica:* regulation by light intensity. *J. Phycol.* 10, 193–9.

Waaland S. D. 1975. Evidence for a species-specific cell fusion hormone in red algae. *Protoplasma* 86, 253–61.

Waaland, S. D. 1978(a) Parasexually produced hybrids between female and male plants of *Griffithsia tenuis* C. Agardh, a red alga. *Planta* 138, 65–8.

Waaland, S. D. 1978(b) In vitro cell fusion in the red alga. *Griffithsia.* In Fogg, G. E. (ed.), *Proceedings of the International Seaweed Symposium, Bangor.* (In press.)

Waaland, S. D., and Cleland, R. 1972. Development in the red alga, *Griffithsia pacifica:* control by internal and external factors. *Planta* (Berlin) 105, 196–204.

Waaland, S. D., and Cleland, R. E. 1974. Cell repair through cell fusion in the red alga *Griffithsia pacifica. Protoplasma* 79, 185–96.

Waaland, S. D., and Waaland, J. R. 1975. Analysis of cell elongation in red algae by fluorescent labelling. *Planta* (Berlin) 126, 127–38.

Waaland, S. D., Nehlsen, W., and Waaland, J. R. 1977. Phototropism in a red alga, *Griffithsia pacifica. Plant and Cell Physiol.* 18, 603–12.

Whittick, A., and South, G. R. 1972. *Olpidiopsis antithamnimnis* n. sp. (Oomycetes, Olpidiopsidaceae), a parasite of *Antithamnion floccosum* (O. F. Müll.) Kleen from Newfoundland. *Arch Mikrobiol.* 82, 353–60.

10: Production and selection of mutants in blue-green algae

CATHERINE L. R. STEVENS AND
S. EDWARD STEVENS, JR.

*Department of Microbiology and Cell Biology,
The Pennsylvania State University, University Park, Pennsylvania 16802*

CONTENTS

I. Introduction

Since the initial paper by Van Baalen (1965) on mutagenesis of blue-green algae, increasingly more investigators have "discovered" blue-green algae as potential genetic systems. The mutant is the tool for the development of genetic systems, and mutagenesis is the technique for the production of mutants. There are a great many mutagens that have been used with varying degrees of success on many living organisms (Fishbein et al. 1970). Of these, only ultraviolet irradiation (Asato and Folsome 1969), methyl methanesulfonate (Shestakov and Khyen 1970), *N*-methyl-*N*-nitrosourea (Stoletov et al. 1965), and *N*-methyl-*N'*-nitro-*N*-nitrosoguanidine (NTG) (Van Baalen 1965) have been demonstrated to be mutagens for blue-green algae. In this chapter we describe procedures for the production of blue-green algal mutants induced by nitrosoguanidine (NTG)

II. Test organisms

The procedures described in this chapter were developed around two blue-green algae, *Agmenellum quadruplicatum* and *Anacystis nidulans,* and have been extended to a third, *Nostoc* sp. strain MAC.

Agmenellum quadruplicatum strain PR-6 (Van Baalen 1962) is a marine, unicellular, blue-green alga. PR-6 is also designated as strain 7002 (Stanier et al. 1971). *Anacystis nidulans* strain TX 20 (Kratz and Myers 1955) is a freshwater, unicellular, blue-green alga that has undergone considerable taxonomic reclassification (Stevens and Myers 1976). TX 20 is also designated as UTEX 625 and UTEX 1550 (University of Texas Culture Collection), strain 6301 (Stanier 1971), strain B 1405/1 (University of Cambridge Culture Centre for Algae and Protozoa) and strain 602 (Institute of Plant Physiology Collection, U.S.S.R.). *Nostoc* sp. strain MAC (Bowyer and Skerman 1968) is a freshwater filamentous blue-green alga. *Agmenellum* and *Nostoc* can be obtained from the authors, or from Dr. C. Van Baalen (University of Texas, Port Aransas).

III. Procedures

A. Growth

Exponential-phase algal cells from a cloned culture are used as starting material. These cells can be obtained from batch-grown cultures or from continuous cultures. The batch-culture facility derived from the original design of Myers (1950) consists of an aquarium tank (20 × 60 × 23 cm) with a Plexiglas top drilled to accommodate 24 culture tubes (22 × 175 mm), a 300 W immersion heater, a thermoregulator (H-B Instrument Co.), and a pump (Gelber Pumps, Inc.). Carbon dioxide (1% in air) is supplied through a brass manifold equipped with 24 angle needle valves. Illumination is provided by two banks of fluorescent lamps (two Westinghouse F24T12 CW/HO). The continuous culture facility followed the design of Myers and Clark (1944). Convenience and reproducibility make the continuous culture facility more desirable for growing cells for subsequent treatment with mutagen. Both methods provide the cultures with CO_2-enriched air supporting a growth rate three times greater than that obtained with air alone. Both facilities also minimize self-shading.

Agmenellum grows optimally at 39°C in medium A (Stevens et al. 1973); *Anacystis,* at 39°C in medium C_s (Stevens and Myers 1976); and *Nostoc,* at 35°C in medium B (Stevens et al. 1973).

B. Mutagenesis

The following experimental approach to the mutagenesis of blue-green algae was developed around the use of NTG; however, the experimental strategy is applicable to most other chemical mutagens.

1. Pretreatment. If a particular mutant is sought, the chances of success may be increased by the use of one of the following pretreatments.

a. The DNA replication point has been shown to be the primary target of NTG-induced mutagenesis (Cerdá-Olmedo et al. 1968). One may take advantage of this fact in synchronizing the growth of the alga by placing a cell suspension in the dark overnight. The cell suspension is then placed in the light, mutagen added, and samples taken over time, as described by Asato and Folsome (1970).

b. If a mutation in a specific biosynthetic pathway is desired, growth of cells before mutagenesis in the presence of pathway intermediates often increases the number of recoverable mutants affected in the pathway.

c. Large numbers of mutants with point mutations may be obtained by starving cells for carbon dioxide or nitrogen prior to NTG mutagenesis. Cells starved in this manner show much higher survival levels following NTG mutagenesis (between 35–50%) than unstarved cells.

2. *Treatment with mutagen.* **Caution:** NTG is considered to be an extremely dangerous compound because of its carcinogenic and mutagenic effects. Every effort should be made to avoid contact with NTG. Mouth pipetting should be avoided. If bodily contact is made, the area should be rinsed with alcohol and then flooded with water. Glassware and other materials coming into contact with NTG should be soaked in strong base (>pH 9.0) or strong acid (<pH 2.0) prior to the normal washing sequence. Acid- or base-labile materials may be soaked in 10% ethanol prior to washing.

a. The cells are harvested by centrifugation, washed once, and resuspended to a concentration of 5×10^7 to 1×10^8 cells/ml.

b. For each experiment, stock NTG is freshly made up in the growth medium or in buffer to a concentration of 3.3 mg/ml. It is then filter-sterilized, by being passed through a 0.45 μm filter (Millipore).

c. A 5-ml sample of cell suspension is mixed with 0.05 ml of the NTG stock and treated for 15 min at 35–39°C while the suspension is being bubbled with 1% CO_2–air. Increased survival can be attained by carrying out the mutagenesis under a yellow safelight consisting of a tungsten lamp covered by a piece of Plexiglas (No. 2422, Rohm and Haas).

d. Following mutagenesis, the cells are recovered by centrifugation (1800*g*, 10 min) in capped sterile centrifuge tubes. Cells are rinsed once to remove the NTG.

e. Alternatively, if posttreatment is not required, cells in NTG may be directly plated, because agar rapidly decomposes NTG (Greenberg 1960).

3. *Posttreatment.* Once mutagenesis has occurred, any further procedures used are designed to maximize recovery of mutants. The posttreatment selected depends on what the investigator is seeking. After the cells have been resuspended in medium or buffer, the investigator may choose to plate out immediately, or allow for an expression period. If one is seeking auxotrophs, a penicillin counterselection procedure can be used.

a. Plating

i. There are several ways to plate the blue-green algae; by spreading, by pour plate, and by overlay. The method of choice depends on the investigator, but the most accurate method for viable cell counts is the overlay technique (see also Chap. 11.III.A.1.). We pipette a 0.1 ml sample of a cell suspension directly onto the surface of an agar plate and add 3 ml of soft agar held at 45°C. The plate is then quickly swirled to spread the cell–soft agar overlay evenly over the plate. This avoids cell loss in the soft agar tube and avoids prolonged exposure of cells to the 45°C heat. A viable cell count before and after mutagenesis can give an accurate kill rate, and if plating is also done on complete versus minimal plates (for auxotrophs) it can provide the mutation frequency.

ii. Plates, sealed with tape, are incubated at ca. 30°C under fluorescent and/or tungsten lamps providing an illumination of 5400–6400 lux (500–600 ft-c). Plates are examined with a dissecting microscope; clones may be visible in 2–7 days.

b. Expression period. This period also allows mutant cells to increase in number, thereby increasing the investigator's chances of recovering the mutant. For example, if one is seeking a uracil auxotroph, then the medium in which the cells are placed would contain uracil (50–100 mg/liter).

i. Usually, 0.5 ml (5×10^7 cells) of the mutagenized cells, in buffer, are inoculated into 10–20 ml of medium and incubated in the light, with CO_2 bubbling, at 35–39°C.

ii. Once the suspension has grown to 2×10^8 cells/ml (approximately 5–6 generations), penicillin counterselection or plating may be done.

c. Penicillin counterselection. To increase the chances for finding an auxotroph, one must increase the mutant/wild-type ratio by the process of penicillin counterselection.

i. The cell suspension, which has allowed expression of the auxotroph (cells grown up in medium plus uracil is useful as an example), is centrifuged, then resuspended in minimal medium (medium minus uracil) and incubated for a period of time equivalent to four generations. This is done to starve the auxotrophs so that they stop dividing.

ii. The suspension is then placed in the dark (with CO_2 bubbling) for 12–24 h, to induce cell synchronization. This is done so that the majority of wild-type cells will begin their cell cycle at the same time, thereby causing penicillin-induced lysis to occur simultaneously (Stevens et al. 1975).

iii. When the cells are returned to the growth facility, a sterile penicillin stock is injected (final concentration of 15.8 units penicillin/ml), and the cells are incubated under normal growth conditions for a period of time equivalent to three generations. Penicillin prevents the continued synthesis of the gram-negative cell wall, and this results in the selective elimination of growing cells. Microscopic examination (400×) of the cell suspension treated with penicillin reveals long coenocytic cells and cell fragments resulting from lysis.

iv. To terminate penicillin treatment, cells are collected by centrifugation and washed. A dilution series is then plated onto complete medium. Plating onto minimal medium will give an estimate of the number of auxotrophs at the same time.

C. Selection of mutants

1. Auxotrophic mutants. This type requires either organic or inorganic, factors for growth.

a. Selection of clones. Clones from the agar plates can be selected randomly or in a manner indicated by preliminary replication results. Furthermore, pigmentation may give some clue to auxotrophic requirements; for example, bleaching of the center of a colony on complete medium may indicate a limiting nutrient condition. Isolation of a clone should always be done with the aid of a microscope to eliminate the possibility of more than one colony's being picked up (dumbbell-shaped colonies are believed to arise from two daughter cells that have not completely separated). Sterile conditions are more difficult to maintain at this point but are very important for auxotrophs with organic requirements. Picking the inoculum from a clone singled out on the agar plate may be done with a sterilized straight needle or sterile wooden toothpicks. This first inoculum should be streaked onto a new, sterile, complete medium plate in such a way that upon incubation the plate will have many single, isolated colonies. Two major phenotypes associated with auxotrophs are: (1) complete auxotrophs, unable to grow without the required nutrient; and (2) leaky auxotrophs which can grow without the required nutrient but at a reduced rate (compared to the parent). For initial studies on a metabolic pathway, complete auxotrophs are more desirable and their genetic lesion(s) is usually easier to define. For detailed studies on regulation of a metabolic pathway, leaky mutants are more desirable. Since the frequency of mutations is still relatively low compared to the background wild-type population, a rapid screening procedure is required to find the mutants in this population.

i. One excellent screening procedure is that devised by Lederberg and Lederberg (1952) for bacteria. It is well suited for unicellular blue-green algae but is less useful for filamentous forms. In this procedure, all the colonies on an agar plate are transferred from complete to minimal medium by replica plating (see also Chap. 11.III.A.4). For each agar plate, there also must be a separate replicate plating device, usually made of velvet or filter paper. We use sterilized velvet circles, which are placed on an automobile piston turned down to fit a petri dish. The complete agar plate is pressed down hard enough on the sterile velvet so that cells adhere, without wetting the velvet. The sterile minimal agar plate is then pressed down on the velvet template; its orientation is marked so that any colonies that grow on complete but not on minimal medium can be found. Lack of growth on the minimal agar plate is only a probable indicator of auxotrophy.

ii. The second screening technique is described by Massey and Mattoni (1965) and can be adapted for mutant screening of filamentous or gliding blue-green algae. This particular technique requires liquid inocula that must be prepared from the initial isolation plates. Inocula are made up in small test tubes and are transferred with a set of Pasteur pipettes to a new set of liquid cultures.

iii. Confirmation assays. Once the initial screening procedures have been done there will be a number of "probable" mutants. For example, in our screening for possible δ-aminolevulinic acid auxotrophs we found that of 260 clones, 39 were probable mutants and 23 more had faulty replication. The probable mutants are transferred from plates to liquid medium containing a substrate that allows growth of a desired mutant. After growth is visible (about 10^7 cells/ml), the cells are washed and transferred into minimal medium. Growth in complete medium but lack of growth in minimal medium is good evidence that an isolate is auxotrophic. A control using the parent should be done to ensure that proper growth conditions have been provided. Some substrates at high concentrations (for example, uracil at >250 mg/liter) will affect the parent's growth rate and/or normal pigmentation; this would suggest toxic effects. Maximum growth rate is determined from a concentration series of inoculum size. The maximum growth rate should be used in maintaining the mutant against reversion. Slow growth in minimal medium (a depressed growth rate and/or a distortion of the normal pigmentation) is indicative of a leaky auxotroph, particularly if it can be returned to normal (wild-type) growth by addition of the substrate.

2. Nitrogen assimilation mutants

a. Selection of clones. Auxotrophic mutants of blue-green algae with genetic lesions affecting inorganic nitrogen metabolism are one of the most extensively studied classes of mutants. These types of mutants are recognized by their color, ranging from yellow-green to golden, on nitrate-containing solidified media (Stevens and Van Baalen 1969; 1970). The change of color from blue-green to yellow-green or golden, results from a decrease in phycocyanin content caused by starvation (Stevens and Van Baalen 1973). The color change is only regarded as a presumptive mutant with an impaired nitrogen pathway.

b. Confirmation of mutants. Isolated colonies may be further differentiated by their growth response to nitrate, nitrite, and ammonia (Stevens and Van Baalen 1970). Four common classes of mutants have been observed, all of which grow normally on ammonia.

i. Mutants with altered nitrite reductase activity. These have nitrate reductase activity and carry out the reduction of nitrate to nitrite. Nitrite accumulates in the medium.

ii. Mutants in which the altered nitrite reductase responds to an increased iron concentration in the medium. Increasing the iron concentration results in a decrease in rate of nitrite production and an increase in growth rate.

iii. Mutants with altered nitrate reductase activity. These do not grow on nitrate but grow normally on nitrite.

iv. Mutants in which the altered nitrate reductase responds to an increased molybdenum concentration in the medium. Increasing the molybdenum concentration results in an increased rate of growth on nitrate.

3. Pigment mutants

a. Selection of clones. Initial selection is made by their color. The normal pigmentation of the parent strains of *A. nidulans* and *A. quadruplicatum* are bluish-green because phycocyanin and chlorophyll *a* are present in a 1:1 ratio. The normal pigmentation of *Nostoc* sp. is brownish-green (grown under fluorescent light as above), with phycoerythrin, phycocyanin, and chlorophyll *a* being present in proportions of 1:1:2 when grown in white light. In working with such a complex mixture of pigments, one must try to visualize the phenotype one will see for the particular lesion sought. For example, the "blue" mutants of *A. nidulans,* with only one third of the parental chlorophyll content, appear more blue than the parent (Stevens and Myers 1976).

b. Confirmation of mutants. The presumed pigment mutants selected from the agar plates should be grown on liquid medium until they reach about 1×10^8 cells/ml. An absorption spectrum of each presumptive mutant is made using whole cells and the Shibata scattering technique (Shibata 1958). The absorption spectrum of the parent under the identical growth conditions serves as control. Comparisons of the mutant/parent optical density ratios at each absorption maximum (pigment peaks), corrected for scattering, will provide the information necessary to judge whether these are pigment mutants. To eliminate leaky auxotrophs, which can appear with pigment distortions, a comparison of growth rates may also be undertaken. However, caution in interpreting growth data must be exercised because these organisms are photoautotrophs; a decrease may result from a reduction in reaction-center chlorophyll rather than phycocyanin.

4. Morphological mutants

a. Selection of clones. Morphological mutants are also selected by visual inspection of clones on agar plates. In unicellular organisms, filamentous mutants often exhibit a diffuse, rough colony morphology unlike the smooth entire parent colony morphology (Padan and Shilo 1969; Brown and Van Baalen 1970).

b. Confirmation of mutants. The development of morphological mutants may often be confirmed by microscopic examination (400×). It should be noted, that different substrates may affect filament formation. For example, prolonged uracil starvation of the uracil auxotroph of *A. nidulans* (ANU 1) results in development of filaments composed of 10–20 cells. Electron micrographs of such mutants provide further evidence of their morphological aberrancy.

D. Maintenance of mutants

The major problem, once a mutant has been isolated, is its tendency to revert. Therefore, mutants must be maintained optimally and recloned.

1. Culturing

a. For short-term maintenance, the mutants may be kept on agar slants or in liquid cultures. The temperature should not drop below 25°C, and illumination should be at least 540 lux (50 ft-c). Furthermore, use of a complex medium may prolong the time between transfers (Stevens and Myers 1976). For auxotrophs the required substrate should be provided in optimum or slightly greater concentrations (Stevens et al. 1975).

b. Long-term maintenance of the mutants by freezing or freeze-drying, as with bacteria, has not yet been reported for blue-green algae. Until a long-term maintenance technique is developed, the number of mutants of these organisms that any one laboratory can retain will be small, because of the time required for transferring slants and confirming mutant status. Long-term maintenance is a particularly difficult problem for pigment mutants because of the relative difficulty in keeping reversions out of the mutant culture.

2. Recloning. This is crucial in long-term maintenance and should be routinely done every three months. After plating, clones must be selected and tested to reconfirm mutant status in the same way as in the original selection. Any additional selection criteria obtained should also be used. Often a particular mutant may have several lesions that may or may not be related. With time these various lesions may undergo reversion mutation with varying frequencies, so that the original mutant may change some of its phenotype but not the characteristics first selected.

IV. References

Asato, Y., and Folsome, C. E. 1969. Mutagenesis of *Anacystis nidulans* by N-methyl-N′-nitro-N-nitrosoguanidine and UV irradiation. *Mutat. Res.* 8, 531–36.

Asato, Y., and Folsome, C. E. 1970. Temporal genetic mapping of the blue-green alga, *Anacystis nidulans. Genetics* 65, 407–19.

Bowyer, J. W., and Skerman, V. B. D. 1968. Production of axenic cultures of soil-borne and endophytic blue-green algae. *J. Gen. Microbiol.* 54, 299–306.

Brown, R. M., Jr., and Van Baalen, C. 1970. Comparative ultrastructure of a filamentous mutant and the wild type of *Agmenellum quadruplicatum. Protoplasma* 70, 87–99.

Cerdá-Olmedo, E., Hanawalt, P. C., and Guerola, N. 1968. Mutagenesis of the replication point by nitrosoguanidine: map and pattern of replication of the *Escherichia coli* chromosome. *J. Mol. Biol.* 33, 705–19.

Fishbein, L., Flamm, W. G., and Falk, H. L. 1970. *Chemical Mutagens.* Academic Press, New York, 364 pp.

Greenberg, J. 1960. A factor in agar which reverses the antibacterial activity of 1-methyl-3-nitro-1-nitrosoguanidine. *Nature* 188, 660.

Kratz, W. A., and Myers, J. 1955. Nutrition and growth of several blue-green algae. *Amer. J. Bot.* 42, 282–87.

Lederberg, J., and Lederberg, E. M. 1952. Replica plating and indirect selection of bacterial mutants. *J. Bacteriol.* 63, 399–406.

Massey, R. L., and Mattoni, R. H. T. 1965. New technique for mass assays of physiological characteristics of unicellular algae. *Appl. Microbiol.* 13, 798–800.

Myers, J. 1950. The culture of algae for physiological research. In Brunel, J., Prescott, G. W., and Tiffaney, L. H. (eds.), *The Culturing of Algae,* pp. 45–51. Charles F. Kettering Foundation, Yellow Springs, Ohio.

Myers, J., and Clark, L. B. 1944. Culture conditions and the development of the photosynthetic mechanism. II. An apparatus for the continuous culture of *Chlorella. J. Gen. Physiol.* 28, 103–12.

Padan, E., and Shilo, M. 1969. Short-trichome mutant of *Plectonema boryanum. J. Bacteriol.* 97, 975–76.

Shestakov, S. V., and Khyen, N. T. 1970. Evidence for genetic transformation in blue-green alga *Anacystis nidulans. Mol. Gen. Genet.* 107, 372–5.

Shibata, K. 1958. Spectrophotometry of intact biological materials. *J. Biochem.* (Tokyo) 45, 599–623.

Stanier, R. Y., Kunisawa, R., Mandel, M., and Cohen-Bazire, G. 1971. Purification and properties of unicellular blue-green algae (Order Chroococcales). *Bacteriol. Rev.* 35, 171–205.

Stevens, C. L. R., and Myers, J. 1976. Characterization of pigment mutants in a blue-green alga, *Anacystis nidulans. J. Phycol.* 12, 99–105.

Stevens, C. L. R., Stevens, S. E., Jr., and Myers, J. 1975. Isolation and initial characterization of a uracil auxotroph of the blue-green alga *Anacystis nidulans. J. Bacteriol.* 124, 247–51.

Stevens, S. E., Jr., and Van Baalen, C. 1969. N-methyl-N′-nitro-N-nitrosoguanidine as a mutagen for blue-green algae: evidence for repair. *J. Phycol.* 5, 136–9.

Stevens, S. E., Jr., and Van Baalen, C. 1970. Growth characteristics of selected mutants of a coccoid blue-green alga. *Arch. Mikrobiol.* 72, 1–8.

Stevens, S. E., Jr., and Van Baalen, C. 1973. Characteristics of nitrate reduction in a mutant of the blue-green alga *Agmenellum quadruplicatum. Plant Physiol.* 51, 350–6.

Stevens, S. E., Jr., Patterson, C. O. P., and Myers, J. 1973. The production of hydrogen peroxide by blue-green algae: a survey. *J. Phycol.* 9, 427–30.

Stoletov, V. N., Zhevner, V. D., Garibyan, D. V., and Shestakov, S. V. 1965. Pigment mutations in *Anacystis nidulans,* induced by nitrosomethylurea. *Genetika* (English translation) 6, 77–83.

Van Baalen, C. 1962. Studies on marine blue-green algae. *Bot. Mar.* 4, 129–39.

Van Baalen, C. 1965. Mutation of the blue-green alga, *Anacystis nidulans. Science* 149, 70.

11: Euglena: mutations, chloroplast "bleaching," and differentiation

HARVARD LYMAN AND KAREN TRAVERSE

Biology Department,
State University of New York at Stony Brook,
Stony Brook, New York 11790

CONTENTS

I. Introduction

The use of wild-type and mutant strains of *Euglena gracilis* for the study of chloroplast synthesis and replication has proven to be a valuable approach (Schiff 1973; Schiff et al. 1971; Schmidt and Lyman 1976). *E. gracilis* may be grown on a variety of defined media (Table 11–1) either as a photoautotroph or as an organotroph. Under organotrophic conditions the chloroplast is a gratuitous organelle. The ease with which either the replication (Table 11–6) or the synthesis (Tables 11–7A and B) of the plastids may be specifically inhibited makes *Euglena* an ideal organism for studying the regulatory mechanisms of chloroplast synthesis and replication, especially the interaction of nuclear and organellar genomes.

When grown in the dark on organotrophic medium for many generations, *E. gracilis* cells contain only proplastids. These precursors to the chloroplasts contain a small amount of protochlorophyll and lack chlorophyll; they have few internal membranes, have only small amounts of 70 S plastid ribosomes, RNAs, and soluble enzymes of carbon dioxide reduction. Upon illumination, a rapid synthesis of chloroplast components occurs. Photosynthetic pigments, components of photosynthetic electron transport and carbon dioxide reduction, lipids, 70 S ribosomes, plastid specific RNAs, aminoacyl-tRNA synthetases, and chloroplast DNA all increase in response to induction by light. The ease with which the synthesis of the chloroplast may be induced in a population of genetically uniform microorganisms has made *Euglena* an attractive system for the study of organelle synthesis.

Because a functional chloroplast is not required for growth, under organotrophic conditions, it has been possible to obtain a number of strains with mutated plastids. A number of such mutants have been described by Schiff et al. (1971). These mutants represent strains having structural and functional lesions of the chloroplast as well as strains that have lost their plastids completely ("bleaching") and in which chloroplast-associated DNA is undetectable. In this chapter *bleaching* is defined as the specific inhibition of chloroplast or proplastid replication that results in permanently aplastidic cells. Thus, the effects seen when agents that reversibly inhibit the synthesis of some chloroplast components are used (as in Tables 11–7A and B) would not be considered bleaching.

II. Growth conditions

A. Aseptic precautions

It is necessary that all media, culture vessels, pipettes, and so forth be sterile when culturing *Euglena.* Standard microbiological techniques for maintaining sterility when transferring cultures or performing other manipulations should be used. The generation time of *Euglena* is generally about 10–24 h, and therefore, *Euglena* is easily overgrown by many other microorganisms. Usually 15 min of autoclaving at 15 lb/in.^{2} is sufficient for sterilization. Large volumes require longer autoclaving times.

B. Media

1. Composition. Many types of defined media for growing *Euglena,* both organotrophic and phototrophic, have been developed. The compositions of some of the most commonly used media are described in Tables 11–1 to 11–3. *Euglena* is quite resistant to trace metal toxicity; however, to allow maximum growth, the medium must contain a chelator (e.g., EDTA or citrate), which reduces the level of trace metal contamination and allows for an initially high concentration of required metals (Cook 1968). *Euglena* grows optimally over a wide pH range (pH 3–8). The use of low-pH media generally eliminates bacterial but not fungal contamination. However, certain substances, for example streptomycin, do not enter the cell readily at low pH, and for this and other reasons neutral and alkaline media have been devised. *Euglena* also grows well on complex media such as nutrient broth or 1–2% proteose peptone (Difco).

2. Solidification of media. All of the media described may be solidified by the addition of agar. Both Difco and Meer agar have proved to be satisfactory. Table 11–4 indicates the concentrations of these two agars used for various purposes. For neutral and alkaline media, the agar may be added directly to the liquid media and then autoclaved. However if agar is autoclaved in the presence of acidic media the agar is hydrolyzed. In this case, equal volumes of double strength medium and double strength agar are prepared and autoclaved separately. After slight cooling (ca. 10 min) the two are mixed.

The use of a spigot bottle greatly simplifies the pouring of agar plates. A spigot bottle approximately twice the size of the final volume

Table 11–1. *Organotrophic media for Euglena as used by various investigators (values in grams/liter).*

	Acidic			Neutral			Alkaline
Compound	Schiff et al. (1971)[k]	Hutner et al. (1956)[c]	Price and Vallee (1962)[d]	Schiff et al. (1971)	Freyssinet et al. (1972)[h] PCb	Freyssinet et al. (1972)[h] CN	Schiff et al. (1971)
KH_2PO_4	0.40	0.30	0.272	—	—	—	—
K_2HPO_4	—	—	—	—	0.015	0.060	0.10
$CaCO_3$	0.20	0.08	—	—	—	—	—
$Ca(NO_3)_2$	—	—	0.0156	—	—	—	—
$CaCl_2 \cdot 2H_2O$	—	—	—	—	—	—	0.05
$(NH_4)_2HPO_4$	0.20	—	—	—	—	—	—
$(NH_4)_2SO_4$	—	—	—	0.02	—	—	—
NH_4Cl	—	—	—	—	0.300	0.200	0.40
$MgSO_4 \cdot 7H_2O$	0.50	0.40	0.40	0.50	—	—	0.50
$FeCl_3 \cdot 6H_2O$	0.005	—	—	—	—	—	0.002
$MnSO_4 \cdot H_2O$	—	—	1.5×10^{-4}	—	—	—	—
$CuSO_4 \cdot 5H_2O$	—	—	2.5×10^{-5}	—	—	—	—
$ZnSO_4 \cdot 7H_2O$	—	—	0.0043	—	—	—	—
$FeSO_4 \cdot 7H_2O$	—	—	0.0100	—	—	—	—
EDTA (disodium salt)	—	—	—	—	—	—	0.50
$K_3C_6H_5O_7 \cdot H_2O$	—	—	—	1.0	—	—	—
L-Glutamic acid	5.0	3.0	—	—	—	—	—
Ammonium glutamate	—	—	3.0	—	—	—	—
L-Glutamic acid (gamma ethyl ester)	—	—	—	1.0	—	—	—
DL-Malic acid	2.0	1.0	0.268	—	—	—	—
Sucrose	—	15.0	15[e]	—	—	—	—
DL-Aspartic acid	—	2.0	—	—	—	—	—
Glycine	—	2.5	—	—	—	—	—
Glycine ethyl ester	—	—	—	1.0	—	—	—
Ammonium succinate[a]	—	0.6	—	—	—	—	—
Sodium acetate·$3H_2O$	—	—	—	1.0	—	—	1.00
L-Asparagine·H_2O	—	—	—	1.5	—	—	—

α-Glycerophosphoric acid (disodium salt)	—	—	—	0.5	—	—	—
Sodium butyrate	—	—	—	—	1.400	0.900	—
n-Butanol	—	—	—	—	—	—	3.00
Thiamine HCl[b] (vitamin B_1)	1.23×10^{-3}	6×10^{-4}	0.0600	0.01	0.002	0.002	0.001
Cyanocobalamin (vitamin B_{12})	2.5×10^{-7}	[c]	1.0×10^{-5}	4.0×10^{-6}	2.0×10^{-6}	2.0×10^{-6}	4.0×10^{-7}
Trace metals[j,k]							
Type (see Table 11–3)	A	B	—	F	C	C	F
Amount	0.5 ml	0.022 g	(none designated)	0.04 g	1 ml	1 ml	0.05 g
pH (final)	3.5	3.6	3.5[f]	6.8[g]	7.0	7.0	8.0[i]

[a] Ammonium succinate may be replaced by an equivalent amount of succinic acid and NH_4HCO_3.

[b] Many investigators prefer to add vitamins B_1 and B_{12} as filter-sterilized stock solutions after the rest of the medium has been autoclaved. This prevents any degradation of the vitamins, especially in neutral and alkaline mediums. However, in our laboratory, we found that these vitamins are present in excess, and autoclaving the vitamins in the media does not result in any appreciable change in the growth characteristic of the organism.

[c] The original medium did not contain vitamin B_{12}; however, most investigators now using this medium add vitamin B_{12} to give a final concentration of ca. 0.005 mg/liter.

[d] For the composition of stock solutions used to prepare this medium, the reader is referred to the original reference by Price and Vallee (1962).

[e] One may use 3.5 ml of 95% ethanol per liter of medium as a carbon source in place of the sucrose.

[f] The final pH is adjusted to 3.5 by the addition of HCl.

[g] The pH is adjusted to 6.8 by the addition of TRIS.

[h] A concentrated stock solution is prepared from all the compounds except the trace metals and adjusted to pH 7.4. This solution is sterilized by filtration. The trace-metal solution is dissolved in an appropriate amount of distilled water and then autoclaved. These two sterile solutions are then combined to give the final medium.

[i] The pH is adjusted to 8.0 by the addition of KOH.

[j] See Table 11–3 for the composition of trace metal mixes.

[k] Trace metal mixtures are dry powders or liquid stock solutions (Table 11–3). The volumes or weights shown refer to the amount of either the dry powder or stock solution needed to give the desired trace metal concentration. This medium may be prepared in quadruple strength.

Table 11–2. *Phototrophic media for Euglena as used by various investigators (values in grams/liter).*

Compound	Acidic			Neutral	
	Lyman and Siegelman (1967)[c]	Hutner et al. (1966)	Cramer and Myers[d] (1952)	Edmunds (1965)[d]	Eisenstadt and Brawerman (1967)
KH_2PO_4	0.3	0.150	1.0	1.0	0.3
$MgSO_4 \cdot 7H_2O$	0.5	0.283	0.409[e]	0.409[e]	0.4
$MgCO_3$	—	0.300	—	—	—
$CaCO_3$	0.06	0.020	—	—	0.08
$CaCl_2 \cdot 2H_2O$	—	—	0.027	0.026	—
$(NH_4)_2SO_4$	1.0	—	—	—	1.0
NH_4HCO_3	—	0.500	—	—	—
$(NH_4)_2HPO_4$	—	—	1.0	1.0	—
EDTA (disodium salt)	0.5	—	—	—	—
HEDTA[a]	—	0.200	—	—	—
$H_3C_6H_5O_7 \cdot H_2O$	—	4.00	—	—	—
$K_3C_6H_5O_7 \cdot H_2O$	—	0.400	—	—	—
$Na_3C_6H_5O_7 \cdot 2H_2O$	—	—	0.52	0.516	—
L-Histidine $HCl \cdot H_2O$	—	1.0	—	—	—
Thiamine·HCl (vitamin B_1)[b]	6×10^{-4}	0.001	1×10^{-5}	1×10^{-4}	6×10^{-4}
Cyanocobalamin (vitamin B_{12})[b]	5×10^{-6}	2×10^{-5}	5×10^{-7}	5×10^{-7}	4×10^{-6}

Trace metals[f]					
Type	F	D	E	G	H
Amount	0.065 g	0.18 g	2 ml	1.67 ml	1 ml
pH (final)	3.5[c]	3.2–3.5	6.8	6.8	—

[a] Hydroxyethyl ethylenediaminetriacetic acid.

[b] Many investigators prefer to add vitamins B_1 and B_{12} as filter sterilized stock solutions after the rest of the medium has been autoclaved.

[c] Adjust to pH 3.5 with H_3PO_4 or H_2SO_4.

[d] Some investigators have devised organotrophic media by adding organic substrates to Cramer and Myers' (1952) medium, or to Edmund's modification of Cramer and Myers' medium. A few of the many such media in the literature are the following:

Substrate	Concentration (*M*)	Reference
Ethanol	0.2	Buetow and Padilla (1963)
Sodium acetate	0.061	Buetow (1962)
	0.02	Buetow and Padilla (1963)
Glucose	0.025	Blum and Wittels (1968)

Other substrates reported to support growth of *Euglena* include succinate, fumarate, malate, and ketoglutarate (Danforth and Wilson 1961). However, many of these substrates are utilized only at particular pHs. For example, malic acid, aspartic acid, and glutamic acid are utilized only below pH 6.

[e] To prevent a precipitate from forming, the $MgSO_4 \cdot 7H_2O$ should be added as a solution after all of the other components of the medium have been added and dissolved. For 1 liter of medium, 1.64 ml of a stock solution (25 g/liter) of $MgSO_4 \cdot 7H_2O$ may be added.

[f] See Table 11-3 for trace metal mixes.

Table 11–3. *Trace metal mixes*

Liquid mixtures[a] (g/liter)		Dry powder mixtures[a] (g)	
Trace metal mix A–metals No. 49 (*Lyman and Siegelman, 1967*)		*Trace metal mix B–metals No. 45* (*Hutner et al. 1956*)	
$Fe(NH_4)_2(SO_4)_2 \cdot 6H_2O$	14.0	$FeSO_4(NH_2)_2SO_4 \cdot 6H_2O$	14.4
$MnSO_4 \cdot H_2O$	62.0	$ZnSO_4 \cdot 7H_2O$	4.4
$ZnSO_4 \cdot 7H_2O$	44.0	$MnSO_4 \cdot H_2O$	1.55
$CuSO_4 \cdot 5H_2O$	8.0	$CuSO_4 \cdot 5H_2O$	0.31
$(NH_4)_6Mo_7O_{24} \cdot 4H_2O$	1.41	$CoSO_4 \cdot 7H_2O$	0.48
$CoSO_4 \cdot 7H_2O$	4.8	H_3BO_3	0.57
$Na_3VO_4 \cdot 16H_2O$	0.92	$(NH_4)_6Mo_7O_{24} \cdot 4H_2O$	0.64
Sulfosalicylic acid	8.0	$Na_3VO_4 \cdot 16H_2O$	0.093
H_3BO_3	1.05		
Trace metal mix C–MEb (*Freyssinet et al. 1972*)		*Trace metal mix D–metals No. 60A* (*Hutner et al. 1966*)	
$MgSO_4 \cdot 7H_2O$	12.0	$Fe(NH_4)_2(SO_4)_2 \cdot 6H_2O$	42.0
$CaCl_2 \cdot 2H_2O$	0.9	$MnSO_4 \cdot H_2O$	15.5
$FeSO_4 \cdot 7H_2O$	0.4	$ZnSO_4 \cdot 7H_2O$	22.0
H_3BO_3	0.2	$(NH_4)_6Mo_7O_{24} \cdot 4H_2O$	3.6
EDTA (disodium salt)	0.5	$CuSO_4$ (anhydrous)	1.0
$ZnSO_4 \cdot 7H_2O$	0.3	$Na_3VO_4 \cdot 16H_2O$	3.7
$MnSO_4 \cdot H_2O$	0.1	$CoSO_4 \cdot 7H_2O$	0.48
H_2SO_4	Bring to 0.01 *N*	H_3BO_3	0.57
		Pentaerythritol	88.8

Trace metal mix E (*Cramer and Myers 1952*)	
$Fe_2(SO_4)_3 \cdot 6H_2O$	1.5
$MnCl_2 \cdot 4H_2O$	0.9
$Co(NO_3)_2 \cdot 6H_2O$	0.65
$ZnSO_4 \cdot 7H_2O$	0.2
H_2MoO_4	0.1
$CuSO_4 \cdot 5H_2O$	0.01

Trace metal mix F (*Schiff et al. 1971*)	
$Fe(NH_4)_2(SO_4)_2 \cdot 6H_2O$	28.0
$MnSO_4 \cdot H_2O$	24.8
$ZnSO_4 \cdot 7H_2O$	52.8
$CuSO_4 \cdot 5H_2O$	0.8
$(NH_4)_6Mo_7O_{24} \cdot 4H_2O$	0.36
$CoSO_4 \cdot 7H_2O$	4.8
$Na_3VO_4 \cdot 16H_2O$	0.37
H_3BO_3	1.14

Trace metal mix G (*Edmunds 1965*)	
$FeSO_4 \cdot 7H_2O$	1.98
$MnCl_2 \cdot 4H_2O$	1.08
$CoCl_2 \cdot 6H_2O$	0.62
$ZnSO_4 \cdot 7H_2O$	0.24
$Na_2MoO_4 \cdot 2H_2O$	0.19
$CuSO_4 \cdot 5H_2O$	0.02

Trace metal mix H (*Eisenstadt and Brawerman 1967*)	
$FeCl_3$	12.0
$ZnSO_4 \cdot 7H_2O$	4.0
$MnSO_4$	1.4
$CuSO_4 \cdot 5H_2O$	0.4
H_3BO_3	0.6

[a] For dry powder mixtures, these trace metal mixes are prepared by grinding the salts together and then weighing out the appropriate amount. Alternatively, a concentrated stock solution may be prepared from the powder. Addition of HCl will generally aid in dissolving high concentrations of these salts.

Table 11–4. *Solid media*

Plating method	Difco agar (g/100 ml)	Meer agar (g/100 ml)
Agar plates	2.0	1.5
Slants	3.0	2.0
Overlay agar	1.5	1.0

is used. A length of tubing is attached to the spigot and clamped shut near the spigot with a screw clamp. A short piece of glass tubing is inserted into the other end of the tubing and this end of the tubing is closed off with a pinch clamp. The desired medium plus the proper amount of agar is added to the bottle along with a magnetic stirring bar. After autoclaving, the medium is allowed to cool, but not to solidify. The bottle is placed on a magnetic stirrer raised several centimeters above the level of the plates. Agar is dispensed into the plates using the pinch clamp to control the flow rate and volume.

C. Incubation, illumination, and aeration

1. Liquid cultures. Either Erlenmeyer or Fernbach flasks are filled with medium to approximately half the volume (this prevents wetting of the plug during agitation of the flask). The flasks may be capped with either foam sponge plugs, gauze-covered cotton plugs, or inverted plastic beakers taped down onto the flasks. After autoclaving and cooling, the flasks are inoculated.

2. Incubation. Euglena grows optimally at 26°C, and this is the most commonly used incubation temperature. Flasks may be incubated in water-bath shakers, in rotary incubators, or on shaking platforms. Cultures may be stirred on a magnetic stirrer, but the culture should be protected from the heat generated by the stirrer by placement of an asbestos pad or other insulation beneath the flask.

3. Illumination conditions. Cook (1966) has reported that the growth of *E. gracilis* strain Z cultured under strict phototrophic conditions is maximal over the range of 6.4×10^3 to 1.9×10^4 erg/cm²/sec (ca. 400–1,200 ft-c; 4,300–13,000 lux) (1 ft-c = ~16.1 erg/cm²/sec). Most investigators culture *Euglena* under light conditions somewhere in the lower half of this range. Cook (1968) also reports that a light intensity of 4.8×10^4 erg/cm²/sec (3,000 ft-c; 32,000 lux) is slightly inhibitory. Cultures growing photoheterotrophically require less light. Under these conditions investigators have used a wide variety of light conditions: from 240 erg/cm²/sec [150 ft-c; 1,600 lux (Epstein and Allaway 1967)] to 1.6×10^4 erg/cm²/sec [1,000 ft-c; 11,000 lux (Gnanam and Kahn 1967)]. Stern et al. (1964) have determined that the optimal light intensity for chloroplast development, or "greening," is approximately 1,600 erg/cm²/sec (100 ft-c; 1,100 lux) for nondividing cells in mannitol media. Intensities of 6.4×10^3 to 1.1×10^4 erg/cm²/sec (400–700 ft-c; 4,300–7,500 lux) were found to inhibit development of photosynthetic capacity in *Euglena.*

To obtain dark-grown cells, cultures are incubated in a dark room or in a dark environmental chamber. Another method is to cover the flasks with black plastic electrician's tape; they should be carefully examined for light leaks before use. Moreover, some care should be used in selecting tape, because not all black tapes are opague. Foil caps should be placed over the usual foam plug to prevent light from striking the edge of the flask and being transmitted to the culture. All manipulations involving dark-grown cells should be performed under a dim, green safelight (Schiff 1972). Light leaks can be detected by placing a strip of film (Kodak Tri-X or Plus X Pan) in a dry darkened vessel for 24 hours and then developing it.

4. Aeration. For maximal growth, phototrophic cultures should be incubated in an atmosphere of 5% CO_2 in air, which may be bubbled through the culture medium. The gas should first be filtered and then moistened. The gas may be filtered by passing it through a disposable filter (e.g., Koby Junior) or a filter constructed of glass wool or cotton. The gas is then moistened by simply being allowed to bubble through an Erlenmeyer flask containing sterile water before it is bubbled through the culture.

D. Other culture methods

Because space does not allow for detailed descriptions of several specialized culture methods, only pertinent references are given here for: synchronous culture (Edmunds 1965); mass culture (Bach 1960; Cook 1968; Lyman and Siegelman 1967); semicontinuous culture (Cook 1968; Edmunds 1965); and continuous culture (Cook 1968).

III. Mutagenesis

Mutants of *Euglena* are easily obtained, and a great variety of mutagenic agents have been employed to produce such mutants. Most of the mutations affecting chloroplasts are assumed to be mutations of the chloroplast genome. Because sexual reproduction in this organism is unknown, genetic analysis of these mutants is impossible. However, nuclear mutations are considered unlikely, because X-ray and ultraviolet killing curves imply that *Euglena* is octaploid (Hill et al. 1966), thus making unlikely an expression of a mutated nuclear genome. Recently, a study of the reassociation kinetics of *Euglena* DNA

(Rawson 1975) indicated that *Euglena* may in fact be diploid. If this is true, many of the mutations now assumed to be chloroplastic may in reality be nuclear.

Mitochondrial mutations are also considered unlikely, because studies indicate that *Euglena* is an obligate aerobe (Schiff et al. 1971), and any mutations adversely affecting mitochondrial function would probably be lethal. In addition, each organism contains many mitochrondria (200–500), and the chances of segregating a mitochondrial mutant are therefore very low.

Chloroplast DNA appears to be very sensitive to many chemical and physical agents. Chloroplast mutations are thought to arise when such an agent has inactivated all plastid genomes but one, and has inflicted some lesion on this remaining genome [UV target analysis indicates that there are about 30 plastid genomes per cell (Schmidt and Lyman 1976)]. This remaining genome then multiplies until the cell has regained its normal complement of plastids.

Table 11–5 describes several methods of mutagenizing *Euglena* cells. Many other agents also mutagenize *Euglena;* several of these agents are discussed in a review by Mego (1968) and in Section IV below.

A. Selective procedures

1. Plating procedure. After treatment with a mutagen, the cells are generally plated for selection and isolation of mutant clones. Sometimes the investigator may prefer to grow the cells for several generations in a liquid medium in the absence of the mutagen to allow time for any mutations to be expressed, or he may choose to grow the cells in a selective liquid medium before plating them onto the solid medium. Plating efficiencies approach 100% when the overlay technique is employed. Overlay agar is prepared from the same medium used to prepare the plates with a reduced agar concentration as indicated in Table 11–4. The overlay agar may be dispensed in 4-ml aliquots into separate, sterile Wasserman tubes (12 × 100 mm). Prior to use, these tubes are heated for a short time in a boiling water bath to melt the agar completely. The tubes are then kept in a 45°C water bath to keep them from solidifying before use. Alternatively, a large volume of overlay agar may be prepared in a flask. If acidic medium is used to prepare the overlay agar, double-strength medium and double-strength agar are prepared and autoclaved in separate flasks and combined aseptically after they have cooled somewhat. The agar is kept molten in a 45°C water bath and aseptically dispensed onto the plates in 4-ml aliquots from a wide-bore 10-ml pipette.

The cell suspension to be plated should be diluted so that only 100–300 cells are delivered to each plate. However, if the plate contains a

Table 11–5. *Mutagenic production in Euglena cells*

Mutagen	Treatment	Viable cells after treatment (%)	Types of mutants obtained[a]
Ultraviolet light	1. Cell suspension, 6 ml, (2×10^6 cells/ml) pipetted into sterile 100×25-mm glass petri dish. 2. Irradiated with 100 erg/mm^2 from a low-pressure mercury arc lamp (15-W germicidal lamp). Cover is removed and cell suspension agitated during irradiation. 3. Irradiation and all subsequent operations carried out under dim green safelight,[c] or a red or yellow incandescent safelight to prevent photoreactivation. 4. Cells are plated.[d] 5. Plates incubated in the dark for 5–7 days and placed in light to allow chloroplast development.	100%	White cells (99%) Motility mutants[e] Yellow mutants[b] Photosynthetic mutants[b]
Nitrosoguanidine[f]	Nitrosoguanidine is extremely carcinogenic–handle with care! Prepare fresh. Filter-sterilize solutions of nitrosoguanidine.		
	Method 1 (Russell and Lyman 1968) 1. Cells suspended to final 10^6 cells/ml in 50 m*M* citrate buffer (pH 5) containing 50 μg/ml nitrosoguanidine. 2. Incubated for 20 min. 3. Cells washed and plated.	95%	White or pale green photosynthetically deficient
	Method 2 (McCalla 1966) 1. Cells suspended (10^5 cells/ml) in acidic organotrophic medium containing 2–8 μg/ml nitrosoguanidine. 2. Incubated for 8–24 h. 3. Cells washed and plated.	1% (8 μg/ml for 24 h)	White cells (80%–2.5 μg/ml for 24 h)
	Method 3 (Schiff et al. 1971) 1. Cells suspended in resting medium[g] containing 25 μg/ml nitrosoguanidine. 2. Incubated for 60 min. 3. Cells washed and plated.		

Table 11–5. (*cont.*)

Mutagen	Treatment	Viable cells after treatment (%)	Types of mutants obtained[a]
Nalidixic acid	1. Stock solution: 5 mg nalidixic acid in 10 ml of 1.0 *N* NaOH brought to 25 ml with distilled H_2O. Sterilized by filtration and stored at −15°C 2. Log phase cells ($1-2 \times 10^4$ cells/ml) are suspended in acidic organotrophic medium containing 50 μg/ml nalidixic acid. 3. Incubated in light for 24 h. 4. Cells washed and plated.	100%	White (99%), nalidixic acid-resistant, pale green, yellow; photosynthetically deficient; light-sensitive, cold-sensitive
Streptomycin[b]	1. Concentrated stock solution of streptomycin is filter-sterilized. 2. Cells suspended in neutral or alkaline organotrophic medium containing 0.05–0.1% streptomycin. 3. Incubated for several generations.[h] 4. Cells washed and plated.	100%	White (95%–2.4 mg/ml streptomycin for 3 generations in the light). Ben Shaul and Ophir (1970). Sm^R
X rays[b]	1. Cells ($1-2 \times 10^6$ cells/ml) are pipetted into polyethylene cap or glass dish covered with aluminum foil 2. Cells exposed to a dose of 30,000–50,000 r. 3. Cells plated.	1%	White, pale green

[a] Particular mutants isolated. Percentage refers to the proportion of cells after treatment which exhibit this phenotype.
[b] Schiff et al. (1971).
[c] For description of a dim green safelight see Schiff (1972).
[d] A description of plating is found in Sect. III.A.I.
[e] Lewin (1960).
[f] *N*-Methyl-*N*′-nitro-*N*-nitrosoguanidine.
[g] Resting medium Section V.B.2.
[h] The cells may be incubated in the light or in the dark; Mego (1968) reports that streptomycin is more effective in the dark.

selective agent and only a small proportion of the cells being plated are expected to grow, larger numbers of cells may be plated. The concentration of the cell suspension is most easily determined by the use of a Coulter Electronic Cell Counter (Parsons 1973) equipped with a 100 μ aperture. Cell concentration may also be determined by counting cells in a hemocytometer after they have been immobilized by addition of a drop of saturated mercuric chloride to the culture medium. Turbidity curves of a cell suspension (measured at 675 nm) also have been used to calculate the cell concentration; however, this method is only useful for cell concentrations of less than 5×10^5 cells/ml. If dilution of the cell suspension is found to be necessary, the most convenient method is to serially dilute the suspension in Wasserman tubes (12 × 100 mm) containing either sterile water or medium. Care must be taken to constantly mix the cells and diluent when performing serial dilutions, because *Euglena* cells rapidly sink to the bottom of the tubes.

To plate, an aliquot (0.1–0.5 ml) of diluted cell suspension is pipetted directly onto the surface of an agar plate. Approximately 4 ml of overlay agar is aseptically poured or pipetted onto the plate. The plate is covered and immediately swirled and shaken gently to spread the cells and overlay agar evenly over the surface of the plate. After the overlay agar has solidified, the plates are inverted and incubated under light. To prevent photoreactivation, plates containing cells that have been exposed to UV light are incubated in the dark for 4 or 5 days before being placed in the light. Colonies are generally visible in 1 week.

2. *Selecting photosynthetic mutants.* A large proportion of pale-green mutants have been found to be deficient in their ability to carry out photosynthesis. Therefore one may select for photosynthetic mutants by screening any pale-green colonies for oxygen evolution and CO_2 fixation. Colonies that have grown up on organotrophic plates may be replicated onto phototrophic media plates (Table 11–2).

Also many nalidixic acid resistant mutants (Nal^R) are also photosynthetic mutants. A functional photosynthetic electron transport system is necessary for nalidixic-acid-induced bleaching to occur (Lyman et al. 1975). Cells with impaired photosynthetic electron transport systems therefore remain green in the presence of nalidixic acid. Thus, one can plate mutagenized cells onto organotrophic agar plates containing 50 μg/ml nalidixic acid and select green colonies. These Nal^R mutants can then be tested for oxygen evolution and CO_2 fixation.

Shneyour and Avron (1975) have described another method of selecting photosynthetically deficient cells. Cells are grown phototrophically to a cell density of 5×10^4 cells/ml at which point DCMU [3-(3,4 dichlorophenyl)-1,1 dimethylurea] in methanol is added to give a concentration of 20 μM DCMU. The cells are then incubated for 7 days. The cells are washed in 50 mM sodium citrate buffer (pH 5.0) and resuspended in the same buffer to give a cell concentration of 10^6 cells/ml. The cells are then mutagenized with nitrosoguanidine according to method I in Table 11–5, washed with phototrophic medium, resuspended in 9 ml of phototrophic medium and incubated phototrophically for 5 days. To this cell suspension, 1.0 ml of a 250 mM sodium arsenate solution is added, and the cells are incubated for 2 more days under phototrophic conditions. Arsenate kills dividing cells 300 times more efficiently than it does nondividing cells. The cells are then washed, diluted, and plated onto organotrophic agar plates and incubated in the light. Green colonies are selected, grown in organotrophic liquid medium and checked for light-dependent oxygen evolution.

A method of detecting photosynthetic *Euglena* mutants by their increased level of fluorescence was devised by Bennoun and Levine (1967) for *Chlamydomonas.* This technique is based upon the observation that chlorophyll fluorescence increases when photosynthetic electron transport is blocked. After treatment with a mutagen, cells are plated and incubated until colonies are plainly visible. The fluorescence is activated by illumination of the colonies with a high-pressure mercury arc lamp in a Zeiss lamp housing. The light is passed through a water filter and then through a Corning glass filter No. 4305, which blocks all wavelengths greater than 640 nm. The intensity of the light at the surface of the plate should be ca. 6,000 erg/cm^2/sec (370 ft-c; 4,000 lux). The enhanced fluorescence of photosynthetically deficient colonies can be detected visually by photography with red-sensitive Polaroid film type 413. The camera lens is fitted with a Corning glass filter No. 2030 to allow only light with a wavelength greater than 640 nm to be transmitted to the film. A neutral density filter (1.2) is also placed over the camera lens so that the exposure time can be adjusted to approximately 5 sec. The colonies are illuminated for 1 min before the picture is taken.

Another selective method, devised by Levine (1960) for *Chlamydomonas,* can be used to detect photosynthetic mutants of *Euglena.* In this technique (carried out in a hood) *Euglena,* after mutagenic treatment, are plated onto organotrophic plates. Then 0.1 ml of a 1 N solution of acetic acid is pipetted into a small watch glass; this is followed by the addition of 0.3 ml of ^{14}C-labeled sodium bicarbonate (2.5 μCi/ml). The petri dish containing the colonies is immediately inverted over the watch glass and illuminated by two 150 W floodlamps for 5 min. A comparable set of colonies should be exposed to ^{14}C in the dark as a control. The colonies are then replicated onto a piece of filter paper (method in III.A.4. and Chap. 10.III.C.1.a). The paper replica must be marked so that colonies on the filter paper replica can be correlated subsequently with colonies growing on the petri dish. The filter paper is exposed to concentrated HCl fumes to drive off any unfixed CO_2. An autoradiograph is then made by exposing Kodak No Screen X-ray film to the filter paper for 1 week, after which the film is developed. Colonies able to fix CO_2 are indicated by a darkening of the film, whereas colonies that are photosynthetically deficient will show no darkening.

More recently, a method for selecting mutants of *Chlamydomonas* with deficient electron transport mechanisms has been described by Schmidt et al. (1977). This method should prove useful for selecting photosynthetic mutants of *Euglena* also. After mutagenesis, cells are grown for several days to allow expression of any mutations. Metronidazole (Flagyl) is added to give a final concentration of 6–12 m*M* to a dilute suspension of cells (incubated in the light to ensure that photosynthetic electron transport rates are high). Metronidazole kills cells that possess an active transport system. After 24 h of metronidazole treatment the cells are plated. Most colonies should be mutants.

3. Selecting drug-resistant mutants. One can select drug-resistant mutants simply by plating cells onto agar plates containing an appropriate concentration of the drug (overlay agar should also contain the drug). Gradient plates (Szybalski and Bryson 1952) can be used to determine the range of drug concentration to which a mutant is resistant. Gradient plates are prepared in the following way: Ten ml of molten agar is pipetted into a sterile petri dish, and the dish is tilted so that one end of the plate is barely covered by agar while the opposite end is covered by a deep layer of agar. The agar is allowed to solidify in this position. The plate is then placed in a level position and 10 ml of agar containing the highest concentration of the drug to be tested is pipetted on top of the other layer. The drug then diffuses downward into the bottom layer to produce a linear concentration gradient. Cells of the mutant are then streaked across the plate along the concentration gradient. One can estimate the concentration at the point where the mutant is no longer resistant to the drug by noting its position on the gradient.

4. Selecting conditional mutants. The two conditional mutants of *Euglena* that have been described (Lyman et al. 1976) are both conditional chloroplast mutations and make less chlorophyll than wild type under the restrictive conditions. However, it is possible that other types of conditional mutants, for instance, those affecting cell viability or an enzyme's function, will be found.

The technique of replica plating can be used to screen for conditional mutants. In this technique, replica plates are made from a master plate containing about 100 colonies. These replica plates are then incubated under various conditions to determine if any of the cells are conditional mutants. In preparing the master plate, one cannot use the overlay technique because the overlay agar interferes with the replica plating. Instead, a 0.1-ml aliquot containing about 100 cells is pipetted onto the surface of the agar plate and evenly spread with a bent glass rod, which is sterilized by dipping in alcohol and flaming. The plates should not have excess water on the agar surface to avoid colony spreading due to the motility of *Euglena*.

A cylindrical wooden block slightly smaller than the inner diameter of the plate is washed in alcohol, and one end is covered by a piece of sterile velvet (or filter paper). The velvet pad is held in place with a sterile rubber band. The velvet-covered block is then pressed firmly and evenly over the agar surface of the master plate. Subsequently, the pad is pressed firmly over the surface of fresh agar plates (again, these plates should contain no excess water). About 4–5 replica plates can be made from one master plate. All plates should be marked at a specific reference point so that colonies on the replica plates can be compared with the corresponding colonies on the master plate. Fifteen minutes after plating, the plates are inverted and incubated under appropriate conditions. The master plates are kept at 4°C for later comparison.

The method of patch plating can also be used to screen for conditional mutants. The bottoms of several petri dishes containing media are marked off into grids, and the spaces of the grid are numbered. The grid spaces should be large enough to accommodate fully grown colonies without any overlapping. Alternatively, a grid may be drawn on a circle of paper and the paper may then be placed beneath the petri dishes. In this case, some reference point must be marked on the plates so that the colonies can be identified at a later time. Presumptive mutants may be transferred to the plates either with a dropper as a small drop of cell suspension or with a sterile toothpick. Each mutant is transferred to a specific space in the grid; if several such plates are made, each plate can be incubated under different conditions. Approximately 12–16 mutants can be compared with the wild type on each plate.

The techniques of replica plating and patch plating should be useful in the isolation of auxotrophic mutants; however, there has been little success in isolating such mutants from *Euglena*.

5. Maintenance of mutant strains. Many mutants are unstable. This is probably due to unequal distribution of chloroplast genomes during cell division. Therefore, after isolation, mutant clones should be grown for many generations on a liquid medium and replated several times to ensure that a stable substrain is obtained.

Mutant strains are conveniently maintained on agar slants (Table 11–4); these should be transferred every fifteen days. If possible, it is best to maintain a mutant strain on a selective medium or under selective conditions to minimize the problem of reversion.

In addition, all mutant strains should be regularly cloned. This is important to ensure that the strain has not changed over time due to the accumulation of spontaneous mutations or revertants.

B. Nomenclature and availability of mutants

1. Nomenclature. A brief description of the nomenclature system used to describe *Euglena* mutants is found in this section, but the reader should refer to the review by Schiff et al. (1971) for details and for a list of the symbols used in this system.

The name of a mutant is based on an easily recognized phenotypic character (e.g., colony color) rather than on the underlying molecular mechanisms. The name consists of a four-letter code in which the first letter describes the phenotype of the mutant – for example, W for white mutants, P for pale green mutants, and Sm^R for streptomycin-resistant mutants. This is followed by a numerical subscript, which indicates the order of isolation of this particular type of mutant (e.g., W_5 would be the fifth white mutant to be isolated). At the present time, it is often difficult for an investigator to discover which number should be used to describe a new mutant. Generally, an investigator uses a number corresponding to the order of isolation in his or her own laboratory, provided that this does not conflict with an already published description of a mutant. The second letter of the code describes the parent strain of the mutant (e.g., *B* for Bacillaris; *Z* for the Z strain). The third symbol indicates the mutagen used to produce the mutant (e.g., *Nal* for nalidixic acid, *Sm* for streptomycin). The fourth letter tells whether the parent culture from which the mutant was isolated was light-grown or dark-grown during mutagenesis.

Thus, Y_9ZNal L refers to the ninth yellow mutant derived from the Z strain after treatment of light-grown cells with nalidixic acid.

2. Availability of mutants. Unfortunately, at the present time there is no central repository from which mutant strains of *Euglena* can be obtained. Several of the earlier mutant strains have been deposited in the University of Texas Culture Collection. Usually, specific mutant strains can be obtained by writing to the investigator who isolated the strain. In addition, the *Chlamydomonas-Euglena* Information Exchange Group attempts to aid investigators interested in obtaining mutant strains of *Euglena.**

* The C.E.I.E.G. may be contacted by writing to either Dr. Harvard Lyman, Biology Department, State University of New York, Stony Brook, N.Y. 11794; or Dr. Nicholas Gillham, Zoology Department, Duke University, Durham, N. C. 27706.

IV. Inhibition of chloroplast replication ("bleaching")

Table 11–6 summarizes methods for the specific inhibition of chloroplast replication. The methods cited (with the exception of nalidixic acid) also inhibit proplastid replication. Nalidixic acid requires a functioning photosynthetic electron transport system for maximum effect; therefore, proplastids are not susceptible (Lyman et al. 1975).

Some inhibitors of either chloroplast or cytoplasmic protein synthesis have been reported ultimately to cause inhibition of chloroplast replication too, but these reports are somewhat variable. This topic is covered extensively in a review by Schmidt and Lyman (1976).

Irreversible plastid loss can also occur following transfer from complex media to defined media under certain conditions (Cook et al. 1974).

V. Chloroplast differentiation ("greening")

The biosynthesis of the chloroplast from its precursor, the proplastid, is defined in this chapter as *chloroplast differentiation* or "greening."

A. Cell conditions prior to assessing differentiation

When one is measuring chloroplast synthesis it is essential to start with fully dark-grown cells. These are cells which have been grown for not less than 12 cell generations in complete darkness. This is to ensure that one is using cells that have only proplastids without any remaining chloroplast structure and low levels of plastid RNAs, lipids, and so forth. Studies of chloroplast synthesis using dark-adapted cells (i.e., grown for 3–4 generations in the dark) do not accurately measure either the synthesis or magnitude of the light-induced differentiation of the chloroplast.

For most purposes, it is convenient to measure chloroplast synthesis in the absence of cell division. Dividing cells are used by investigators interested in the partitioning of resources by the cell as it divides and synthesizes its chloroplasts, but this approach can lead to ambiguities in interpretation of results, if one is interested in the regulatory mechanisms of chloroplast synthesis. The following approach provides uniform, nondividing cells that will complete chloroplast synthesis in 72–96 h.

B. Methods

All manipulations of dark-grown cells must be done under a safelight that will not induce chloroplast synthesis. The light described by Schiff (1972) is best suited for this purpose.

1. Dark-grown cells, near or at the end of the logarithmic phase of growth, are used.

Table 11–6. *Specific inhibition of chloroplast replication (bleaching)*

Agent[a,b]	Method[c] (treatment)
Ultraviolet light	A 6-ml cell suspension (2×10^6 cells/ml) is placed into a sterile 100 × 25-mm petri dish. Exposure to 100 erg/mm^2 from low-pressure 15 W germicidal lamp. Protect from photoreactivation (see Table 11-5).
Growth at elevated temperature	Cells are grown at 34°C. After 10 generations ca. 100% of the cells give rise to white colonies when plated. Dark-grown cells bleach at a slightly faster rate: 8 generations for 100% bleaching. Treatment effective only if cells divide at 34°C. At temperatures higher than 34°C, cell division is inhibited. At 32°C, cells must be in the light if bleaching is to occur (Uzzo and Lyman 1969).
Nalidixic acid	Cells are grown in the light for at least 2 generations in 50 μg/ml nalidixic acid (Table 11-5). Cells unable to photosynthesize are highly resistant (Lyman et al. 1975).
Streptomycin	Cells are grown in a medium of neutral or alkaline pH (see Table 11-1) for several generations. Streptomycin concentrations of 0.05 to 0.1% are used.

[a] Specific inhibition of chloroplast replication is easily assayed by plating. A cell that when plated gives rise to a colorless colony when plated is unable to replicate its plastids. Plating also determines whether the agent kills any of the cells. Alternatively, chloroplast counts can be made directly by means of a light microscope whose light source is equipped with blue-colored glass filter (Oriel Optical Co. No. 40). Chloroplasts strongly absorb wavelengths between 430 and 460 nm, and so, can be easily distinguished from other particles in the cell. Paramylum granules often refract green light when viewed under white light and can be mistaken for chloroplasts by the unwary (Gross et al. 1956). Fluorescence microscopy has also been used to count chloroplasts within cells (Epstein and Schiff 1961). If chloroplasts are lost from *Euglena* cells there is a concomittant loss of chloroplast DNA. This can be detected in $CsCl_2$ gradients of DNA extracted from whole cells (Lyman et al. 1975).

[b] The agents described in this table have been shown specifically to inhibit chloroplast or proplastid replication with no effect on cell viability under the specified conditions.

[c] Treatment will result in inhibition of chloroplast replication in approximately 100% of cells under the indicated conditions.

2. Cells are washed either in autotrophic medium (Table 11–1) or in resting medium (1% mannitol w/v, 0.01 *M* K_2HPO_4, 0.01 *M* KH_2PO_4, 0.01 *M* $MgCl_2$).

3. They are resuspended in either of the above media and incubated on a shaker in darkness for 24 h to allow any residual cell divisions to occur.

4. Then they are placed in the light (II. C.3) and agitated.

5. Cell number and chloroplast parameters (Tables 11–7A and B) are determined during the initial 72–96 h. A parallel culture is kept in darkness as a control. It is also advisable to have parallel cultures of a permanently aplastidic strain (Schiff et al. 1971; Schmidt and Lyman 1976) as additional controls. The use of such strains as controls allows the investigator to determine the existence of light-induced phenomena that occur in the absence of the plastid and its genome.

C. *Parameters of chloroplast synthesis*

The biosynthesis of the chloroplast involves the formation of many substances. An accurate assessment of chloroplast synthesis should involve the analysis of more than a single constituent. While the rate and amount of chlorophyll synthesis is a good indication of the rate of synthesis of the entire plastid, the kinetics of appearance of some constituents is quite different from that of chlorophyll. This is especially true of mutant strains, particularly pale green or yellow strains, which may synthesize a small amount of chlorophyll but normal (wild-type) amounts of other compounds (i.e., enzymes of CO_2 reduction) (Schmidt and Lyman 1976). Methods for the determination of lamellar components (pigments, pigment complexes, lipids) and electron-transfer components are summarized in Table 11–7A. The soluble chloroplast enzyme methods are summarized in Table 11–7B. Of course many other chloroplast constituents show light-induced increases. Most of these are listed in a recent review (see Table 13.5 in Schmidt and Lyman 1976).

Tables 11–7A and B do not list every chloroplast component that increases during chloroplast differentiation. Many of these are listed in Schmidt and Lyman (1976) along with information on their inhibition. The plastid constituents in this reference include plastid-associated DNA, chloroplast ribosomes, various species of RNA, aminoacyl-tRNA synthetases, aminolaevulinate dehydratase, ferredoxin NADP reductase, glycolate DCPIP oxidoreductase, and phosphoglycolate phosphatase.

A procedure is given by Nicolas and Nigon (1974) for the assay of chloroplast-associated 3′, 5′ cyclic AMP receptor protein. Egan et al. (1975) have described an assay for a light-induced DNase. Ernst-Fonberg et al. (1974) have characterized the light-induced synthesis of an acyl carrier protein-dependent fatty acid synthetase. Light-induced and chloroplast associated mRNAs have been described by Verdier (1975) and Sagher et al. (1976), respectively. Chloroplast DNA polymerase can be assayed by the methods of McLennan and Keir (1975).

Table 11–7A. *Methods for measuring chloroplast synthesis and inhibition: lamellar and electron-transfer components and lipids*

Lamellar components	Method of determination	Inhibitors[a]
Chlorophyll (total chlorophylls *a* and *b*)	80% Acetone extracts of whole cells. From absorption at 645, 652, 663 nm, concentrations are calculated from equations of Arnon,[c] MacKinney,[c] or Jeffrey and Humphrey (1975).	Act.D: 10–45 μg/ml CAP: 0.06–2.0 mg/ml CHE: 0.003–20 μg/ml 5-FU: 25–500 m*M* Mit.: 30 μg/ml Myxin: 8 μg/ml Rif: 250 μg/ml SM: 0.15–2.0 mg/ml
Carotenoids (total)	80% Acetone extracts of whole cells. From absorption at 480 nm, 645 nm, and 663 nm concentrations are calculated (Liaaen-Jensen and Jensen 1971). Further fractionation possible according to Gross et al (1975).	Act.D: 0.13–10 μg/ml CAP: 1–2.0 mg/ml SM: 0.5 mg/ml SAN 9789 (Vaisberg and Schiff (1976): 20–100 μg/ml
Protochlorophyll (-ide)	Dark-grown cells are extracted at 0–4°C with 90% acetone containing a small amount of $MgCO_3$. Mixed with $\frac{1}{2}$–$\frac{3}{4}$ volume of cold diethyl ether, then with cold distilled water for separation. Dried in vacuo and resuspend in diethyl ether. Extinctions of pchl(-ide) in diethyl ether are 662 nm (0.20), 638 nm (4.27), and 623 nm (39.9) (Cohen and Schiff 1976).	
Chlorophyll proteins (CP I and CP II)	Isolated chloroplasts (see below) are washed in 50 m*M* pH 8.5 borate buffer and lysed by the addition of SDBS[d] to final 50 parts SDBS per part chlorophyll. Lysate is centrifuged at 100,000*g*, 30 min. Chlorophyll proteins are separated by electrophoresis from supernatant on 7% polyacrylamide gels with 0.1% SDBS in 50 m*M* pH 8.5 borate buffer at 6 mA/gel,	(Bishop et al. 1973 have measured the effect of CAP and CHE on the activity of photosystems I and II).

Table 11–7A. (*cont.*)

Lamellar components	Method of determination	Inhibitors[a]
	ca. 15 min (Genge et al. 1974). For rapid estimate of chlorophyll protein complexes, whole cells are broken in French press (3,000 psi); whole cells are removed by low-speed centrifugation and a crude chloroplast fraction is sedimented at 40,000*g* for 30 min. Plastid membrane polypeptides can be analyzed by the methods of Bingham and Schiff (1976), Genge et al. (1974), or Vasconceles et al. (1976).	
Photosynthetic electron transfer components:		
Ferredoxin	Cells washed with 0.9% NaCl in 0.01 *M* TRIS–HCl (pH 7.5–8.0) are resuspended in the same medium, frozen and thawed three times, then vigorously shaken for several min. They are centrifuged at 1000*g*, 10 min, and supernatant is placed on a 1 × 10-cm DEAE–cellulose column, washed in cold with 40 ml 0.13 *M* NaCl in the above TRIS buffer and 40 ml 0.15 *M* NaCl in the same buffer. Ferredoxin and cytochrome 552, (see below) remaining as red band on top, are eluted with 0.8 *M* NaCl in TRIS buffer; the eluate is dialyzed overnight against 0.005 *M* EDTA in 0.001 *M* TRIS, pH 8. Activity may be assayed by the methods described by Buchanan and Arnon (1971).	
Cytochrome 552 (*Euglena* cytochrome f)	Extracted simultaneously with ferredoxin (see above). DEAE–cellulose column is washed with 0.1 *M* phosphate buffer, pH 7.4, until pink cytochrome band reaches bottom; final elution is in buffer with 0.1 *M* NaCl. Cytochrome 552 may also be prepared from acetone powders (Mitsui 1971). It is reduced by ascorbic acid, sodium dithionate, and sodium borohydride. The extinction coefficients of the reduced cytochrome are 552 nm (29.7); 523 nm (18.6); and 416 nm (157.4).	CAP: 1.0 mg/ml CHE: 1.5 μg/ml SM: 0.5 mg/ml

Cytochrome b (cytochrome 561, *Euglena* cytochrome b_6)	Detected in extracts of acetone powders by its reduced absorption maxima at 561 nm, 530 nm, and 432 nm (Perini et al. 1964).	CAP: 1.0 mg/ml CHE: 15 μg/ml
Plastoquinone A	Cells in 50 m*M* phosphate buffer, pH 7.2, are mixed in a solution of acetone, methanol, and isopropyl alcohol (1 : 1 : 1 v/v) and sonicated. In dim light, the broken cells are mixed with petroleum ether (30–60°C) and heptane (1 : 1 v/v) in a separatory funnel. The green lipid phase is evaporated to dryness in vacuo and redissolved in heptane. The heptane extract may be chromatographed on silica gel GHR plates with chloroform–heptane (80:20). Quinones are located by leucomethylene blue and may be eluted with ethanol for spectrophotometric assay (Schwelitz et al. 1972).	
Chloroplast lipids	Harvested cells, or chloroplasts, are suspended in 7 volumes of methanol and 14 volumes chloroform are added. These are mixed 30 min and centrifuged to remove cellular residue. The residue is extracted twice in 2 volumes methanol and 4 volumes chloroform. The combined extracts are dialyzed overnight in the cold. The chloroform layer is removed then dried with anhydrous sodium sulfate. The chloroplast-associated monogalactosyl diacylglycerol and digalactosyl diacylglycerol may be estimated by thin-layer chromatography on silica gel H plates using chloroform : methanol : acetic acid : ether (35 : 15 : 5 : 45) as solvent and detecting the spots with diphenylamine–aniline reagent. Other lipids can be analyzed using the procedures outlined by Bishop et al. (1973).	CAP: 0.25–1.0 mg/ml CHE: 2.5–10 μg/ml

Note: Footnotes *a* to *d* appear at end of Table 11–7B.

Table 11–7B. *Methods for measuring chloroplast synthesis and inhibition: enzymes*

Soluble enzyme	Method of determination	Inhibitors[a]
NADP-dependent glyceraldehyde 3-phosphate dehydrogenase, EC 1.2.1.13 (D-glyceraldehdye-3-phosphate; NADP oxidoreductase phosphorylating)	The enzymatic conversion of 1,3-diphosphoglyceric acid to 3-phosphoglyceraldehyde in the presence of NADPH is measured according to the method described by Latzko and Gibbs (1969). A 1.0-ml reaction mixture is made up in replicate 1-ml cuvettes as follows: TRIS–HCl, pH 8.4, 100 μmol; $MgCl_2$, 10 μmol; ATP, 5 μmol; NADPH, 0.2 μmol; 3-phosphoglyceric acid kinase (ATP: 3-phospho-D-glycerate-1-phosphotransferase, EC 2.7.2.3), 0.2 units; glutathione, 5 μmol; 3-phosphoglycerate, 4 μmol; cell-free extract, 0.02 ml. Glutathione solubilized in 0.05 *M* TRIS–HCl buffer, pH 7.8, is added to the reaction mixture immediately before extracts are removed from the centrifuge. The extracts are incubated in the reaction mixture for 4 min at room temperature before the reaction is initiated by the addition of 3-phosphoglycerate. The cuvettes are placed immediately in a recording spectrophotometer. The rate of NADPH oxidation is measured at 340 nm for 10 min. Specific acitvity is calculated after protein determinations as moles NADPH oxidized/h/mg protein.	Act. D: 134 μg/ml CAP: 1–2 mg/ml CHE: 15 μg/ml[b] Rif: 250 μg/ml SM: 0.5 mg/ml
3-Phosphoglyceric acid kinase, EC 2.7.2.3 (ATP; 3-phospho-D-glycerate-1-phosphotransferase)	The enzymatic conversion of 3-phosphoglycerate to 1,3-diphosphoglycerate is measured according to the method of Latzko and Gibbs (1969). A 1.0-ml reaction mixture is made in replicate 1-ml cuvettes as follows: TRIS–HCl, pH 7.4, 100 μmol; $MgCl_2$, 10 μmol; glutathione, 5 μmol; ATP, 5 μmol; NADH, 0.2 mol; glyceraldehyde-3-phosphate dehydrogenase (D-glyceraldehyde-3-phosphate; NAD oxidoreductase phosphorylating, EC 1.2.1.12), 0.2 units; 3-phosphoglycerate, 4 μmol; cell-free extract, 0.02 ml. Glutathione dissolved in 0.05 *M* TRIS–HCl buffer, pH 7.8, is added to the reaction mixture immediately before extracts are removed from the centrifuge. The extracts are incubated in the reaction mixture for 4 min at room tem-	

perature before the reaction is initiated by the addition of 3-phosphoglycerate. The reaction is assayed in the same way as glyceraldehyde-3-phosphate dehydrogenase.

Enzyme	Assay	Concentrations
Class 1 fructose 1,6-diphosphate aldolase (EC 4.1.2.7) (ketose-1-phosphate aldehydelyase)	The enzymatic cleavage of fructose 1,6-diphosphate to yield dihydroxyacetone phosphate and glyceraldehyde 3-phosphate in the presence of EDTA is measured by the method of Wu and Racker (1959). The EDTA severely inhibits the non-chloroplast, metal-requiring "class II" fructose 1,6-diphosphate aldolase also present in the cell-free extracts. 1.0-ml reaction mixtures are made up in duplicate 1-ml cuvettes: TRIS-HCl, pH 7.6, 40 μmol; NADH, 0.2 mol; disodium EDTA, 1 μmol; α-glycerophosphate dehydrogenase-triosephosphate isomerase [L-glycerol-3-phosphate, NAD oxidoreductase (EC 1.1.1.8), D-glyceraldehyde-3-phosphate ketol-isomerase (EC 5.3.1.1)], 10 μg; cell-free extract, 0.05 ml; fructose 1,6-diphosphate, 2 μmol. The extracts are incubated in the reaction mixture for 10 min at room temperature before the reaction is initiated by the addition of fructose 1,6-diphosphate. The cuvettes are covered with parafilm and inverted three times upon addition of substrate. The cuvettes are immediately placed in a recording spectrophotometer and the rate of NADH oxidation is measured at 340 nm for 20 min. Specific activity is calculated as moles fructose 1,6-diphosphate cleaved/h/mg protein. (The protein content of the cell-free extracts must be determined). The stoichiometry of the coupled reaction is 1 mole fructose 1,6-diphosphate cleaved/2 moles NADH oxidized.	CAP: 0.5–1.5 mg/ml CHE: 3.0 μg/ml
Ribulose 1,5-diphosphate carboxylase. EC 4.1.1.39, (3-phospho-D-glycerate carboxy-lyase dimerizing)	Ribulose 1,5-diphosphate carboxylase is measured by a modification of the method of Weissbach et al. (1956). This enzyme catalyzes the formation of 2 moles of 3-phosphoglycerate from one mole of ribulose 1,5-diphosphate and 1 mole of carbon dioxide. The following 1.0-ml reaction mixture is made up in replicate 5-ml shell vials: TRIS–HCl, pH 7.8, 200 μmol; $MgCl_2$, 10 μmol; EDTA, 0.06 mol; glutathione, 6 μmol; $NaH^{14}CO_3$, 20 μmol; ribulose 1,5-diphosphate, 0.4 μmol; cell-free extract, 0.2	CAP: 1.0–2.0 mg/ml CHE: 4–15 μg/ml SM: 0.5 mg/ml

Table 11–7B. (*cont.*)

Soluble enzyme	Method of determination	Inhibitors[a]
	ml. Glutathione solubilized in 0.05 *M* TRIS–HCl buffer, pH 7.8, is added to the reaction mixture immediately before extracts are removed from the centrifuge. The reaction is initiated by the addition of ribulose 1,5-diphosphate. At 2-min intervals, 0.1-ml aliquots of the reaction mixture are dispensed into 5-ml shell vials containing 0.5 ml of 6 *M* HCl to arrest the reaction and drive off most of the unfixed bicarbonate. The reaction is carried out for 15–20 min. To each vial, 1 ml of acetone is added as a carrier for the removal of unfixed ^{14}C-bicarbonate. The vials are dried overnight at 80°C; then, 5 ml of Bray's scintillation fluid is added to each vial for counting. Specific activity is calculated after protein determinations as moles CO_2 fixed/h/mg protein.	

Note: For all enzyme measurements, 1×10^8 cells are harvested by centrifugation at 3020*g* for 5 min at 4°C. The cells are washed with 20 ml pH 7.8 0.05 *M* TRIS–HCl buffer and pelleted. Cells are then resuspended in 12 ml of 0.05 *M* pH 7.8 TRIS–HCl buffer containing 0.005 *M* β-mercaptoethanol and transferred to a 15-ml Branson sonication rosette in an ice water bath. All cells and 90% of the chloroplasts are disrupted by sonication for 2 min. We use a MSE 150-W ultrasonic disintegrator operated in the low-power mode with a 8.4 micron amplitude at setting 4. The disrupted cell solution is centrifuged for 15 min at 27,550*g*, 4°C. The uppermost portion of the cell-free extracts is decanted into chilled glass tubes and used immediately for enzyme assays.

[a] Abbreviations: Act.D., actinomycin D; CAP, chloramphenicol; CHE, cycloheximid; 5FU, 5-fluorouracil; Mit, mitomycin; Rif, rifampicin; SM, streptomycin; SAN 9789, 4-chloro-5-(methylamino)-2-(α,α,α,-trifluro-*m*-tolyl)-3(2H) pyridazinone.

[b] In the reference by Schmidt and Lyman (1976) there is an error in Table 13.6 which omits the word *cycloheximide* in the description of CHE inhibition of this enzyme.

[c] The MacKinney equations are Chl a (mg/liter) = 12.7(OD 663) − 2.69(OD 645); Chl b (mg/liter) = 22.9(OD 645) − 4.68(OD 663); Chl a + b (mg/liter) = 8.02(OD 663) + 20.2(OD 645). The Arnon equation for total chlorophyll as modified by Bruinsma (1961) is (OD 652)/36 = mg Chl a + b/ml acetone extract.

[d] SDBS: Sodium dodecyl benzene sulfonate.

D. Inhibition of chloroplast differentiation

Specific inhibition of chloroplast synthesis is best accomplished using nondividing dark-grown cells as described in Section V.B. The agent to be tested is added, and the cells are illuminated and agitated. Because the cells are not dividing, the only events occurring are those related to chloroplast synthesis. Synthesis is monitored for 72–96 h, by which time control cultures are fully green.

It is very important to distinguish chloroplast synthesis in nondividing cells from that in dividing cells, especially with respect to the use of inhibitors. For example, if nondividing, dark-grown cells are illuminated at 34°C, normal chloroplast synthesis occurs. However, if dividing dark-grown cells are illuminated at 34°C, the synthesis of some chloroplast components (i.e., chlorophyll, carotenoids) stops before the plastids stop dividing. (Growth at 34°C is usually assumed to result in specific inhibition of plastid replications as is shown in Table 11–6). This indicates that a condition or agent assumed to affect plastid replication only can also affect plastid synthesis under the proper conditions. The interpretation of results obtained from methods using inhibitors of either plastid replication or synthesis must take this into account. Analysis of much of the work with inhibitors indicates that the effects of many of these may be felt in chloroplast gene expression and that this in turn results in the observed effect on replication and/or synthesis. In this sense, growth at 34°C should properly be described as a restrictive condition that probably affects the expression of several chloroplast genes. It is possible that some of the agents that have been thought of as specific inhibitors of chloroplast replication may also be similar in their action.

Tables 11–7A and B also list inhibitors reported to affect the synthesis of the indicated chloroplast components. Three extensive reviews describe the rationales underlying these and other inhibitor studies (Ebringer 1972; Mego 1968; Schmidt and Lyman 1976). The two most common inhibitors used are chloramphenicol and cycloheximide. They are used because of their specific inhibition of protein synthesis on chloroplast (70 S) and cytoplasmic (80 S) ribosomes, respectively. In general, these inhibitors have proved useful in determining where a particular protein is synthesized, but one must bear in mind that inhibition of a regulatory protein coded in the chloroplast might affect the synthesis on a nonplastid ribosome. One must also consider the possibility of nonspecific effects of these inhibitors (McMahon 1975; Hoxmark and Nordby 1977). For example, in the light, *Euglena* can reduce chloramphenicol to aminochloramphenicol, which is without effect as an inhibitor of protein synthesis (Vaisberg et al. 1976).

In using inhibitors to block chloroplast differentiation, it is advisable to use parallel cultures in the dark and aplastidic mutants in light and darkness as controls. It is also essential that aliquots of cells be plated with inhibitors during the course of experiments to determine if any or all of the cells are becoming bleached (i.e., to determine whether plastid replication is being affected as well as synthesis). The effects of many inhibitors of chloroplast differentiation are reversible (that is, when the agent is washed out, normal synthesis occurs), but some apparently affect plastid replication too (Ebringer 1972; Schmidt and Lyman 1976). Proper interpretation of results requires that inhibitors of chloroplast synthesis be checked for their effect on chloroplast–proplastid replication by plating and for reversability by removal of the agent. In Tables 11–7A and B, a range of inhibitor concentrations is indicated. This represents the lowest and highest concentrations reported. For particular concentrations, the reader is referred to a recent review, which lists most of the intermediate concentrations used and also the relevant references (Schmidt and Lyman 1976).

Chloroplast differentiation is also inhibited by certain metabolizable carbon sources (App et al. 1963; Garlaschi et al. 1974; and references in Murray et al. 1970). Ethanol can inhibit chlorophyll synthesis by 30–50%. Glycolate (43.7 m*M*) produces about a 50%, and serine (31.4 m*M*) about a 60% inhibition of chlorophyll synthesis; glycine and acetate will also inhibit chlorophyll synthesis.

E. Isolation of chloroplasts

Most of the assays of chloroplast components can be performed with whole cells, but it is sometimes desirable to work with purified chloroplasts. The method of Eisenstadt and Brawerman (1967), modified by Graham and Smillie (1971), is simplest.

1. Cells are harvested and washed in the cold with 10 m*M* TRIS–HCl, pH 7.6, washed again in STM [sucrose 10% (w/v); 10 m*M* TRIS–HCl, pH 7.6; 4 m*M* $MgCl_2$; 1 m*M* 2-mercaptoethanol], and resuspended in 3–4 volumes STM.
2. The suspension is passed through a chilled French pressure cell (American Instruments) at 3,000–4,000 psi, and breakage is checked microscopically.
3. The broken cell suspension is diluted with an equal volume of STM and centrifuged for 10 min at 500*g*.
4. The pellet is resuspended in STM (100 ml per 10 ml of original packed cell volume), allowed to stand 10 min, and then filtered through four layers of gauze.
5. The filtrate is centrifuged for 10 min at 500*g* and the pellet resuspended in STM (20–30 ml per 10 ml of original packed volume).

6. The suspension is mixed thoroughly with 2 volumes 75% (w/v) sucrose and centrifuged at 23,000*g* for 10 min in a swinging bucket rotor. (An angle head rotor may be used but the surface layer of chloroplasts is found along one side of the tube and must be scraped off.)

7. The floating surface layer of plastids is removed, diluted with STM (50–100 ml), and centrifuged for 5 min at 3000*g*. The plastids may be repurified by this flotation technique.

Chloroplasts isolated in this manner appear intact but will usually be found to have lost most of their cytochrome 552 and ferredoxin. Many soluble enzymes of the Calvin cycle and other soluble components may also be lost. DNA, RNA, ribosomes, and membrane components are retained. Other methods for chloroplast isolation are described by Brown and Haselkorn (1972), Davis and Merrett (1973), Preston et al. (1972), Price (1978), and Salisbury et al. (1975).

VI. References

App, A. A., and Jagendorf, A. 1963. Repression of chloroplast development in *Euglena gracilis* by substrates. *J. Protozool.* 10, 340–3.

Bach, M. K. 1960. Mass culture of *Euglena gracilis. J. Protozool.* 7, 50–2.

Bennoun, P. and Levine, R. P. 1967. Detecting mutants that have impaired photosynthesis by their increased level of fluorescence. *Plant Physiol.* 42, 1284–7.

Bingham, S., and Schiff, J. A. 1976. Cellular origin of plastid membrane polypeptides in *Euglena.* In Bucher Th. et al. (eds.), *Genetics and Biogenesis of Chloroplasts and Mitochondria,* pp. 79–86. North Holland, Amsterdam.

Bishop, D. G., Bain, J. M., and Smillie, R. M. 1973. The effect of antibiotics on the ultrastructure and photochemical activity of a developing chloroplast. *J. Exp. Bot.* 24, 361–76.

Blum, J., and Wittels, B. 1968. Mannose as a metabolite and an inhibitor of metabolism in *Euglena. J. Biol. Chem.* 243, 200–10.

Brown, R., and Haselkorn, R. 1972. The isolation of *Euglena gracilis* chloroplasts uncontaminated by nuclear DNA. *Biochim. Biophys. Acta* 259, 1–4.

Bruinsma, J. 1961. Comment on spectrophotometric determination of chlorophyll. *Biochim. Biophys. Acta.* 52, 576–8.

Buchanan, B. B., and Arnon, D. I. 1971. Ferredoxins from photosynthetic bacteria, algae, and higher plants. In San Pietro, A. (ed.), *Methods in Enzymology,* vol. 23A, pp. 413–40. Academic Press, New York.

Buetow, D. E. 1962. Differential effects of temperature on the growth of *Euglena gracilis. Exp. Cell Res.* 27, 137–42.

Buetow, D. E., and Padilla, G. M. 1963. Growth of *Astasia longa* on ethanol. I. Effects of ethanol on generation time, population density and biochemical profile. *J. Protozool.* 10, 121–3.

Cohen, C. E., and Schiff, J. A. 1976. Events surrounding the early development of *Euglena* chloroplasts – XI Protochlorophyll(ide) and its photoconversion. *Photochem. Photobiol.* 24, 555–66.

Cook, J. R. 1966. The synthesis of cytoplasmic DNA in synchronized *Euglena*. *J. Cell Biol.* 29, 369–73.

Cook, J. R. 1968. The cultivation and growth of *Euglena*. In Buetow, D. E. (ed.), *The Biology of Euglena*. vol. 1, pp. 243–314. Academic Press, New York.

Cook, J., Harris, P., and Nachtwey, D. 1974. Irreversible plastid loss in *Euglena gracilis* under physiological conditions. *Plant Physiol.* 53, 284–90.

Cramer, M., and Myers, J. 1952. Growth and photosynthetic characteristics of *Euglena gracilis*. *Arch. Mikrobiol.* 17, 384–402.

Danforth, W. F., and Wilson, B. W. 1961. The endogenous metabolism of *Euglena gracilis*. *J. Gen. Microbiol.* 24, 95–105.

Davis, B., and Merrett, M. J. 1973. Malate dehydrogenase isoenzymes in division-synchronized cultures of *Euglena*. *Plant Physiol.* 51, 1127–32.

Ebringer, L. 1972. Are plastids derived from prokaryotic microorganisms? Action of antibiotics on chloroplasts of *Euglena gracilis*. *J. Gen. Microbiol.* 71, 35–52.

Edmunds, L. N. 1965. Studies on synchronously dividing cultures of *Euglena gracilis* Klebs (strain Z). I. Attainment and characterization of rhythmic cell division. *J. Cell Comp. Physiol.* 66, 147–58.

Egan, J. M., Dorsky, D., and Schiff, J. A. 1975. Events surrounding the early development of *Euglena* chloroplasts. VI. Action spectra for the formation of chlorophyll, lag elimination in chlorophyll synthesis, and appearance of TPN-dependent triose phosphate dehydrogenase and alkaline DNase activities. *Plant Physiol.* 56, 318–23.

Eisenstadt, J. M., and Brawerman, G. 1967. Isolation of chloroplasts from *Euglena gracilis*. In Grossman, L., and K. Moldave (eds.), *Methods in Enzymology* vol. 12A, pp. 476–8. Academic Press, New York.

Epstein, H. T., and Allaway, E. 1967. Properties of selectively starved *Euglena*. *Biochim. Biophys. Acta* 142, 195–207.

Epstein, H. T., and Schiff, J. 1961. Studies of chloroplast development in *Euglena*. 4. Electron and fluorescence microscopy of the proplastid and its development into a mature chloroplast. *J. Protozool.* 8, 427–32.

Ernst-Fonberg, M., Dubinskas, F., and Jonak, Z. 1974. Comparison of two fatty acid synthetases from *Euglena gracilis* var. *bacillaris*. *Arch. Biochem. Biophys. 165,* 646–55.

Freyssinet, G., Heizmann, P., Verdier, G., Trabuchet, G., and Nigon, V. 1972. Influence des conditions nutritionelles sur la réponse à l'éclairement chez les euglènes etiolées. *Physiol. Vég.* 10, 421–42.

Garlaschi, F., Garlaschi, A., Lombardi, A., and Forti, G. 1974. Effect of ethanol in the metabolism of *Euglena gracilis*. *Plant Sci. Lett.* 2, 29–39.

Genge, S., Pilger, D., and Hiller, R. G. 1974. The relationship between chlorophyll *b* and pigment–protein complex II. *Biochim. Biophys. Acta* 347, 22–30.

Gnanam, A., and Kahn, J. S. 1967. Biochemical studies on the induction of chloroplast development in *Euglena gracilis*. I. Nucleic acid metabolism during induction. *Biochim Biophys. Acta* 142, 475–85.

Graham, D., and Smillie, R. M. 1971. Chloroplasts (and lamellae): algal preparations. In San Pietro, A. (ed.), *Methods in Enzymology* vol. 23A, pp. 228–48. Academic Press, New York.

Gross, J., Wirtschafter, S., Bernstein, E., and James, T. 1956. Monochromatic microscopy of *Euglena. Trans. Am. Microscop. Soc.* 75, 480–3.

Gross, J. A., Stroz, R. J., and Britton, G. 1975. The carotenoid hydrocarbons of *Euglena gracilis* and derived mutants. *Plant Physiol.* 55, 175–7.

Hill, H. Z., Schiff, J. A., and Espstein, H. T. 1966. Studies of chloroplast development in *Euglena.* XIII. Variation of UV sensitivity with extent of chloroplast development. *Biophys. J.* 6, 125–33.

Hoxmark, R. C., and Nordby, O. 1977. A warning against using chloramphenicol in the light. *Plant Sci. Lett.* 8, 113–18.

Hutner, S. H., Bach, M. K., and Ross, G. I. M. 1956. A sugar-containing basal medium for vitamin B_{12}-assay with *Euglena;* Application to body fluids. *J. Protozool.* 3, 101–12.

Hutner, S. H., Zahalsky, A. C., Aaronson, S., Baker, H., and Frank, O. 1966. Culture media for *Euglena gracilis.* In Prescott, D. M. (ed.), *Methods in Cell Physiology* vol. 2, pp. 217–28. Academic Press, New York.

Jeffrey, S. W., and Humphrey, G. F. 1975. New spectrophotometric equations for determining chlorophylls a_1 b_1 c_1 and c_2 in higher plants, algae and natural phytoplankton. *Physiol. Pflanzen* (BPP) 167, 191–4.

Latzko, E., and Gibbs, M. 1969. Enzyme activities of the carbon reduction cycle in some photosynthetic organisms. *Plant Physiol.* 44, 295–300.

Levine, R. P. 1960. A screening technique for photosynthetic mutants in unicellular algae. *Nature* 188, 339–40.

Lewin, R. A. 1960. A device for obtaining mutants with impaired motility. *Can. J. Microbiol.* 6, 21–5.

Liaaen-Jensen, S., and Jensen, A. 1971. Quantitative determination of carotenoids in photosynthetic tissues. In San Pietro, A. (ed.), *Methods in Enzymology,* vol. 23A, pp. 586–602. Academic Press, New York.

Lyman, H., and Siegelman, H. W. 1967. Large-scale autotrophic culture of *Euglena gracilis. J. Protozool.* 14, 297–9.

Lyman, H., Jupp, A. S., and Larrinua, I. 1975. Action of nalidixic acid on chloroplast replication in *Euglena gracilis. Plant Physiol.* 55, 390–2.

Lyman, H., Alberte, R., and Thornber, J. P. 1976. Photosynthetic mutants of *Euglena:* chloroplast lamellar characteristics. *Plant Physiol.* 57, 73 (Suppl.).

McCalla, D. R. 1966. Action of some analogs of nitrosoguanidine on the chloroplasts of *Euglena gracilis. J. Protozool.* 13, 472–4.

McLennan, A. and Keir, H. 1975. Subcellular location and growth stage dependence on the DNA polymerases of *Euglena gracilis. Biochim. Biophys. Acta* 407, 253–62

McMahon, D. 1975. Cycloheximide is not a specific inhibitor of protein synthesis in vivo. *Plant Physiol.* 55, 815–21.

Mego, J. L. 1968. Inhibitors of the chloroplast system in *Euglena.* In Buetow, D. E. (ed.), *The Biology of Euglena,* vol. 2, pp. 351–81. Academic Press, New York.

Mitsui, A. 1971. *Euglena* cytochromes. In San Pietro, A. (ed.), *Methods in Enzymology,* vol. 23A, pp. 368–71. Academic Press, New York.

Murray, D. J., Giovanelli, J., and Smillie, R. 1970. Photoassimilation of glycolate, glycine and serine by *Euglena gracilis. J. Protozool.* 17, 99–104.

Nicolas, P., and Nigon, V. 1974. Chloroplast and non-chloroplast adenosine

3′,5′-cyclic-monophosphate-receptor proteins in *Euglena gracilis. FEBS Letters* 49, 254–9.

Ophir, I., and Ben-Shaul, Y. 1973. Separation and ultrastructure of proplastids from dark-grown *Euglena* cells. *Plant Physiol.* 51, 1109–16.

Parsons, T. R. 1973. Coulter counter for phytoplankton. In Stein, J. R. (ed.), *Handbook of Phycological Methods: Culture Methods and Growth Measurements,* pp. 346–58. Cambridge University Press, Cambridge.

Perini, F., Kamen, M. D., and Schiff, J. A. 1964. Iron-containing proteins in *Euglena.* I. Detection and characterization. *Biochim. Biophys. Acta* 88, 74–90.

Preston, J., Parenti, F., and Eisenstadt, J. 1972. Studies on the isolation and purification of chloroplasts from *Euglena gracilis. Planta* (Berlin) 107, 351–67.

Price, C. A. 1978. Chloroplasts from *Euglena gracilis.* In Hellebust, J. A. and Craigie, J. S. (eds.), *Handbook of Phycological Methods: Physiological and Biochemical Methods,* pp. 5–13. Cambridge University Press, Cambridge.

Price, C. A., and Vallee, B. L. 1962. *Euglena gracilis,* a test organism for study of zinc. *Plant Physiol.* 37, 428–33.

Rawson, J. 1975. The characterization of *Euglena gracilis* DNA by its reassociation kinetics. *Biochim. Biophys. Acta* 402, 171–8.

Russell, G. K., and Lyman, H. 1968. Isolation of mutants of *Euglena gracilis* with impaired photosynthesis. *Plant Physiol.* 43, 1284–90.

Sagher, D., Grosfeld, H., and Edelman, M. 1976. Large subunit ribulose biphosphate carboxylase messenger RNA from *Euglena* chloroplasts. *Proc. Nat. Acad. Sci. U.S.* 73, 722–6.

Salisbury, J., Vasconcelos, A. C., and Floyd, G. L. 1975. Isolation of intact chloroplasts of *Euglena gracilis* by isopycnic sedimentation in gradients of silica. *Plant Physiol* 56, 399–403.

Schiff, J. A. 1972. A green safelight for the study of chloroplast development and other photomorphogenetic phenomena. In San Pietro, A. (ed.), *Methods in Enzymology,* vol. 24, pp. 321–2. Academic Press, New York.

Schiff, J. A. 1973. The development, inheritance and origin of the plastid in *Euglena.* In Abercrombie, M., and Brachet, J. (eds.), *Advances in Morphogenesis,* vol. 10, pp. 265–309. Academic Press, New York.

Schiff, J. A., Lyman, H., and Russell, G. K. 1971. Isolation of mutants from *Euglena gracilis.* In San Pietro, A. (ed.), *Methods in Enzymology* vol. 23A, pp. 143–62. Academic Press, New York.

Schmidt, G. W., and Lyman, H. 1976. Inheritance and synthesis of chloroplasts and mitrochondria of *Euglena gracilis.* In Lewin, R. (ed.), *The Genetics of Algae,* pp. 257–99. Blackwell Scientific, Oxford.

Schmidt, G., Matlin, K., and Chua, N. 1977. A rapid procedure for selective enrichment of photosynthetic electron transport mutants. *Proc. Nat. Acad. Sci. U.S.* 74, 610–14.

Schwelitz, F., Dilley, R., and Crane, F. 1972. Biochemical and biophysical characteristics of a photosynthetic mutant of *Euglena gracilis* blocked in photosystem II. *Plant Physiol.* 50, 161–5.

Shneyour, A., and Avron, M. 1975. A method for producing, selecting, and isolating photosynthetic mutants of *Euglena gracilis. Plant Physiol.* 55, 142–4.

Stern, A. J., Schiff, J. A., and Epstein, H. T. 1964. Studies of chloroplast development in *Euglena.* VI. Light intensity as a controlling factor in development. *Plant Physiol.* 39, 226–31.

Szybalski, W., and Bryson, V. 1952. Genetic studies of microbial cross-resistance to toxic agents. I. Cross-resistance of *E. coli* to 15 antibiotics. *J. Bacteriol.* 64, 489–99.

Uzzo, A., and Lyman, H. 1969. Light dependence of temperature-induced bleaching in *Euglena gracilis. Biochim. Biophys. Acta* 180, 573–5.

Vaisberg, A. J., and Schiff, J. A. 1976. Events surrounding the early development of *Euglena* chloroplasts. VII. Inhibition of carotenoid biosynthesis by the herbicide SAN 9789 (4-chloro-5-(methylamino)-2-(α,α,α-trifluro-*m*-tolyl)-3(2H) pyridazinone) and its developmental consequences. *Plant Physiol.* 57, 260–269.

Vaisberg, A. J., Schiff, J. A., Li, L., and Freedman, Z. 1976. Events surrounding the early development of *Euglena* chloroplasts. VIII. Photo control of the source of reducing power for chloramphenicol reduction by the ferredoxin-NADP reductase system. *Plant Physiol.* 57, 594–601.

Vasconceles, A., Mendiola-Morgenthaler, L. R., Floyd, G., and Salisbury, J. L. 1976. Fractionation and analysis of polypeptides of *Euglena gracilis* chloroplasts. *Plant Physiol.* 58, 87–90.

Verdier, G. 1975. Synthesis and translation site of light-induced RNAs in etiolated *Euglena gracilis. Biochim. Biophys. Acta* 407, 91–8.

Weissbach, A. B., Horecker, L., and Hurwitz, J. 1956. The enzymatic formation of phosphoglyceric acid from ribulose diphosphate and carbon dioxide. *J. Biol. Chem.* 218, 795–810.

Wu, R., and Racker, E. 1959. Regulatory mechanisms in carbohydrate metabolism. III. Limiting factors in glycolysis of ascites tumor cells. *J. Biol. Chem.* 234, 1029–35.

12: Crypthecodinium: sexual reproduction and mutagenesis

ROBERT C. TUTTLE

Department of Biology,
University of California at San Diego, La Jolla, California 92093

CONTENTS

I. Introduction

The dinoflagellates are a biologically successful group, ranging from free-living autotrophic species to parasitic species living in invertebrates and other phytoplankton. To understand their mechanisms for adaptation to various ecological niches, it is important that these organisms be susceptible to genetic manipulation. For this, one important characteristic is the production of mutants and another is that these organisms can be induced to undergo sexual reproduction.

The genetics of most classes of algae has been investigated in detail (Lewin 1976), but dinoflagellate genetics has been studied only in two laboratories, those of Allen et al. (1975); and those of Himes and Beam (1975). Since dinoflagellates ferment sugars without acid production and are resistant to penicillin as well as being difficult to replica plate, the time-saving approaches devised for selection of bacterial mutants have not been applicable to the Pyrrophyta. Mutants have been found only by direct inspection or by laborious replica-picking of survivors onto enriched or minimal growth media.

This chapter deals with the saprophytic organism *Crypthecodinium cohnii* and describes successful methods for inducing sexual reproduction and for production of mutants. Alternate methods in addition to specific details on mutagenesis for *Euglena* and Cyanophyta are given in Chaps. 10 and 11. Sexual reproduction and genetic recombination were only recently discovered in this species (Beam and Himes 1974; Tuttle and Loeblich 1974). Tuttle and Loeblich (1974) produced three classes of *Crypthecodinium* mutants: (1) pigment-deficient strains, (2) purine and pyrimidine-requiring auxotrophs, and (3) slow-growing prototrophs. Using 5 albino strains, they reported the first demonstration of genetic recombination in pigment-deficient mutants. With UV light, Himes and Beam (1975) produced impaired-motility mutants; in these they showed genetic recombination as well as segregation.

II. Organism, equipment, and growth conditions

A. Organism

Crypthecodinium cohnii can be obtained from the University of Texas Culture Collection and the Culture Centre of Algae and Protozoa.

B. Special equipment

Two types of microscopes are needed: an inverted microscope, such as the Leitz Diavert with 40× and 100× magnification; and a stereoscopic microscope, such as a Bausch and Lomb with a 15× ocular and 2× objective lense (60–90× magnification). All media are autoclaved, and aseptic techniques are used in culturing and in all mutagenic treatments.

C. Growth conditions

Stock cultures of *C. cohnii* are grown in 50 ml liquid medium (250-ml flasks) incubated at 27°C in the dark with shaking. The heterotrophic growth medium (MLH) contains: 340 m*M* NaCl; 30 m*M* $MgSO_4$; 8 m*M* $CaCl_2$; 9 m*M* KCl; 0.8 m*M* disodium glycerophosphate; 1.5 m*M* $(NH_4)_2SO_4$; 15 m*M* sodium acetate; 0.8 m*M* 1-histidine·HCl; $8 \times 10^{-9}M$ biotin; $3 \times 10^{-6}M$ thiamine·HCl; $8 \times 10^{-10}M$ vitamin B-12, 20 m*M* *D*-glucose, 10 m*M* betaine·HCl, 8 m*M* *N*-morpholino ethane sulfonic acid, F metal mix [100 × stock contains: 6 m*M* nitrilotriacetic acid, 0.08 m*M* 5-sulfosalicylic acid, 0.2 m*M* $Fe(NH_4)_2(SO_4)_2$, 30 m*M* NaOH; pH is adjusted to 3.8 with HCl]. The final pH of the medium is 6.6, adjusted with NaOH (Tuttle and Loeblich 1975).

III. Demonstrations of sexual reproduction

A. Method of gamete induction

The conditions that have been found to induce sexuality in dinoflagellates are those of environmental stress; these include nitrogen and phosphorus depletion. *C. cohnii* cells are grown to mid-log phase (ca. 1×10^6 cells/ml) under standard conditions. To induce mating, the cells are diluted 1:25 by being inoculated into fresh NPM medium (containing 1/10 the nitrogen and phosphorus source levels of medium MLH). A 5-ml aliquot is transferred into a 12-ml conical Pyrex centrifuge tube and allowed to stand for 10 min. Since fusing gametes swim poorly and settle, 0.1-ml samples are taken from the bottom of the tube and added to 0.3 ml of 1/10 NPM medium in a sterile glass depression slide. Under an inverted microscope (40–100×), pairs of fusing gametes are drawn up in a 20-μl micropipette (previously prepared and loaded with medium). They are transferred to successive

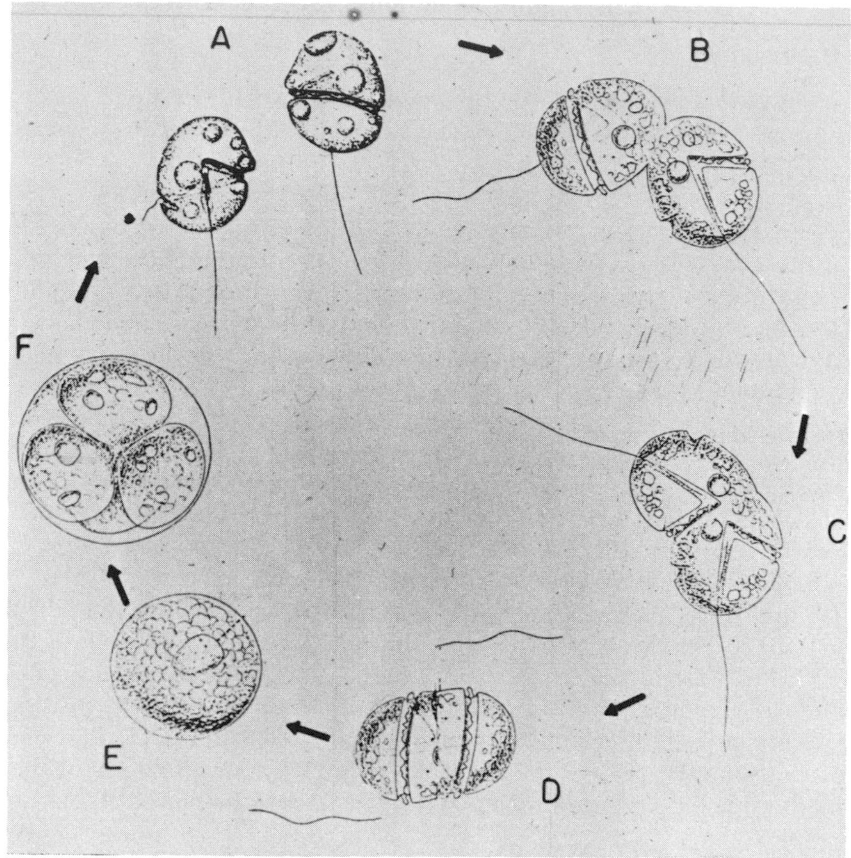

Fig. 12–1. Sexual life cycle of *Crypthecodinium cohnii.* Vegetative motile cells (A) function directly as gametes. The gametes fuse pairwise (B) by their epiconal regions. Partial cytological fusion (C) precedes nuclear fusion. In the process of gamete fusion, the swimming of the copulating pair becomes increasingly erratic and feeble, until the flagella are shed (D), forming a nonmotile zygote (E). The zygote matures, grows to twice its original size, undergoes meiosis, and germinates to produce 4 (or rarely 8) motile daughter cells (F–A).

drops of media (in depression slides) until one pair of fusing gametes remains per drop. For production of zygotes, they are then incubated in a sterile glass 22 × 100-mm petri dish (humidified by 2 layers of water-saturated paper towels) for 5 h at 27°C. For zygote germination, the fusing pairs of gametes are further incubated for an additional 15–24 h at the same temperature under the same conditions.

Somatic fusion, zygote maturation, and germination are followed by microscopic examination (ca. 200×). The *C. cohnii* sexual cycle is illustrated in Fig. 12-1. Most of the fusing pairs are isogamous. Since sexual reproduction occurs in cultures arising from a single cell, no restrictive mating types (+, −) exist in *C. cohnii.* Upon staining, the nuclear fusion process can be observed with a microscope (400–1,100×). Fusing gametes and zygotes, are transferred to 0.1 ml of 0.5% methyl green dye in 0.1 *M* acetic acid (pH 4.4) on a microscope slide. The cells are stained for 10–30 min, excess stain is removed by a pipette, and the cells are washed twice on the slide with distilled water. All nuclei stain a deep blue; a zygote is recognized by the single large nucleus, which persists until meiosis.

B. Results and application

Genetic recombination of chromosomal genes is the final proof of *C. cohnii* sexuality and is easy to demonstrate (Tuttle and Loeblich 1974). Segregation analysis of *C. cohnii,* however, is difficult and has been shown only with the impaired-flagella mutants of a specially selected, durable strain by Himes and Beam (1975). For dinoflagellates, which do not grow on agar, all segregation analysis manipulations must be done in liquid media.

The taxonomy of controversial dinoflagellate genera can be resolved by use of the information from the sexual cycle. In addition to *C. cohnii,* sexual reproduction has also been shown to occur in the fresh water form *Peridinium willei* (Pfiester 1976) and in *Ceratium,* where von Stosch (1973) helped in the clarification and correct interpretation of previous observations of the sexual and asexual cycles of several species of this genus. A comparison of conditions that repress or induce sexuality and the presence or absence of mating types will be useful for clarifying family groups. Attempts at "interspecies" matings using single cells of two (assumedly distinct) species will serve to define species limits among the dinoflagellates.

IV. Mutagenic procedures

A. Preparation of C. cohnii for mutagenesis

Cultures are grown to mid-log phase, with a cell density of 5×10^5 to 1×10^6 cells/ml (determined with a hemocytometer or Coulter Counter). Cells from a 10-ml aliquot are collected on a 8-μm pore filter (Millipore, MF-SCWP) in a glass filter holder. The cells are washed free of organic media components with a rinse (10 ml) of a M-salt solution [340 m*M* NaCl, 30 m*M* $MgSO_4$, 8 m*M* $CaCl_2$, 9 m*M* KCl, and 8 m*M* *N*-morpholino ethane sulfonic acid; pH 6.0].

B. Ultraviolet light as mutagenic agent

Cells prepared as above are suspended in a watch glass (5-cm diameter) with 5 ml M-salts to a final concentration of 1.5×10^4 cells/ml. They are then irradiated for ca. 4 min with ultraviolet light at 200 erg/cm^2/sec (15 W germicidal lamp placed 60 cm above the cell suspension, intensity is determined with a YSI-Kettering radiometer). This exposure produces the desired 90–95% mortality rate. Individual source lamps should be calibrated, and a mortality curve (Withers and Tuttle 1979) should be determined. After irradiation, the cells are diluted to 10^3 cells/ml and plated out on MLH agar (1% w/v) plates and incubated at 27°C. A typical dose–mortality curve is shown in Fig. 12-2. Survivors are diluted (to 10^5 cells/ml) and plated on solidified agar and allowed to undergo two divisions before being tested on selection media (IV.D).

C. Chemical mutagenesis

The mutagenic agent *N*-methyl-*N'*-nitro-nitrosoguanidine (NTG) is a dangerous carcinogen and must be handled with caution. Contact with the skin or pipetting by mouth must be avoided, and manipulations must be done in a fume hood while wearing inert gloves (Van-Lab Poly Scientific). NTG-contaminated glassware is cleaned in 0.1 *N* NaOH, which destroys the mutagenic acid. The NTG stock solution (10mg/ml) is freshly prepared for each experiment by dissolving it in a small volume of dimethylsulfoxide (DMSO) and diluting with sterile water to the final concentration. Cells prepared as in IV.A above and suspended in 10 ml M-salts are treated with NTG (50 μg/ml final concentration) for 6 min. This should result in a 5–10% survival of cells as previously determined (Tuttle and Loeblich 1974). To remove the mutagenic agent, cells are washed and then suspended in MLH growth medium and incubated (27°C) to allow them to undergo two division cycles (in ca. 16 h).

D. Selection of mutants using growth inhibitors

To test *C. cohnii* sensitivity to growth inhibition compounds (Table 12-1) (mutagenically treated) cells are plated on a series of MLH agar plates (10^5 cells/plate). Then, sterile 1-cm filter paper discs saturated with stock solutions of the test compounds are placed on the plates and the samples are incubated under standard conditions. Only toxic compounds produce a zone of inhibition around the filter discs. These compounds are further tested in a series of concentrations in liquid media to find a minimal concentration that inhibits growth. To select resistant mutants, a concentration 5 or 10 times the minimal inhibitory level is used. Toxicity tests can also be done in liquid media

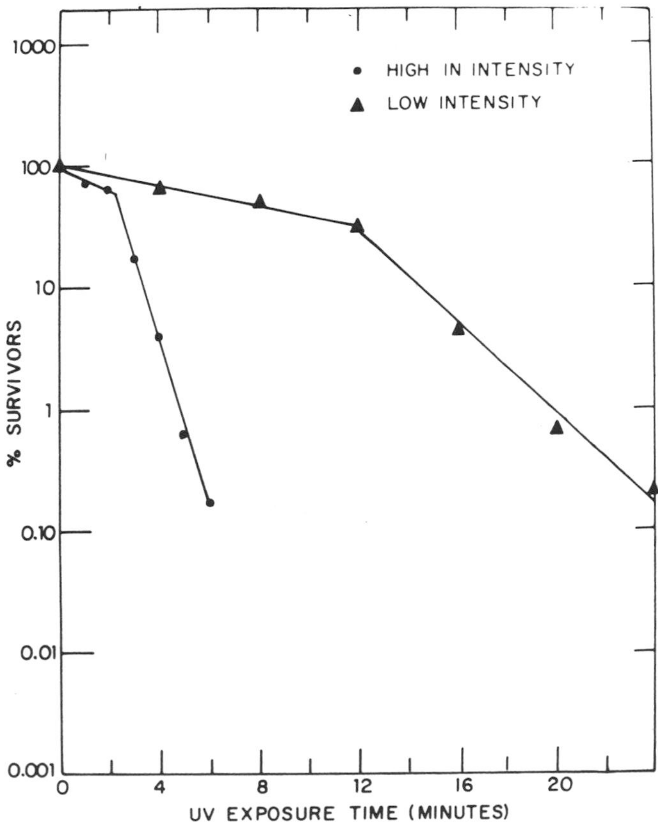

Fig. 12-2. Mortality kinetics of *C. cohnii* at two ultraviolet light intensities. High intensity, 200 erg/cm²/sec; low intensity, 50 erg/cm²/sec.

with double strength stock solutions and double strength growth media mixed 1:1.

E. Mutants recovered from C. cohnii mutagenesis

Typical results of NTG and UV mutagenesis are given in Table 12-2. Three types of mutants were found. Wild-type *C. cohnii* makes yellow-orange colonies (due to the presence of carotenoids) when grown in low light, 0.2 W/m². Many distinct types of carotenoid-deficient cream colored or albino strains (car⁻) were obtained by scoring the petri plates visually after 14 days incubation. Strains that required the nucleotide bases adenine, guanine, or cytosine (nuc⁻) were isolated by plating NTG-treated cells on solid medium MLH containing 50μg/ml of the compounds. Colonies were replica-picked with sterile wooden toothpicks and put on a new set of plates with or without the various

Table 12–1. *Inhibitors for the selection of mutants from Crythecodinium cohnii*

Compound	Final concentration[a]	Process affected
Sulfonamide	6 mg and 10 μg dTMP[b]/ml	Thymine biosynthesis
2-Valine	80 μg/ml	Valine synthesis regulation
Fucidic acid	800 μg/ml in 2 m*M* EDTA	DNA replication
Rifampicin	200 μg (in methanol)/ml	RNA polymerase
Phosphonomycin	2 mg/ml	Glycerophosphate transport
Kasugamycin	1 mg/ml	Protein synthesis
5-Methyl tryptophan	200 μg/ml	Tryptophan biosynthesis
L-Canavanine	100 μg/ml	Arginine biosynthesis
Methyl glyoxal	1 m*M*	Glyoxalate metabolism
Sodium azide	2 m*M*	Electron transport poison

[a] Filter sterilized as 10-fold concentrated stock solutions.
[b] Deoxythymidine-5′-monophosphate (Wickner 1974).

Table 12–2. *Experimental results of screening NTG or UV treated C. cohnii for mutant survivors*

Mutagen	Dose	Colonies screened (No.)	Recovered mutants (No.)	Mutants survivors (%)
Nitrosoguanidine	100 μg/ml (10 min)	441	2 car^{-} [a]	0.45
	50 μg/ml (6 min)	31,500	33 car^{-}	0.10
	50 μg/ml (6 min)	10,228	5 car^{-}	0.049
	20 μg/ml (9 min)	5,000	4 car^{-}	0.08
	50 μg/ml (6 min)	1,176	2 ade^{-} [b]	0.17
	50 μg/ml (6 min)	10,228[c]	20 nuc^{-} [c]	0.20
	20 μg/ml (5 min)	4,162	11 nuc^{-}	0.26
	50 μg/ml (5 min)	10,228	5 slo^{-} [d]	0.049
Ultraviolet light	200 erg/cm^2/sec	6,250	2 car^{-}	0.032

[a] Contain no visible carotenoids.
[b] Require adenine for growth.
[c] Require adenine, cytosine, guanine (100 μg/ml) for growth.
[d] Slow growth, 10–40% rate of wild type, in medium MLH.

bases. Nuc^{-} strains did not grow on minimal medium but grew on the media containing the required nucleotide base. Slow-growing prototrophs (slo^{-}) were found while screening for nuc^{-} strains. Slo^{-} mutants grew on minimal medium MLH, but at a rate significantly slower than wild type, which is a characteristic of mitochondrial mutants.

F. Alternate methods of mutagenesis

For direct positive selection of resistance mutants, crystals of NTG can be placed on medium MLH plates (1% agar) with 10^3–10^4 *C. cohnii* cells and an appropriate concentration of inhibitory compound. Although it is a potent mutagen of low toxicity to microbes, NTG has been shown to make multiple mutations and chromosomal alterations with high frequency. It is a health hazard and very dangerous for the researcher as it quickly breaks down to generate the carcinogenic gas, diazomethane. NTG should only be used when other chemical mutagens, such as ethyl methane sulfonate, methyl methane sulfonate, and 2-aminopurine are not effective. These are stable as liquids and can be used in the above protocols without most of the problems encountered with NTG.

V. Acknowledgments

I wish to thank Alfred Loeblich in whose laboratory these experiments were conducted. Discussions with Paul Levine and J. Woodland Hastings on algal mutagenesis and genetics were indispensable in the completion of this study.

VI. References

Allen, J. R., Roberts, T. M., Loeblich, A. R. III, and Klotz, L. C. 1975. Characterization of the DNA from the dinoflagellate *Crypthecodinium cohnii* and implications for nuclear organization. *Cell* 6, 161–9.

Beam, C. A., and Himes, M. 1974. Evidence for sexual fusion and recombination in the dinoflagellate *Crypthecodinium cohnii. Nature* 250, 435–6.

Himes, M., and Beam, C. A. 1975. Genetic analysis in the dinoflagellate *Crypthecodinium (Gyrodinium) cohnii*: evidence for unusual meiosis. *Proc. Nat. Acad. Sci. U.S.* 72, 4546–69.

Lewin, R. A. (ed.). 1976. *The Genetics of Algae.* Blackwell Scientific, Oxford. 360 pp.

Pfiester, L. 1976. Sexual reproduction of *Peridinium willei. J. Phycol.* 12, 234–8.

Stosch, H. A. von. 1973. Observations on vegetative reproduction and sexual life cycles of two freshwater dinoflagellates, *Gymodinium pseudopalustre* Schiller and *Woloszynskia apiculata* sp. nov. *Brit. Phycol. J.* 8, 105–34.

Tuttle, R. C., and Loeblich, A. R., III. 1974. Genetic recombination in the dinoflagellate *Crypthecodinium cohnii. Science* 185, 1061–2.

Tuttle, R. C., and Loeblich, A. R., III. 1975. An optimal growth medium for the growth of the dinoflagellate *Crypthecodinium cohnii. Phycologia* 14, 1–8.

Wickner, R. B. 1974. Mutants of *Saccharomyces cerevisiae* that incorporate deoxythymidine-5′-monophosphate into deoxyribonucleic acid in vivo. *J. Bacteriol.* 117, 252–60.

Withers, N. W., and Tuttle, R. C. 1979. Effect of visible and ultraviolet light on carotene deficient mutants of *Crypthecodinium. J. Protozool.* 26, 135–8.

13: Protoplast and spheroplast production

MARINA ADAMICH*

Department of Biological Sciences, University of California, Santa Barbara, California 93106

BARBARA B. HEMMINGSEN

Department of Microbiology, San Diego State University, San Diego, California 92182

CONTENTS

* Current address: Chemistry Department, School of Medicine, University of California at San Diego, La Jolla, California 92093

I. Introduction

A. Objectives

The area of algal protoplast and spheroplast production has grown rapidly within the past 4 years chiefly because such "wall-less" cells are potentially useful as starting material for the isolation of organelles and other particulate systems, and for studies of cell-wall synthesis, intermediary metabolism, cell-surface phenomena, and the incorporation of foreign genomes. Methods developed for the removal of cell walls in the preparation of algal protoplasts and spheroplasts include exposure of cells to autolysine, to hypo- and hyperosmotic conditions, to detergent, to various external enzymes of nonalgal origin, and to other physiological conditions not yet fully understood. It is the aim of this chapter to describe those methods that offer the investigator high yields of healthy protoplasts or spheroplasts in the shortest time possible.

B. Definition of terms

Algal cells from which cell walls have been removed have been referred to as protoplasts, naked cells, cytoplasts, spheroplasts, cell-wall-free algae, and naked protoplasts. To reduce some of the confusion resulting from the intermixing of these terms, we will follow, in part, the convention used with bacteria (Brenner et al. 1958; Kaback 1971), yeast (Streiblová 1968), and higher plants (Cocking 1972) and define algal protoplasts and spheroplasts as follows:

Protoplast: an algal cell of any genotype from which the cell wall and all coatings external to the plasmalemma are removed but which retains all cellular components.

Spheroplast: an algal cell of any genotype from which all or part of the cell wall is removed but which retains coating(s) external to the plasmalemma and all cellular components.

Cytoplasts are discussed in Chap. 14. Neither of the above definitions is suitable to describe such "wall-less" cytoplasmic droplets since they do not contain a full complement of cytoplasmic components.

To apply these terms correctly, it is necessary to establish whether cell layers external to the plasmalemma remain. This is best determined by transmission, although scanning electron microscopy, the use of various cell wall stains, and possibly increased sensitivity of treated cells to osmotic stress may give some indication of the presence or absence of all or part of a cell wall. Differences in osmotic sensitivities sometimes reported for protoplasts and spheroplasts are omitted from the definitions because of the lack of a standard method for such determinations.

C. Evaluation criteria of the protoplast and spheroplast state

One or more of the following criteria may be used to assess the degree of structural and physiological integrity of the protoplasts and spheroplasts:

1. Viability: orderly increase of all cell components followed by division with or without cell wall regeneration.
2. Cell wall regeneration with or without growth and division.
3. Cytoplasmic fine structure similar to that of the normal cell.
4. Retention of normal membrane permeability (Gaff and OKong'O-Ogola 1971; Stadelmann and Kinzel 1972; Larkin 1976).
5. Retention of normal levels of respiratory and/or photosynthetic activity.
6. Motility where applicable.

II. Organisms and methods

Algae from which protoplasts and spheroplasts have been prepared are listed in Table 13–1, together with strain number, source, method used, and reference(s).

A. Cyanophyceae

The blue-green algae have a cell wall similar to that of gram-negative bacteria (Drews 1973). This wall contains peptidoglycan, a lysozyme-sensitive heteropolymer that confers shape and osmotic protection on the cell, and other material not sensitive to lysozyme. It is likely that all lysozyme-treated blue-green algae are spheroplasts; indeed, some have been shown by electron microscopy to retain cell wall material (Jensen and Sicko 1971; Lindsey et al. 1971). Spheroplasts can be prepared from various species (Table 13–1) using the basic method developed by Biggins (1967; 1971) as described here and variations of it.

1. Enzymatic preparation of spheroplasts. Phormidium luridum var. *olivaceae* (Table 13–1, Fig. 13–1a) is cultured in liquid medium C (Kratz and Myers 1955) supplied with 4% CO_2 in air (v/v) under fluorescent lamps. The cells are harvested by centrifugation (500*g*, 10 min) dur-

Table 13–1. *Test organisms and methods*

Alga	Strain	Culture sources	Method	Reference
Cyanophyceae[a]–unicellular				
Anacystis nidulans	TX-20	Not given	Enzymatic	Lindsey et al. (1971)
A. nidulans	625	UTEX[b]	Enzymatic	Sigalat and de Kouchkovsky (1975a, b); Cosner (1978)
Gloeocapsa alpicola	Not given	Not given	Enzymatic	Grodzinski and Colman (1973)
Coccochloris peniocystis	Not given	Not given	Enzymatic	Grodzinski and Colman (1973)
Microcystis aeruginosa	Not given	Not given	Enzymatic	Vance and Ward (1969)
Cyanophyceae–filamentous				
Oscillatoria amoena[c] (Kützing) Gomont	Not given	Not given	Enzymatic	Fuhs (1958)
O. tenuis	Not given	UTEX	Enzymatic	Crespi et al. (1962)
O. formosa[c]	Not given	UTEX	Enzymatic	Crespi et al. (1962)
Fremyella diplosiphon	Not given	UTEX	Enzymatic	Crespi et al. (1962)
Plectonema calothricoides	Not given	UTEX	Enzymatic	Crespi et al. (1962)
Synechococcus lividus[c]	Not given	From Dr. D. L. Dyer	Enzymatic	Crespi et al. (1962)
Plectonema boryanum (Gomont)	594	UTEX	Enzymatic	Sofrová et al. (1974); Kessel et al. (1973)
Anabaena flos-aquae	Not given	Not given	Enzymatic	Vance and Ward (1969)
A. variabilis	Not given	Not given	Enzymatic	Gusev et al. (1970)
A. ambigua Rao	1403/7	Cambridge Culture Coll.[d]	Enzymatic	Bhattacharya and Talpasayi (1973)
Cylindrospermum sp.	LB 942	UTEX	Enzymatic	Jensen and Sicko (1971)
Calothrix parietina[e] (Näg.)	See footnote	Cambridge Culture Coll.[d]	Enzymatic	Kessell et al. (1973)
Phormidium luridum	Not given	UTEX	Enzymatic	Crespi et al. (1962)
P. luridum var. *olivaceae* Boresch	Not given	UTEX	Enzymatic	Biggins (1967)

Chlorophyceae				
Chlamydomonas reinhardii	11-32a	Sammlung von Algenkult.[f]	Autolysine	Schlösser et al. (1976)
	11-32b	Sammlung von Algenkult.[f]	Autolysine	Schlösser et al. (1976)
	11-32c	Sammlung von Algenkult.[f]	Autolysine	Schlösser et al. (1976)
	11-32aM	Sammlung von Algenkult.[f]	Autolysine	Schlösser et al. (1976)
	73.72	Sammlung von Algenkult.[f]	Autolysine	Schlösser et al. (1976)
C. incerta	7.73	Sammlung von Algenkult.[f]	Autolysine	Schlösser et al. (1976)
C. smithii	54.72	Sammlung von Algenkult.[f]	Autolysine	Schlösser et al. (1976)
C. globosa	81.72	Sammlung von Algenkult.[f]	Autolysine	Schlösser et al. (1976)
C. spec.	11.31	Sammlung von Algenkult.[f]	Autolysine	Schlösser et al. (1976)
Chlorella saccharophila	211-9a	Sammlung von Algenkult.[f]	Enzymatic	Braun and Aach (1975); Feyen (1977)
C. ellipsoidea	211-1b	Sammlung von Algenkult.[f]	Enzymatic	Braun and Aach (1975); Feyen (1977)
C. vulgaris	Not given	Carolina Biol. Supply Co.[i]	Enzymatic	Berliner (1977)
Uronema gigas	Not given	Not deposited	Enzymatic	Gabriel (1970)
Cosmarium lundellii (Delp.)[g]	Not given	Laboratoire d'Hydrologie[h]	Osmotic	Chardard (1972)
C. turpinii	Not given	Carolina Biol. Supply Co.[i]	Enzymatic	Berliner and Wenc (1976a; b)
Klebsormidium (*Hormidium*) *flaccidium*	LB 1958	UTEX	Enzymatic	Marchant and Fowke (1977)
Micrasterias thomasiana	Not given	Carolina Biol. Supply Co.[i]	Enzymatic	Berliner and Wenc (1976a; b)
M. angulosa	Not given	From Dr. D. H. Tippit	Enzymatic	Berliner and Wenc (1976a; b)
M. denticulata	Not given	From Dr. D. H. Tippit	Enzymatic	Berliner and Wenc (1976a; b)
Mougeotia sp.	Not given	Not deposited	Enzymatic	Marchant and Fowke (1977)
Spirogyra sp.	Not given	Not deposited	Enzymatic	Ohiwa (1977)
Stigeoclonium sp.	Not given	Not deposited	Enzymatic	Marchant and Fowke (1977)
Ulothrix fimbriata	638	UTEX	Enzymatic	Marchant and Fowke (1977)
Zygnema extenue Jao	Not given	UTEX	Enzymatic	Ohiwa (1977)

Table 13–1. (*cont.*)

Alga	Strain	Culture sources	Method	Reference
Euglenophyceae				
Euglena gracilis (Klebs)	Z (Hutner)	Not given	Enzymatic	Price and Bourke (1966)
Bacillariophyceae				
Nitzschia alba Lewin and Lewin	LT-P-1	Not deposited	Other	Hemmingsen (1971)
Pyrrophyceae				
Gonyaulax polyedra Stein	70	Algal Culture Collection[j]	Osmotic	Adamich and Sweeney (1976)
Rhodophyceae				
Porphyridium cruentum Naegeli	No. 161	UTEX	Enzymatic	Clément-Metral (1976)

[a] Fulco et al. (1967) found that several species of marine and freshwater blue-green algae were sensitive to lysozyme but only after 2–12 days of treatment. No other workers report such long exposure times.
[b] UTEX, formerly the Indiana University Culture Collection, is now located at the Department of Botany, University of Texas, Austin, Texas 78712.
[c] Little effect on filaments after 24 h exposure to lysozyme; all other species were considerably affected after a much shorter exposure.
[d] Culture Centre of Algae and Protozoa (Cambridge), 36 Storeys Way, Cambridge, England CB3 ODT.
[e] Received as *Tolypothrix tenuis* (Kützing) No. B1482/3 and reclassified.
[f] Sammlung von Algenkulturen, Universität Göttingen, 18 Nikolausbergerweg, Göttingen, Federal Republic of Germany.
[g] This culture may not be available presently.
[h] Laboratoire d'Hydrologie de Gif-sur-Yvette, 9119s, Ersone, France.
[i] Carolina Biological Supply Company, Burlington, North Carolina 27215.
[j] Algal Culture Collection, University of California at Santa Barbara, Santa Barbara, California 93106.

ing the late exponential phase of growth, and washed once with a solution of 0.5 *M* mannitol in 0.03 *M* potassium phosphate buffer, pH 6.8, and resuspended in this solution to a concentration of about 5% cells (w/v). Solid lysozyme (Worthington Biochemical Co.) is added to a final concentration of 0.05%, and the preparation is maintained at 35°C for 2 h with occassional swirling. A 70% yield of spheroplasts (Fig. 13–1b) is usually obtained. Spheroplasts may be separated from remaining filaments by passing the preparation through glass wool and collecting the spheroplasts by centrifugation (500*g*, 4 min). *P. luridum* spheroplasts are stable for several hours when maintained at 0–4°C.

2. *Evaluation.* Spheroplasts of *P. luridum* suspended in mannitol–buffer solution are capable of endogenous respiration and photoassimilation of CO_2 at rates comparable to those of untreated cells. The Hill activity of spheroplasts is best maintained using 0.2 *M* to 0.5 *M* KCl (Binder et al. 1976). While *P. luridum* spheroplasts appear to remain healthy after treatment with lysozyme, spheroplasts of *Anacystis nidulans* (Lindsey et al. 1971) and of *Anabaena variabilis* (Baulina et al. 1975) are reported to sustain some fine-structural damage.

B. Chlorophyceae

1. *Autolytic technique.* Schlösser et al. (1976) have prepared protoplasts from vegetative cells of a number of species and strains of *Chlamydomonas* (Table 13–1) using the cell wall-lytic factor, "gamete-autolysine," described by Claes (1971). This heat-labile, enzymelike factor is released into the medium by *C. reinhardii* gametes only at the time of their fusion. It will dissolve the cell walls of zoospores, sporangia, gametes, and the vegetative cell, but not the wall of the zygote.

a. Preparation of protoplasts. Gamete-autolysine is prepared by the fusion of gametes from compatible mating types as follows: Two clonal strains, *C. reinhardii* 11–32b and 11–33c (Table 13–1), are cultured in Kuhl's medium (Kuhl 1962) and aerated with a mixture of air and 2% CO_2. The cultures are maintained on a 12:12 h LD cycle (20,000 lux; 1,800 ft-c) at 34°C and diluted to 1.5×10^6 cells/ml at the end of each dark period. Gamete formation is initiated, 9 h after the onset of the light cycle, by sedimenting the cells (1,500*g*, several min) and resuspending them in gamete-formation medium (10^{-3} *M* $MgSO_4$, 10^{-4} *M* $CaCl_2$, 5×10^{-3} *M* sodium phosphate, pH 6.0) to 50% of the original volume of the growth medium. Six hours after the onset of the following dark cycle, the cells are again sedimented and resuspended in fresh gamete-formation medium to 20% of the original volume of growth medium. Of this cell suspension 60-ml aliquots are placed into 2-cm diameter glass dishes and illuminated for 3 h

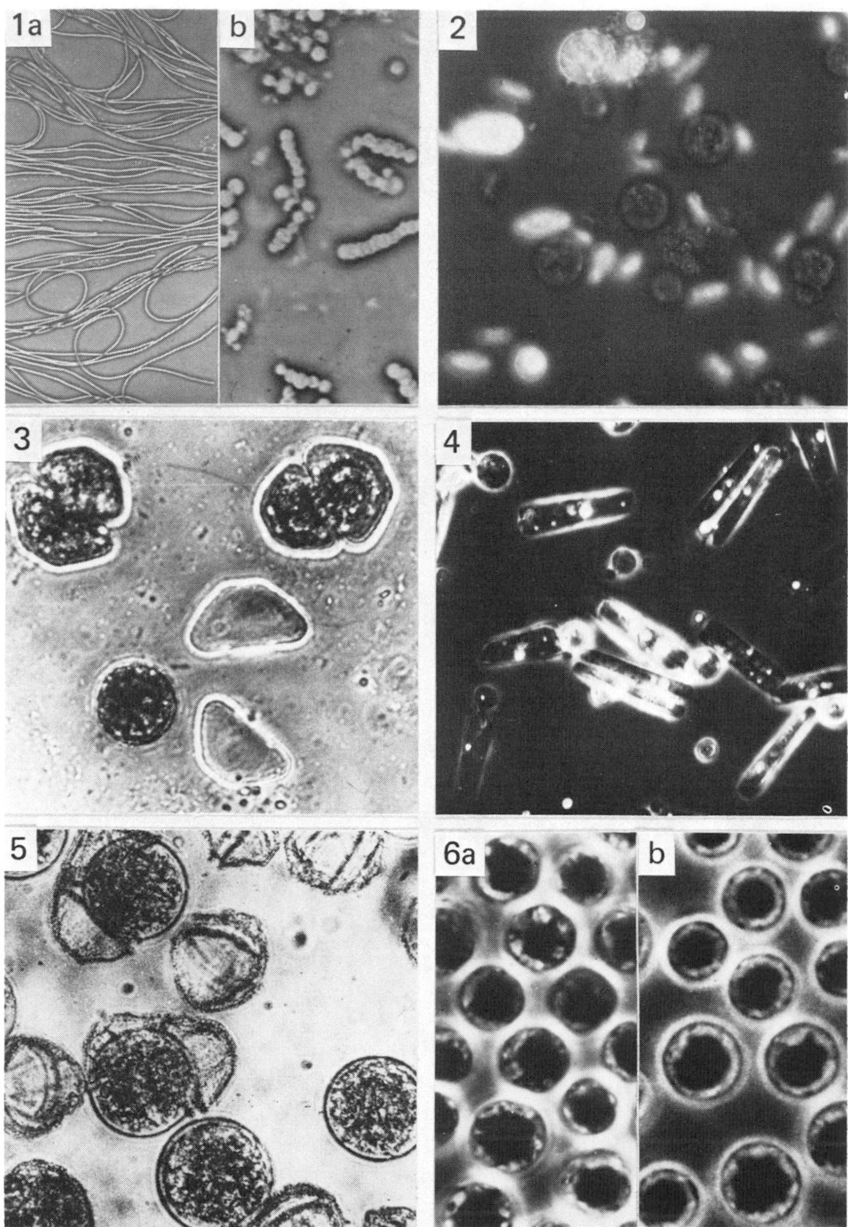

Figs. 13–1 to 13–6. 13–1: Light micrographs of *Phormidium luridum;* (a) untreated filaments (×175) and (b) spheroplasts prepared by treatment with lysozyme in mannitol–buffer (×1,050). (Courtesy of J. Biggins.) 13–2: Fluorescent-light micrograph of *Chlorella saccharophila* (Krüger) Nadson treated with cellulase in hyperosomotic medium and stained with Calcofluor-White. Normal cells (ellipsoid) and spheroplasts (spherical) fluoresce, but protoplasts (spherical) do not (×700). (Courtesy of H. G. Aach, S.

(41,000 lux; 375 ft-c) at 25°C. During this time gametes are formed. At the end of 3 hours, the gamete suspensions are mixed. Gamete fusion will occur immediately. The suspension is centrifuged (15,000*g*, 20 min) 1.5 h following the mixing of the gametes and the supernatant containing gamete-autolysine is removed for the subsequent preparation of protoplasts. The gamete-autolysine may be maintained at 3°C for a few days without the loss of activity, or concentrated by ammonium sulfate precipitation, dialyzed, and frozen.

To prepare protoplasts, *Chlamydomonas reinhardii* 11–32b or *C. smithii* (Table 13–1) is cultured as described above and, 11 hours after the onset of the light period, an aliquot is centrifuged (1,500*g*, 3 min), the supernatant decanted, and 1 ml gamete-autolysine solution added for every 9×10^6 cells. During the next 20 min, protoplasts ecdyse from their cell walls through an apparent opening in the wall.

b. Evaluation. The protoplasts remain motile, and a new cell wall is regenerated within 2 h after resuspension in fresh growth medium. This wall, and successive walls, may be removed repeatedly by following the above procedure without apparent damage to the cell.

2. *Enzymatic technique.* The cell walls of some *Chlorella* species are resistant to enzymatic digestion due to the presence of sporopollenin (Atkinson et al. 1972). Braun and Aach (1975) and Feyen (1977) have reported a method employing hyperosmotic conditions and cellulase in the preparation of spheroplasts and protoplasts of *Chlorella saccharophila* and *C. ellipsoidea* (Table 13–1), two species that apparently lack sporopollenin. In the presence of Calcofluor White, a purported cellulose stain (Chap. 9), some cells appear to retain fragments of cell wall (spheroplasts), while others do not (protoplasts). However, both spheroplasts and protoplasts are osmotically fragile and lyse readily in distilled water.

a. Preparation of spheroplasts. Cultures of *C. saccharophila* or *C. ellipsoidea* (Table 13–1) are grown in mineral nutrient medium (Ruppel 1962) to a density of 10^7 cells/ml under continuous illumination

(continued from facing page)
Bartsch, and V. Feyen.) 13–3; Light micrograph of *Cosmarium turpinii* treated with cellulysin in hyperosmotic medium. In the normal cell, the two hemicells are joined at the isthmus, and the cellular contents occupy both hemispheres. Following treatment, the hemicells separate at the isthmus and the cell contents ecdyse (not shown) to form a single, round, dense protoplast (×210). (Courtesy of M. Berliner.) 13–4: Light micrograph of *Nitzschia alba* after exposure to medium rich in organic compounds and deficient in divalent cations. Protoplasts (spherical) ecdyse from the separated girdle bands (×500). 13–5: Light micrograph of *Gonyaulax polyedra* in hypotonic solution containing 0.013% Liquinox detergent. Spheroplasts ecdyse from the separated girdle region (×600). 13–6: Light micrographs of *Porphyridium cruentum;* (a) untreated cells, and (b) spheroplasts prepared by treatment with a mixture of digestive enzymes (×1,344). (Courtesy of J. D. Clément-Metral.)

(7,000–10,000 lux; 630–930 ft-c) at 25–27°C. Air, filtered through sterile glass wool, should be bubbled through the medium (ca. 4–6 ml/min). A 10-ml aliquot of cell suspension is harvested by centrifugation (ca. 800*g*, 30 sec), the supernatant decanted, 1 g sterile sea sand added, and the mixture vortexed for 30–60 sec. A 10 ml solution containing 0.3 *M* sorbitol, 0.3 *M* mannitol, and 4% Onozuka SS-cellulase (All Japan Biochemicals, Ltd.) is then added to the cell–sand mixture, and the mixture incubated with gentle shaking in a shaker bath under normal culture conditions but without aeration. Protoplasts and spheroplasts (Fig. 13–2) form within 2–3 h with yields of approximately 80%.

b. Evaluation. Information is not yet available regarding the physiological state of these protoplasts and spheroplasts.

c. Preparation of protoplasts. Chardard (1972) and Berliner and Wenc (1976a, b) have reported the preparation of protoplasts from a number of desmids (Table 13–1) by the exposure of cells to hyperosmotic conditions in the presence or absence of hydrolytic enzymes. The hyperosomotic method of Berliner and Wenc (1976b) employing exo-enzymes is described here.

Cosmarium turpinii (Table 13–1) is cultured in 250-ml Erlenmeyer flasks containing 50 ml FWV growth medium (Lee and Loeblich 1971). The cultures are maintained at 21°C on a rotary shaker, on a 15:9 h LD schedule. Illumination is supplied by two 15 W Daylight fluorescent bulbs at a distance of about 20 cm.

In the middle of the dark period, 3 ml of the culture is placed into a 60 × 15-mm plastic petri dish (Falcon) and the cells are allowed to settle overnight (M. Berliner, personal communication). Next, 3 ml of growth medium containing 0.8 *M* D-mannitol is added to the dish, followed by 0.65 ml of a 20% w/v Cellulysin solution. Cellulysin (CalBiochem, La Jolla) is completely dissolved in FWV medium containing 0.4 *M* D-mannitol. After a brief mixing, the cell suspension is incubated undisturbed for 3–4 h under stock culture conditions. During this time, one protoplast is formed at the isthmus from the contents of two hemicells (Fig. 13–3). Protoplast ecdysis is complete within 4 h, with a yield of 90%. Protoplasts will not form during the cell division period of *C. turpinii*.

d. Evaluation. The protoplasts exclude trypan blue dye for some time but do not regenerate a cell wall, grow, or divide. In addition, pigment discoloration and morphological alterations can be observed after a few days in the incubation medium. The protoplasts are sensitive to hypotonic conditions and to physical manipulations.

C. *Bacillariophyceae*

Many diatoms have been observed to release protoplasts (for examples see Hendey 1945; and von Stosch 1965), but the precise conditions and yields have not been specified. Protoplasts of the nonphotosynthetic marine diatom *Nitzschia alba* can be prepared in large numbers by suspending exponentially growing cells in a medium deficient in divalent cations and rich in organic compounds (Hemmingsen 1971). Electron microscopic examination of thin sections of the protoplasts shows the complete absence of wall layers external to the plasmalemma.

1. Preparation of protoplasts. Nitzschia alba (Table 13–1) is grown axenically in 1-liter amounts of growth medium plus glucose (Hemmingsen 1971; Azam and Volcani 1974) in 2-liter Erlenmeyer flasks and incubated in darkness or light at 30°C with vigorous agitation provided by a magnetic stirring bar. Cells in the exponential phase of growth are aseptically harvested by centrifugation (10,000*g*, 10 min) and the pellets drained carefully. The pellets are resuspended with a minimum of agitation in 50 ml sterile protoplast-producing medium [1% Bactotryptone, 0.5% Bacto-yeast (Difco) in 3% NaCl prepared with glass-distilled water; pH is adjusted to 6.2 with HCl before autoclaving]. The cell suspension is then transferred to a sterile 125-ml Erlenmeyer flask and incubated at 30°C in darkness or light with gentle agitation from either a magnetic stirring bar or a rotary shaker. The final concentration of cells should be between $1-5 \times 10^6$ cells/ml. Within an hour after transfer to the protoplast-producing medium, the cells become nonmotile but appear otherwise normal. Five hours later, nearly every cell is expanded: the girdle bands on one side of the cell are apparently weakened, the two halves of the cell wall separate, and the protoplasts begin to extrude partially from the cell wall (Fig. 13–4). By 10 h, a substantial number have ecdysed and by 12 h, 50–60% of the total population is composed of free protoplasts.

2. Evaluation. Determinations of viability and cell wall regeneration are not yet possible because of the difficulty in separating protoplasts from normal cells, cell walls, and debris.

Certain precautions are advised in the use of this method: (a) pipetting of cells seriously hampers their ability to form protoplasts; (b) gentle but not vigorous agitation of the medium is necessary; (c) the addition of Ca^{++} (to 12 m*M*) or Mg^{++} (to 27 m*M*) suppresses protoplast formation and allows growth and division of the cells; thus, it is essential to drain sedimented cells to minimize the introduction of growth medium into the protoplast-producing medium.

D. Pyrrophyceae

Adamich and Sweeney (1976) have prepared spheroplasts of the marine dinoflagellate *Gonyaulax polyedra* by suspending cells in a hypotonic solution containing a small amount of detergent. The spheroplasts retain a carbohydrate coating external to the plasmalemma

(Loeblich 1970; Sweeney 1976). They are stable to handling and to osmotic pressure changes, but will lyse in distilled water within minutes.

1. Preparation of spheroplasts. Gonyaulax polyedra (Table 13–1) is cultured in a liter of half-strength f medium (f/2) (Guillard and Ryther 1962) contained in a 2.8 liter Fernbach flask on an alternating 12 h LD cycle (1,500 $\mu W/cm^2$) at 21°C. A cell density of approximately 10^4 cells/ml is reached in several weeks. Between 06 and 12 circadian time (Chap. 19), a 25-ml aliquot of the laboratory culture, not to exceed a final cell concentration of 10^5 cells, is transferred to a 40-ml conical centrifuge tube and centrifuged (ca. 40*g*, 30 sec). Higher speeds and longer centrifugation times may be required for older cultures (6 weeks) and cultures sampled at 0 circadian time (Adamich 1976). The supernatant is removed by aspiration to approximately 0.1 ml above the packed pellet, and the cells are resuspended in 4 ml L-solution [0.013% v/v Liquinox, a mixture of nonionic, cationic, and anionic detergents (Alconox), 7 m*M* NaCl, 1 m*M* sodium phosphate, pH 8.1, with a determined value of 14 milliosmoles]. The cell suspension is hand-agitated gently for 3 min, the cells are sedimented, and the supernatant is removed to approximately 0.5 ml above the packed pellet, which is left undisturbed for 15 min. Spheroplast formation can be followed in the light microscope by mixing a few drops of the packed cells with a drop of Evan's blue dye (0.25% w/v) (Gaff and Okong'O-Ogola 1971) and covering the preparation with a No. 0 coverslip; the microscope light should be used sparingly. The cell walls are stained blue, but the spheroplasts are not stained except when the spheroplast membrane loses its permeability characteristics.

Ecdysis of the spheroplasts (Fig. 13–5) from the girdle region occurs approximately 15 min following the removal of excess L-solution above the pellet, and yields of 100% are routine. Ecdysis is inhibited if EDTA is present in either the growth medium or L-solution. Once ecdysis is complete (15–20 min), the spheroplasts may be washed with f/2 medium. Cell walls and other debris can be removed by careful layering of the cell suspension onto 5 ml of 40% (w/v) Ficoll (Pharmacia Fine Chemicals) which has been further purified and recrystallized (Uvnäs and Thon 1961).

2. Evaluation. Spheroplasts exclude Evan's blue dye (indicating an intact plasmalemma) and regain motility, bioluminescence, and a cell wall within 1–2 h after resuspension in fresh growth medium. They also display a circadian rhythm in bioluminescence and evolve oxygen at rates comparable to the natural population (Adamich and Prézelin, unpublished observations). A cell wall is not synthesized when spheroplasts are maintained as a packed pellet.

E. Rhodophyceae

Clément-Metral (1974; 1976) has reported a method for preparing spheroplasts from the unicellular red alga *Porphyridium cruentum* (Table 13–1) by utilizing a mixture of digestive enzymes. The spheroplasts, but not normal cells, lyse in 0.05 *M* acetate buffer, pH 5, and they no longer stain with Calcofluor White, which stains the wall of intact cells. Light and scanning electron microscopy of the spheroplasts suggest the disappearance of the mucilaginous cell covering, but it is not clear whether all layers external to the plasmalemma are removed.

1. Preparation of spheroplasts. An axenic culture of *Porphyridium cruentum* (Table 13–1, Fig. 13–6a) is grown in artificial seawater medium (Jones et al. 1963) at 21°C under continuous illumination (2,000 lux; 190 ft-c) with forced aeration (2.5% CO_2 in air) and mechanical agitation. Cells are harvested at the end of the early log phase of growth (ca. 2×10^6 cells/ml) when their sheaths are thinnest and washed in 0.6 *M* NaCl. One gram (wet weight) of cells is suspended in 20 ml 0.6 *M* NaCl containing 0.4% (w/v) Macerozyme pectinase and 4% (w/v) Onozuka cellulase (All Japan Biochemicals, Ltd.) (final pH 5.5), and the mixture is incubated for 2 h at 37°C with gentle shaking. The yield of spheroplasts (Fig. 13–6b) is reported to be from 88% to 100%. Spheroplasts may be washed free of enzymes in 0.6 M NaCl (12,000*g*, 5 min). They are stable for 24 h provided they are kept as a pellet at 5°C.

2. Evaluation. The spheroplasts evolve oxygen at a rate comparable to that of intact cells when illuminated in saturating white light in 0.6 *M* NaCl, 0.1 *M* $NaHCO_3$, pH 8.0. Production of spheroplasts appears to be dependent on enzyme lot, because not all laboratories have been successful using the above procedure (E. Gantt, personal communication).

III. Closing remarks

Some difficulty may be encountered in preparing protoplasts or spheroplasts of the various algae by the methods cited because information which may be critical – the algal strain, culture conditions, cell age, time of sampling in the cell cycle, and time of day – is not available. In addition, impurities in the mixtures of digestive enzymes and the often unavoidable inconsistencies among batches of enzyme mixtures and detergent blends may cause problems.

Ideally, protoplasts and spheroplasts should be prepared under aseptic conditions using axenic cultures, and their exposure to exogenous enzymes should be minimized to reduce endocytosis of foreign matter (e.g., bacteria, viruses, exo-enzymes) that could conceivably lead to structural artifacts, loss in viability, and irreproducible results. This is especially important if algal protoplasts are to be used in genetic research, mutant isolation, and cell fusion studies.

IV. Acknowledgments

We wish to thank the authors of the methods presented here for helpful comments, criticisms, and unpublished details.

V. References

Adamich, M. 1976. "Membrane Circadian Rhythms in *Gonyaulax polyedra.*" Ph.D. dissertation, pp. 1–122. University of California at Santa Barbara, Santa Barbara, California.

Adamich, M., and Sweeney, B. M. 1976. The preparation and characterization of *Gonyaulax* spheroplasts. *Planta* 130, 1–5.

Atkinson, A. W., Jr., Gunning, B. E. S., and John, P. C. L. 1972. Sporopollenin in the cell wall of *Chlorella* and other algae: ultrastructure, chemistry, and incorporation of ^{14}C-acetate, studied in synchronous cultures. *Planta* 107, 1–32.

Azam, F., and Volcani, B. E. 1974. Role of silicon in diatom metabolism. VI. Active transport of germanic acid in the heterotrophic diatom *Nitzschia alba. Arch. Mikrobiol.* 101, 1–8.

Baulina, O. I., Korzhenevskaya, T. G., Nikitina, K. A., and Gusev, M. V. 1975. An electron-microscopic and biochemical study of the spheroplasts of the blue-green alga *Anabaena variabilis. Microbiology* 44, 109–12.

Berliner, M. D. 1977. Protoplast induction in *Chlorella vulgaris. Plant Sci. Lett.* 9, 201–4.

Berliner, M. D., and Wenc, K. A. 1976a. Osmotic pressure effects and protoplast formation in *Cosmarium turpinii. Microbios Letters* 2, 39–45.

Berliner, M. D., and Wenc, K. A. 1976b. Protoplast induction in *Micrasterias* and *Cosmarium. Protoplasma* 89, 389–93.

Bhattacharya, N. C., and Talpasayi, E. R. S. 1973. Action of lysozyme on *Anabaena ambigua* Rao. *Arch Mikrobiol.* 90, 157–9.

Biggins, J. 1967. Preparation of metabolically active protoplasts from the blue-green alga, *Phormidium luridum. Plant Physiol.* 42, 1442–6.

Biggins, J. 1971. Protoplasts of algal cells. In San Pietro, A. (ed.), *Methods in Enzymology,* vol. 23A, pp. 209–11. Academic Press, New York.

Binder, A., Tel-or, E., and Avron, M. 1976. Photosynthetic activities of membrane preparations of the blue-green alga *Phormidium luridum. Eur. J. Biochem.* 67, 187–96.

Braun, E., and Aach, H. G. 1975. Enzymatic degradation of the cell wall of *Chlorella. Planta* 126, 181–5.

Brenner, S., Dark, F. A., Gerhardt, P., Jeynes, M. H., Kandler, O., Kellenberger, E., Kleineberger-Nobel, E., McQuillen, K., Rubio-Huertos, M., Salton, M. R. J., Strange, R. E., Tomcsik, J., and Weibull, C. 1958. Bacterial protoplasts. *Nature* 181, 1713–14.

Chardard, R. 1972. Production de protoplastes d'Algue par un procédé physique. *Comp. Rend. Acad. Sc. Paris Série D.* 274, 1015–18.

Claes, H. 1971. Autolyse der Zellwand bei den Gameten von *Chlamydomonas reinhardii. Arch. Mikrobiol.* 78, 180–8.

Clément-Metral, J. D. 1974. Preparation of metabolically active protoplasts from the red alga *Porphyridium cruentum.* In Avron, M. (ed.), *Proceedings Third International Congress on Photosynthesis,* vol. 3, pp. 2067–71, Elsevier. Amsterdam.

Clément-Metral, J. D. 1976. Preparation and some properties of protoplasts from the red alga *Porphyridium cruentum. Journal de Microscopie et de Biologie Cellulaire* 26, 167–72.

Cocking, E. C. 1972. Plant cell protoplast–isolation and development. *Annu. Rev. Plant Physiol.* 23, 29–50.

Cosner, J. C. 1978. Phycobilisomes in spheroplasts of *Anacystis nidulans. J. Bact.* 135, 1137–40.

Crespi, H. L., Mandeville, S. E., and Katz, J. J. 1962. The action of lysozyme on several blue-green algae. *Biochem. Biophys. Res. Commun.* 9, 569–73.

Drews, G. 1973. Fine structure and chemical composition of the cell envelopes. In Carr, N. G. and Whitton, B. A. (eds.), *The Biology of Blue-Green Algae,* pp. 99–116. Blackwell Scientific, Oxford.

Feyen, V. 1977. "Versuche zur somatischen Hybridisation von *Chlorella saccharophila.*" Ph.D. thesis, pp. 1–101. Rheinisch-Westfälische Technische Hochschule, Aachen, Germany.

Fuhs, G. W. 1958. Enzymatischer Abbau der Membranen von *Oscillatoria amoena* (Kütz.) Gomont mit Lysozym. *Arch. Mikrobiol.* 29, 51–2.

Fulco, L., Karfunkel, P., and Aaronson, S. 1967. Effect of lysozyme (muramidase) on marine and freshwater blue-green algae. *J. Phycol.* 3, 51–2.

Gabriel, M. 1970. Formation, growth, and regeneration of protoplasts of the green alga, *Uronema gigas. Protoplasma* 70, 135–8.

Gaff, D. F., and OKong'O-Ogola, O. 1971. The use of nonpermeating pigments for testing the survival of cells. *J. Exp. Bot.* 22, 756–8.

Grodzinski, B., and Colman, B. 1973. Loss of photosynthetic activity in two blue-green algae as a result of osmotic stress. *J. Bacteriol.* 115, 456–8.

Guillard, R. R. L., and Ryther, J. H. 1962. Studies of marine planktonic diatoms. I. *Cyclotella nana* Hustedt, and *Detonula confervacea* (Cleve) Gran. *Can. J. Microbiol.* 8, 229–39.

Gusev, M. V., Nikitina, K. A., and Korzhenevskaya, T. G. 1970. Metabolically active spheroplasts of blue-green algae. *Microbiology* 39, 752–7.

Hemmingsen, B. B. 1971. "A mono-silicic acid stimulated adenosinetriphosphatase from protoplasts of the apochlorotic diatom *Nitzschia alba.*" Ph.D. dissertation, pp. 1–119. University of California, San Diego.

Hendey, N. I. 1945. Extra frustular diatoms. *J. Roy. Microscop. Soc.* 65, 34–9.

Jensen, T. E., and Sicko, L. M. 1971. The effect of lysozyme on cell wall morphology in a blue-green alga. *Cylindrospermum* sp. *J. Gen. Microbiol.* 68, 71–5.

Jones, R. F., Speer, H. L., and Kury, W. 1963. Studies on the growth of the red alga *Porphyridium cruentum. Physiol. Plant.* 16, 636–43.

Kaback, H. R. 1971. Bacterial membranes. In Jakoby, W. B. (ed.), *Methods in Enzymology,* vol. 22, pp. 99–120. Academic Press, New York.

Kessel, M., MacColl, R., Berns, D. S., and Edwards, M. R. 1973. Electron microscope and physical characterization of C-phycocyanin from fresh extracts of two blue-green algae. *Can. J. Microbiol.* 19, 831–6.

Kratz, W. A., and Myers, J. 1955. Nutrition and growth of several blue-green algae. *Am. J. Bot.* 42, 282–7.

Kuhl, A. 1962. Zur Physiologie der Speicherung kondensierter anorganischer Phosphate in *Chlorella.* In *Beiträge zur Physiologie und Morphologie der Algen,* pp. 157–64. Fischer Verlag, Stuttgart.

Larkin, P. J. 1976. Purification and viability determinations of plant protoplasts. *Planta* 128, 213–16.

Lee, R. F., and Loeblich, A. R., III. 1971. Distribution of 21:6 hydrocarbon and its relationship to 22:6 fatty acids in algae. *Phytochemistry* 10, 593–602.

Lindsey, J. K., Vance, B. D., Keeter, J. S., and Scholes, V. E. 1971. Spheroplast formation and associated ultrastructural changes in a synchronous culture of *Anacystis nidulans* treated with lysozyme. *J. Phycol.* 7, 65–71.

Loeblich, A. R., III. 1970. The amphiesma or dinoflagellate cell covering. In Yochelson, E. L. (ed.), *Proceedings of the North American Paleontological Convention,* vol. 2, pp. 867–929. Allen Press, Lawrence, Kansas.

Marchant, H. J., and Fowke, L. C. 1977. Preparation, culture, and regeneration of protoplasts from filamentous green algae. *Can. J. Bot.* 55, 3080–6.

Ohiwa, T. 1977. Preparation and culture of *Spirogyra* and *Zygnema* protoplasts. *Cell Structure and Function* 2, 249–55.

Price, C. A., and Bourke, M. E. 1966. "Spheroplasts" prepared from *Euglena gracilis* by proteolysis. *J. Protozool.* 13, 474–7.

Ruppel, H.-G. 1962. Untersuchungen über die Zusammensetzung von *Chlorella* bei Synchronisation im Licht–Dunkel-Wechsel. *Flora* 152, 113–38.

Schlösser, U. G., Sachs, H., and Robinson, D. G. 1976. Isolation of protoplasts by means of a "species-specific" autolysine in *Chlamydomonas. Protoplasma* 88, 51–64.

Sigalat, C., and de Kouchkovsky, Y. 1975a. Preparation and properties of photosynthetic fragments of the unicellular blue-green alga *Anacystis nidulans.* In Avron, M. (ed.), *Proceedings Third International Congress on Photosynthesis,* vol. 1, pp. 621–7. Elsevier, Amsterdam.

Sigalat, C., and de Kouchkovsky, Y. 1975b. Fractionnement et caractérisation de l'appareil photosynthétique de l'algue bleue unicellulair *Anacystis nidulans.* I. Obtention de fractions membranaires par "lyse osmotique" et analyse pigmentaire. *Physiol. Vég.* 13, 243–58.

Sofrová, D., Slechta, V., and Leblová, S. 1974. Photochemical reactions related to photosystem II of the blue-green algae *Plectonema boryanum. Photosynthetica* 8, 34–9.

Stadelmann, E. H., and Kinzel, H. 1972. Vital staining of plant cells. In Prescott, D. M. (ed.), *Methods in Cell Physiology,* vol. 5, pp. 325–72. Academic Press, New York.

Stosch, H. A. von. 1965. Manipulierung der Zellgrösse von Diatomeen im Experiment. *Phycologia* 5, 21–44.

Streiblová, E. 1968. Surface structure of yeast protoplasts. *J. Bacteriol* 95, 700–7.

Sweeney, B. M. 1976. Freeze-fracture studies of the thecal membranes of *Gonyaulax polyedra:* circadian changes in the particles of one membrane face. *J. Cell Biol.* 68, 451–61.

Uvnäs, B., and Thon, I.-L. 1961. Evidence for enzymatic histamine release from isolated rat mast cells. *Exp. Cell. Res.* 23, 45–57.

Vance, B. D., and Ward, H. B. 1969. Preparation of metabolically active protoplasts of blue-green algae. *J. Phycol.* 5, 1–3.

14: Cytoplasts from coenocytic algal cells

YOLANDE KERSEY

Department of Biological Sciences,
Stanford University, Stanford, California 94305

CONTENTS

I. Introduction

Membrane-bound cytoplasmic droplets, commonly referred to as cytoplasts, are easily obtained from the large coenocytic cells of some green algae. Cytoplasts from species of Characeae have been studied extensively for over a century (references in Kuroda 1964; Kersey 1972) with respect to motility and to a lesser extent to membrane properties; studies on *Acetabularia* cytoplasts have been concerned with metabolic processes, differentiation, and nucleocytoplasmic relations (Gibor 1965; Werz 1968). Cytoplasts from coenocytic algae such as *Bryopsis* can fuse spontaneously and subsequently differentiate to form a new organism (Tatewaki 1970; Kobayashi et al. 1976).

Cytoplasts from coenocytic algae are useful for experimental purposes in that they are immediately obtainable in relatively large quantities from living cells without the necessity of enzymatic digestion of cell walls. In cytoplasts, the absence of the cell wall offers greatly improved visibility of the cell contents while life processes continue for hours or days, more or less unabated. These properties have been utilized extensively in studies of motile organelles and fibrils in cytoplasts from Characean algae – studies that have aided in the elucidation of the mechanism of cytoplasmic streaming in green plants. (For a discussion of streaming, see, for example, Kersey et al. 1976; Kersey and Wessells 1976; and Chap. 15 of this volume.)

This chapter contains descriptions of techniques used to obtain cytoplasts and examples of research using cytoplasts obtained from coenocytic green algae. Specific methods used to study cytoplasmic streaming and membrane phenomena are covered in Chap. 15; Chap. 5 deals with *Acetabularia* regeneration.

II. Cytoplast preparation from Characeae

A. Cell sap as ambient medium

An internodal cell (of *Chara* or *Nitella*) is cut at one end with a razor blade or scissors, then held vertically and the uncut end clasped between a pair of forceps or fingers, which are drawn down to expel the cell contents onto a microscope slide. The preparation can be protected from evaporation by expressing petroleum jelly (from a No. 22 hypodermic needle) to form a ring somewhat larger than the drop of cell contents, and then pressing on a coverslip to form a sealed chamber. (The syringe is heated in a 60°C oven and loaded with molten jelly at the same temperature, then allowed to cool before use.)

The cytoplasmic droplets are most easily examined using phase or dark-field microscopy, or differential interference optics (Kersey 1972) (see Chap. 20.V.B and Chap. 21). Under these conditions, the motile mechanism of the intact cell remains partially organized, and chloroplast movement can be observed. Furthermore, motile fibrils are often visible; these link the chloroplasts in intact cells and are presumed to be formed from the ectoplasmic fibrils associated with streaming. Individual fibrils are near the limit of resolution of light microscopy, and, when motile, are visible only by virtue of attached microsomes. Details of movement can best be observed on film taken with a high-speed cine camera (e.g., LoCam, Redlake Instruments). Within a few hours the preparation ages; this is marked by coagulation and cessation of organellar motion.

B. Cytoplasts in synthetic cell sap

Artificial cell sap (ACS) of about the same ionic composition and tonicity as the natural cell sap is useful in some manipulations, such as isolation of endoplasm, where large volumes are required, and also in application of drugs. A simple artificial cell sap suitable for these purposes is described by Kamiya and Kuroda (1957), and consists of KNO_3 (0.08 *M*), NaCl (0.05 *M*), and $Ca(NO_3)_2$ (0.004 *M*). In working with several genera and species of Characeae, I found that the viability of some cytoplasts could be increased considerably by empirically changing the concentration of ACS. Survival time of cytoplasts is somewhat less than in natural cell sap; however, Kuroda (1964) maintained preparations up to three days in enriched ACS containing glucose and other ingredients.

C. Isolation of endoplasmic droplets

Relatively large membrane-bound endoplasmic droplets can be obtained from Characean internodes by a method devised by Kamiya and Kuroda (1957), Kuroda (1964), and Kuroda and Kamiya (1975a). In their procedure, a cell is placed in a small vertical chamber containing artificial cell sap and kept under a negative hydrostatic pressure (see Chap. 15.III.B.1).The cell is amputated near the bottom end which extends into a small cuvette filled with artificial cell sap. The cytoplasm flows out into the cuvette gradually over a period of 20–40 min, and the cell does not collapse from turgor loss because of the negative hydrostatic pressure applied externally. The endoplasm contains very few chloroplasts because most remain fixed in the ectoplasmic gel, but the chloroplasts as well as nuclei in the droplet rotate on their own axes. At first there is no mass streaming, but after a few hours, motile fibrils appear, the site of origin of which is as yet unclear (Kuroda, 1964).

D. Production of functional droplets without limiting membranes

Under certain conditions, the cytoplast membrane can be removed, and the cytoplasmic mass remain more or less intact so that some vital functions continue largely undisturbed. This allows the cytoplasm to be accessible to large molecules, which normally would not penetrate the limiting membrane. Kuroda and Kamiya (1975b) devised a medium that allows chloroplast rotation to continue in vitro after removal of the surface membrane of *Nitella* cytoplasts with a microneedle. The solutions consists of 80 m*M* KNO_3, 2 m*M* NaCl, 1 m*M* $Mg(NO_3)_2$, 1 m*M*$Ca(NO_3)_2$, 30 m*M* EGTA, 1 m*M* ATP, 2 m*M* DTT, 160 m*M* sorbitol, 3% w/v Ficoll, and 5 m*M* PIPES buffer, pH 7.0.

The gradual transition from an ambient cytoplasmic matrix to a cytoplasm-free defined medium seems to be important in maintaining the vital processes of isolated organelles in vitro. For example, Ben Shaul et al. (1974) observed that a peripheral layer of cytoplasm seems to be necessary to maintain the integrity of *Codium* cytoplasts during isolation; however, photosynthetic reactions occur just as well in chloroplasts that are subsequently washed free from the cytoplasmic layer. Similarly, in the experiment discussed above, Kuroda and Kamiya found that the cytoplasmic matrix initially serves a protective function but does not appear to be necessary per se for chloroplast motility.

III. Cytoplast preparation from Acetabularia

In his experiments, Gibor (1965) used the apex (2–3 cm long) of *Acetabularia* stalk cells prior to cap development. The enucleate fragments are cut into sections with iridectomy scissors in a petri dish containing a shallow layer of sterile sea water slightly diluted (9:1) with autoclaved water. The cytoplasm is allowed to flow from the cell for several hours, during which time it divides into many cytoplasts of various sizes. With incubation in a special culture medium at 20°C and illumination at 300 ft-c (12:12 h LD cycle), about 10% of the cytoplasts can be expected to survive for 2–3 weeks. The culture medium consists of 3 parts natural sea water with 1 part of a solution containing KCl (1%), NaCl (0.1%), yeast extract (0.05%), serum albumin

(0.05%), TRIS buffer pH 7.8 (0.05 *M*), and raffinose (0.25 *M*). At "death" of the cytoplasts, the surface membrane disappears, leaving a clump of chloroplasts.

An alternate method of cytoplast production (Kamiya and Kuroda 1966) is accomplished by centrifuging the cell contents to one end at about 1200*g* for 4 min, then cutting the stalk with scissors and expressing the cell contents into liquid paraffin. Droplets prepared in this manner survive several days in vitro. Cytoplasmic motility can be studied by time lapse photography in such preparations.

IV. Cytoplast production from Bryopsis

According to Tatewaki (1970), cytoplasts from *Bryopsis* can be obtained by cutting the plant into several pieces about 1–2 cm long and then crushing the pieces between two glass slides. The cytoplasts are then transferred to a watch glass or petri dish containing sterile sea water or sea water enriched by the method of Provasoli (1968). Cytoplasts produced from *Bryopsis* in this manner are joined by fine membranous strands and tend to fuse within 10–30 min to form irregularly shaped masses. Fusion can be accelerated by centrifugation. During the first week of culture in enriched medium, the spherical masses elongate, and the chloroplasts become distributed in the peripheral cytoplasm around a large vacuole. After ca. 1–3 days, chloroplasts become distributed over the surface layer of cytoplasm, enclosing a large central vacuole. The spherical mass soon elongates to form a filament, which eventually branches and produces gametes (Tatewaki 1970, 1973; Kobayashi et al. 1976). By one month, a mature frond is formed.

V. Concluding remarks

The development of new plants from cytoplasts such as occurs in *Bryopsis* species suggests that these organisms may be suitable for studies of somatic hybrids formed by cytoplasm fusion. Fused cytoplasmic masses would be of interest in studies of differentiation, whether or not nuclear fusion, leading to true somatic hybrids, occurred. Cytoplasts from coenocytic algae might lend themselves well to such studies because they are easily obtained mechanically without the lengthy and possibly deleterious effects of enzyme treatment necessary to obtain cytoplasts from small uninucleate plant cells. Observation and manipulation of such cytoplasts in the process of differentiation will no doubt yield clues to factors affecting normal growth and development. The requirement for nuclei could yield information as to which proteins are coded for at various stages of development.

Among the tasks to be completed to further enhance the value of algal cytoplasts in such research are (1) to improve culture conditions so as to allow differentiation to proceed in cytoplasts from a wider variety of plants, (2) to improve methods of cytoplast membrane fusion.

VI. Acknowledgments

For helpful information and comments, I am grateful to Drs. P. B. Green, P. Hepler, R. Nuttall, M. Adamich, B. Hemmingsen, V. Proctor, J. West, M. Tatewaki, A. Gibor, M. Tazawa, and especially Drs. K. Kuroda and N. Kamiya of Osaka University.

VII. References

Ben-Shaul, Y., Schonfeld, M. and Neumann, J. 1974. Photosynthetic reactions and structural studies in the marine alga *Codium vermilara.* In *Proceedings Third International Congress on Photosynthesis,* M. Avron (ed.),vol. 2, pp. 533. Elsevier, Amsterdam.

Gibor, A. 1965. Surviving cytoplasts in vitro. *Proc. Nat. Acad. Sci. U.S.* 54, 1527–31.

Kamiya, N., and Kuroda, K. 1957. Cell operation in *Nitella.* I. Cell amputation and effusion of the endoplasm. *Proc. Japan Acad.* 33, 149–52.

Kamiya, N., and Kuroda, K. 1966. Some observations on protoplasmic streaming in *Acetabularia. Bot. Mag.* Tokyo 79, 706–13.

Kersey, Y. M. 1972. "Observations on the Streaming Cytoplasm and Motion in Cytoplasts of Characean Algae: Role of Microfilaments." Ph.D. dissertation University of California at Irvine, Irvine, California.

Kersey, Y. M. 1974, Correlation of polarity of actin filaments with protoplasmic streaming in Characean algae. *J. Cell Biol.* 63, 2, Pt. 2: 165a.

Kersey, Y. M., and Wessells, N. K. 1976. Localization of actin filaments in internodal cells of Characean algae: a scanning and transmission electron microscope study. *J. Cell Biol.* 68, 264–75.

Kersey, Y. M., Hepler, P. K., Palevitz, B. A., and Wessells, N. K. 1976. Polarity of actin filaments in Characean algae. *Proc. Nat. Acad. Sci. U.S.* 73, 165–7.

Kobayashi, K., Saikawa, M., Hori, T., Tatewaki, M., Enomoto, S., and Wada, S. 1976. Biology of coenocytic green algae. *The Cell* (Tokyo) 8, 360–77 (in Japanese).

Kuroda, K. 1964. Behavior of naked cytoplasmic drops isolated from plant cells. In Allen, R. D. and Kamiya, N. (eds.), *Primitive Motile Systems in Cell Biology,* pp. 31–41. Academic Press, New York.

Kuroda, K., and Kamiya, N. 1975a. Surgical operations in giant plant cells and their applications. *Seitai no Kagaku* 26, 59–66 (in Japanese).

Kuroda, K., and Kamiya, N. 1975b. Active movement of *Nitella* chloroplasts in vitro. *Proc. Japan Acad.* 51, 774–7.

Provasoli, L. 1968. Media and prospects for cultivation of marine algae. In Watanabe, A., and Hattori, A. (eds.), *Cultures and Collections of Algae.* Proc.

U.S.–Japan Conference Hakone, Sept. 1966. Jap. Soc. Plant Physiol., pp. 63–75.

Tatewaki, M., 1973. The coenocytic alga *Bryopsis* and its protoplasts. *Kagaku to Seibutsu* 11, 665–8 (in Japanese).

Tatewaki, M., and Nagata, K. 1970. Surviving protoplasts in vitro and their development in *Bryopsis. J. Phycol.* 6, 401–3.

Werz, G. 1968. Differenzierung und Zellwandbildung in isoliertem Cytoplasma aus *Acetabularia. Protoplasma* 65, 349–57.

15: Cytoplasmic streaming and membrane phenomena in cells of Characeae

MASASHI TAZAWA

*Department of Botany, Faculty of Science,
University of Tokyo, Hongo, Tokyo 113, Japan*

CONTENTS

I. Introduction

Full-grown internodal cells of *Chara* and *Nitella* are 5–10 cm in length and 0.3–1 mm in diameter. Because of their giant-cell dimensions, they are very good material for quantitative studies on the transport of water, electrolytes, and nonelectrolytes in a single cell. Cells of Characeae are excitable and elicit a large action potential amounting to sometimes more than 150 mV. Insertion of microneedles into the cell is easy. Therefore, many electrophysiological studies have been done on characean cells. Another physiological activity that is especially pronounced in *Nitella* and *Chara* internodes is a very rapid cytoplasmic streaming; the rate reaches 60–80 μm/sec at 20–25°C. It is therefore not surprising that a larger part of the important work on cytoplasmic streaming in plant cells has been done using Characeae (Kamiya 1959). Chapter 14 provides greater detail on cell structure of Characeae and covers experimental methods using cytoplasts from coenocytic algae.

This chapter is an introduction to the methods and techniques for studying cytoplasmic streaming and membrane phenomena. Taking advantage of very simple operational techniques, such as amputation, ligation, centrifugation, intracellular perfusion, and strong illumination, we can modify the cell in different ways. The methods described in this chapter are simple and accessible not only technically but also economically. *Chara* and *Nitella* are distributed fairly widely on the earth and cultured fairly easily in the laboratory.

II. Preparation and conditioning of the material

In the following experiments full-grown internodes of Characeae are used; their sizes vary greatly according to species and culture conditions. After internodes have been isolated (as well as ecorticated, in some species) from neighboring cells, they should be allowed to recover (from several hours to as much as one day) in pond water or in artificial pond water (APW). The cytoplasmic streaming stops transiently on isolation because of mechanical shock. Complete recovery of the streaming rate often requires several hours. The normal streaming rate is therefore a good criterion for the healthy state of the cell. The APW used by many workers contains 0.1–1.0 m*M* NaCl, 0.1 m*M* KCl, and 0.1–1.0 m*M* $CaCl_2$.

III. Methods

A. Transcellular osmosis

1. Measurement of osmotic water permeability. Osmotic water permeability can be measured using the osmometer illustrated in Fig. 15-1 (Kamiya and Tazawa 1956; Tazawa and Nishizaki 1956; Dainty and Hope 1959). The main part of the osmometer is composed of the volumeter which has a glass capillary (E); a chamber (A), which accommodates the cell part *a*; and a branched tube (T) with a cock (K). Before the cell is set, the cock is opened and the volumeter is filled with APW. An internode, the surface of which has been blotted with a piece of filter paper, is brought into a groove engraved on the small glass cock (c). To make it water-tight, the cock loaded with the cell is inserted into the chamber with petroleum jelly. An air bubble (d) is introduced from the end of the capillary by slanting the volumeter, while the large cock remains open. After the air bubble (d) has been introduced into the middle of the capillary, the cock is closed. Two large bent-glass tubes (B and C) are joined to the volumeter through rubber stoppers (R_1 and R_2). Both tubes are filled with APW, and the cell is then ready for the transcellular osmosis (TCO). TCO is induced by replacing APW in tube B with APW to which an osmotic agent has been added. Impermeable nonelectrolytes such as sorbitol or mannitol are commonly used as osmotica. Immediately after the replacement the bubble (d) moves from left to right, reflecting transcellular water movement. The bubble movement is followed by the use of an eyepiece micrometer in a horizontal microscope. Since the volumeter is very sensitive to temperature change, it is necessary to conduct the experiment in a constant temperature bath.

When the cell is equally partitioned into A and B, the transcellular water permeability (L_p') can be calculated by the following equation (Tazawa and Kamiya 1966):

$$L_p' = \frac{(dv/dt)_{t\to 0}}{S\Delta\pi/2} \tag{1}$$

where $(dv/dt)_{t\to 0}$ is the initial rate of water flow (within 10 sec), $S/2$ is the surface area of the cell part in A or B, and $\Delta\pi$ is the difference in osmotic pressure between the media in B and A. Since in TCO water passes the membrane twice, L_p should be half the osmotic water per-

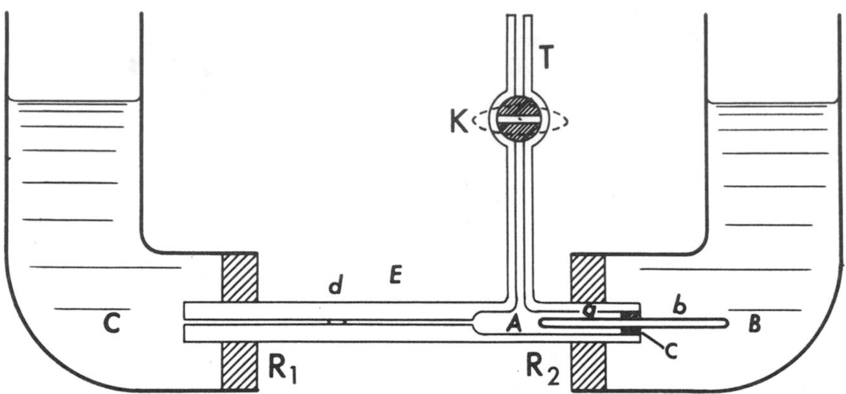

Fig. 15–1. Osmometer for measuring transcellular osmosis. For further explanation see the text. (From Tazawa and Kamiya 1966.)

meability of the membrane (L_p) when the membrane exerts the same resistance to endosmosis and exosmosis. Actually, L_p for endosmosis ($L_{p\,\mathrm{en}}$) is larger than that for exosmosis ($L_{p\,\mathrm{ex}}$). With the increase in $\Delta\pi$, $L_{p\,\mathrm{en}}$ increases, while $L_{p\,\mathrm{ex}}$ decreases (Kiyosawa and Tazawa 1973). To see the effect of a substance on L_p', the APW in chamber A and B is replaced with APW supplemented with the substance. The replacement of the solution in chamber A is done by opening the cock, discarding APW in A and introducing the solution to A through the branch tube T.

The diameter of the capillary is normally ca. 0.5 mm, which is small enough to follow the TCO precisely when the osmotic gradient is not too small. The osmometer shown in Fig. 15-1 can also be used for measuring the flow of water induced by the electrical current supplied through the Ag–AgCl electrodes in chambers A and B (Fensom and Dainty 1963). In this case, because the electroosmotic flow is very small, the diameter of the capillary should be ca. 0.1 mm. To ensure smooth movement of the air bubble in such a fine capillary, use of a dilute detergent solution in the capillary is recommended.

2. *Preparation of cell fragments having higher or lower osmotic pressures.* An internode is placed in a vessel divided into two pools A and B (Fig. 15–2). When one part of the cell (*a*) is bathed in APW and the other part (*b*) is bathed in APW to which mannitol or sorbitol is added, TCO is induced (Fig. 15–2,I). After 15–20 min of TCO, the transcellular water movement almost stops. At this time the osmotic pressures of the cell parts *a* (π_a) and *b* (π_b) in the new equilibrium state are expressed by the following equations:

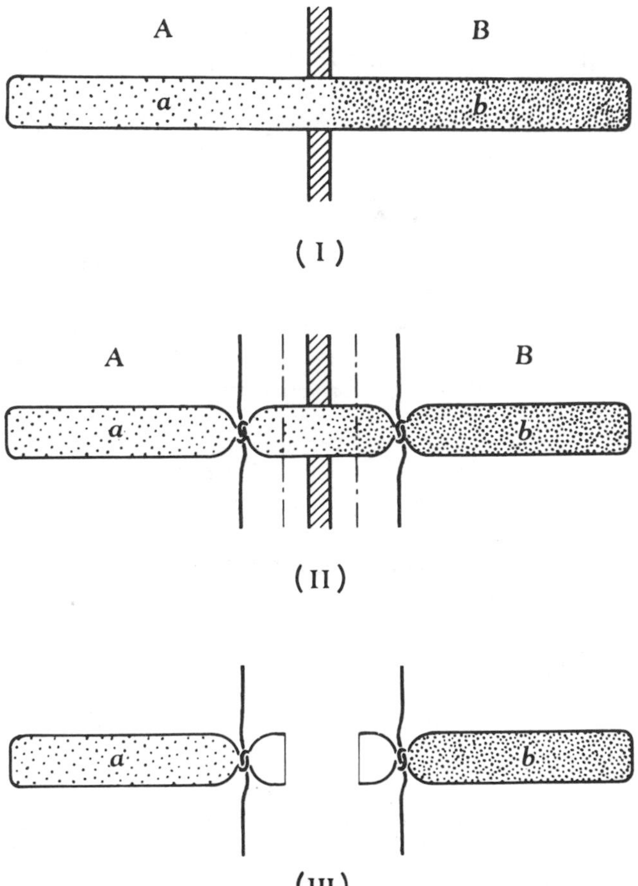

Fig. 15–2. Cell fragment preparation with higher and lower osmotic pressures than the normal osmotic pressure. (I) As the result of transcellular osmosis the cell sap concentration is lowered on side A and heightened on side B. (II) The cell is ligated at two loci. (III) The cell is cut in the middle. (From Kamiya and Kuroda 1956.)

$$\pi_a = \pi_i - \frac{V_b}{V_a + V_b} \Delta\pi \tag{2}$$

$$\pi_b = \pi_i + \frac{V_a}{V_a + V_b} \Delta\pi \tag{3}$$

where π_i is the original osmotic pressure before TCO, V_a and V_b are volumes of a and b and $\Delta\pi$ is the osmotic-pressure gradient between the solutions in A and B. The cell is ligated at the two loci near the partition wall with strips of silk or synthetic thread (Fig. 15–2,II). By amputating the cell at the middle cell fragment, two fragments are produced, one having lower (a) and the other (b) having higher osmotic pressures than the normal one (Fig. 15–2,III).

Cells having abnormal osmotic pressures are excellent material for studies on osmotic and ionic regulations (Kamiya and Kuroda 1956; Nakagawa et al. 1974).

B. Internal perfusion

The composition of the cell sap can be easily modified by replacing the cell sap with artificial solutions containing Ca^{++} by means of vacuolar perfusion (Tazawa 1964). In order to change the composition of the cytoplasm, however, the perfusion medium should contain a Ca^{++} chelating agent that is effective in removing the tonoplast (Tazawa et al. 1976.)

1. Preparation of the cell having artificial cell sap. Before vacuolar perfusion, cells are first stained at ca. pH 7.0 with neutral red, which stains the cell vacuole. Four strips of silk thread (l_1, l_2, l_3, l_4) are placed on the Plexiglas perfusion bench as shown in Fig. 15–3. The top surface of the bench is coated with petroleum jelly. An internode, which has been blotted with a piece of filter paper, is placed on the perfusion bench. Two drops (d_1, d_2) of the perfusion medium are placed on the bench to cover both cell ends. The middle part of the cell is exposed to the air. After loss of turgor, which can be recognized by a slight decrease in cell diameter, both cell ends are amputated. The bench is slanted a little so that the perfusion medium flows into the vacuole and the red cell sap is pushed out from the other end. This should be completed within 1–2 min with a cell length of ca. 5 cm. Observing that no red sap is coming out, the cell is tied off at two loci with l_1 and l_4. The sorbitol solution, being hypotonic by 0.1–0.15 *M* to the perfusion medium, is placed onto the cell to restore the turgor. Then, the cell is tied off again with l_2 and l_3. After removing the small marginal cell fragments, a cell fragment containing the artificial cell sap is obtained. Cells showing vigorous cytoplasmic streaming are then ready for experiments such as membrane potential measurement by means of ordinary microelectrode method (Tazawa and Kishimoto 1964).

The perfusion media must contain Ca^{++} (1 m*M* or more) to maintain the tonoplast intact. For longer survival of the perfused cells, the tonicities of the perfusion media should be adjusted to be isotonic or hypertonic to the natural cell sap. The best survival is obtained when the ionic composition is similar to that of the natural cell sap. For the ionic composition, Hope and Walker (1975, Table 5.4) and Tazawa et al. (1974) should be consulted. Modification of the composition of the vacuolar fluid provides great advantages for studying ionic and os-

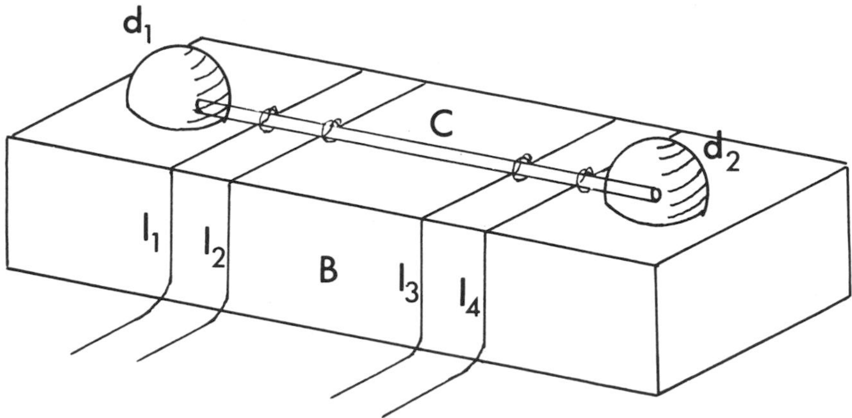

Fig. 15–3. Perfusion bench made of Plexiglas. Internode (C), strips of thread (l_1, l_2, l_3, l_4), and drops (d_1, d_2) of perfusion medium.

motic regulation (Nakagawa et al. 1974) and electrical properties of the tonoplast (Tazawa and Kishimoto 1964; Kikuyama and Tazawa 1976a, b).

2. *Open-vacuole method.* This method (Tazawa et al. 1975) is used for measuring the potential difference between the inside and outside of the cell without having to insert a microelectrode into the cell. An internode (N) whose vacuole has been stained with neutral red is placed on a Plexiglas vessel illustrated in Fig. 15–4. The vessel has three pools (A, B, and C). A and C are connected through a connecting glass or rubber tubing (T) which serves to eliminate the difference in water level between the two pools. A, C, and T are filled with the perfusion medium while B remains empty. The vessel is coated with a thin layer of petroleum jelly to prevent the medium from dispersing readily along the surface of the internode. When the cell has lost turgidity due to evaporation, both cell ends are cut. After closing the valve (V) or pinching off the rubber tubing, one either adds a small amount of the perfusion medium to A or soaks up the solution in B with a piece of filter paper. The perfusion medium in A then flows into the cell, and the red cell sap comes out of the cell opening in B. After checking that all the red sap has come out, one opens the valve or pinchcock.

To measure the vacuolar potential and the membrane resistance of the cell part in B, it is filled with a solution that is isotonic or slightly hypotonic to the perfusion medium. The electrodes consist of polyethylene tubing filled with 100 m*M* KCl agar (2%) and connected to a Ag–AgCl wire through a 3 *M* KCl solution. One electrode is placed in

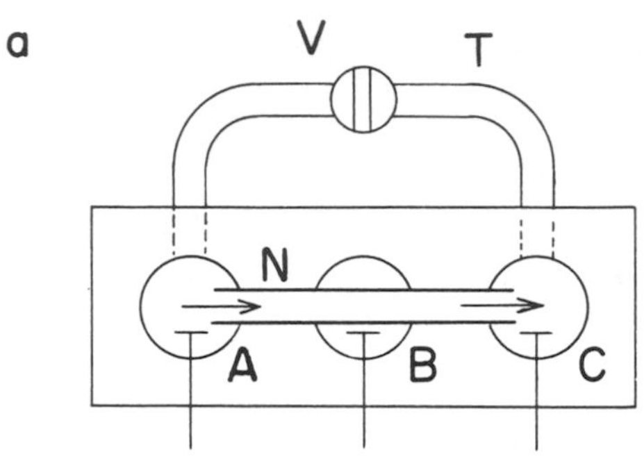

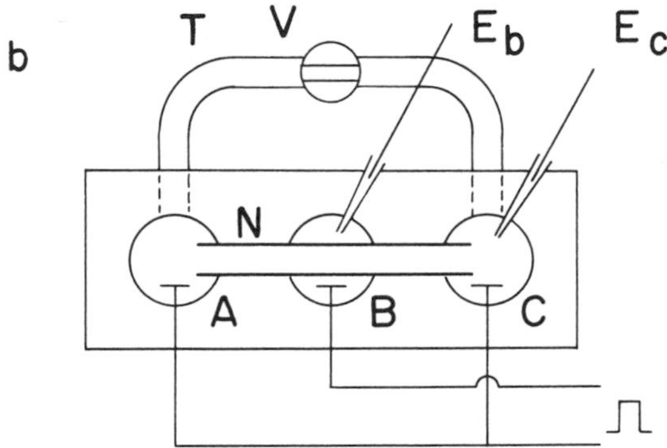

Fig. 15–4. Vessel for measuring electric potential of the cell (N) by the open-vacuole method. a. The vacuole of the cell is perfused with the medium from A to C by giving a pressure difference between A and C. The valve (V) is closed during perfusion. b. After the cell sap has been completely replaced with the artificial medium, V is opened and reservoir B is filled with an isotonic medium. Tips of two agar electrodes (E_b and E_c) are immersed in pools B and C to measure the vacuolar potential of the cell in B. The electric current is supplied through an Ag–AgCl wire. (From Kikuyama and Tazawa 1976a.)

A and the other in B. Since the vacuole is open to A and C, the potential difference between the two electrodes under zero current should represent the vacuolar potential of the cell bathed in B. The electric current pulse for stimulation is given to the cell through Ag–AgCl wire between B and the end pools. Characteristics of the tonoplast of *Nitella* internodes have been studied in detail by the open-vacuole method by Kikuyama and Tazawa (1976a, b).

3. Measurement of motive force of cytoplasmic streaming. The perfusion vessel shown in Fig. 15–5 is like that shown in Fig. 15–4 except that it has only two pools (A_1 and A_2). Two pieces of narrow glass slides (g_1,

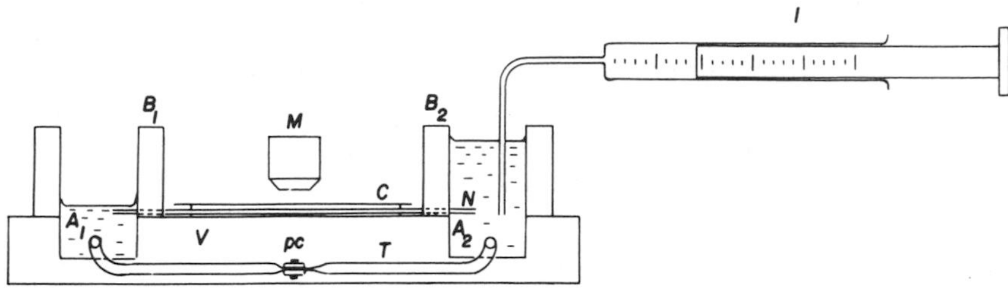

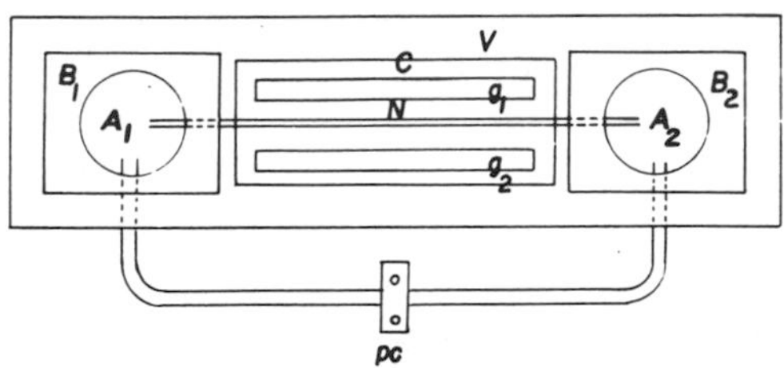

Fig. 15–5. Perfusion vessel for measuring the motive force of cytoplasmic streaming, as seen from the side (top) and from above (bottom). Vessel (V), Plexiglas blocks with cylindrical hole (B_1, B_2) tube (T) connecting pools (A_1 and A_2), pinchcock (pc), internodal cell (N), pieces of glass slides (g_1, g_2), cover slip (c), injection syringe (I). (From Tazawa 1968.)

g_2) are cemented to the top of the vessel between the two pools. An internode (N) is loaded on the vessel. In this case the cell is not stained with neutral red. To produce a difference in water level between the pools, two supplemental Plexiglas blocks (B_1, B_2), each with a cylindrical hole 1 cm in diameter, are attached to the vessel. A small groove, big enough for the internode to be fitted into, is engraved on the bottom of each block. A pair of blocks are placed onto the vessel with petroleum jelly so that each hole comes just over the pool and the cell is held in the groove of each block. This procedure must be done before the cell loses its turgor, because in the turgorless stage the cell is easily compressed by adding the petroleum jelly.

The next step is to put an isotonic perfusion medium in the pools and in the connecting tube. The middle part of the cell is exposed to the air until the cell loses turgor through evaporation. Liquid paraffin is introduced on the vessel between g_1 and g_2, and a cover slip (c) is put on them. Thus, the middle part of the cell is embedded in liquid paraffin. This procedure is necessary not only for preventing further evaporation but for obtaining good optics. After both cell ends are cut off, the vessel is put under the microscope for observation of cytoplasmic streaming.

To induce perfusion, the connecting tube is closed by a pinchcock (pc), and the injection syringe (I) containing the perfusion fluid is fixed to a micromanipulator in a horizontal position. The tip of the bent injection needle is immersed in the perfusion medium in one pool. One supplies the perfusion fluid to the pool from the injection syringe while observing cytoplasmic streaming, which is opposite to the direction of perfusion. As soon as cytoplasmic streaming stops, no more perfusion fluid is added and the pinchcock is opened so that the water levels in the two pools are rapidly equalized. Addition of perfusion medium and stoppage of streaming normally takes about 5 sec. The motive force F of the cytoplasmic streaming is calculated by the following formula:

$$F = \frac{\Delta p \times r}{2L} \tag{4}$$

where Δp is the pressure difference between the two pools, r and L are the radius and the length of the cell, respectively (Tazawa 1968).

Two other methods are also available for measuring the motive force (Kamiya and Kuroda 1958; 1973).

4. Preparation of cells lacking tonoplasts. Using the technique of vacuolar perfusion described in III.B.1 above, but using a perfusion medium containing a Ca^{++}-chelating agent, tonoplast-free cells are prepared. The composition of the media used most commonly for *Chara australis* is as follows: 5 m*M* EGTA [ethyleneglycol-bis-(β-aminoethyl ether) *N,N'*-tetraacetic acid], 5 m*M* TRIS–maleate (pH 7.0), 17 m*M* KOH, 6 m*M* $MgCl_2$, 1 m*M* ATP, and 300 m*M* sorbitol (Tazawa et al. 1976). Loss of the tonoplast is indicated by absence of the clear contour between the endoplasm and the vacuole, or the appearance of endoplasm fragments in the vacuolar space. The rate of cytoplasmic streaming is decreased only slightly by the disappearance of the tonoplast (Tazawa et al. 1976). It should be noted that ATP and Mg^{++} are essential for maintenance of cytoplasmic streaming (Williamson 1975; Tazawa et al. 1976; Shimmen 1978) normal resting potential, and membrane excitability (Shimmen and Tazawa 1977).

Tonoplast-free cells can be subjected to electrical measurements by means of either the conventional microelectrode method (Shimmen et al. 1976) or the open-vacuole method (Shimmen and Tazawa 1977). In case of the open-vacuole method chambers A and C can be filled with solutions of different compositions. For example, the solution in A can contain ATP, but it is omitted from the solution in C. Intracellular composition can then easily be modified by changing the direction of perfusion.

The mechanism of cytoplasmic streaming is easily studied by using tonoplast-free cells, because the chemical environment surrounding the motile fibrils existing at the sol-gel interface (Kamitsubo 1972a; Nagai and Rebhun 1966) can be easily modified. Moreover, possible components of the motile system such as myosin, actin, and others can also be introduced into the tonoplast-free cells by internal perfusion.

Using the perfusion vessel designed for measuring the motive force (Fig. 15–5) most of the endoplasm can be swept away by very rapid perfusion (Williamson 1975). The motile fibrils, however, remain attached to the chloroplasts on the cortical gel layer and the small endoplasmic particles along the fibrils can be observed by differential interference microscopy (Chap. 21).

The tonoplast-free system opens the way to the modeling of cytoplasmic streaming. In this connection, rotation movement of chloroplasts in the isolated endoplasmic drop (the envelope of which is removed by EGTA treatment) is also useful for the study of cytoplasmic movement (Kuroda and Kamiya 1975).

C. Centrifugation

In general, centrifugation of plant cells moves the endoplasm with nuclei and chloroplasts to the centrifugal end. This improves the optical condition of the locus where chloroplasts are removed. To control the extent of chloroplast removal in characean cells, it is practical to use a simple centrifuge microscope (Brown 1940; Hayashi 1957; Kamitsubo 1972a). Fig. 15–6 shows a schematic view of the Brown-type centrifuge microscope. A small internode (s), 0.5–1.5 cm in length, is introduced in a small glass vessel (c) with APW. The vessel is then attached to the front of the rotor (R) (ca. 4 cm in diameter). The light beam coming through the condenser (C) is reflected by a small mirror (m), attached to the rotor, and enters into the objective (ob) of the microscope. Since the mirror forms an angle of 45° to the rotor axis, the virtual image (s′) of the sample (s) is formed at the center of the axis and is therefore observed as a resting image. Detachment of most of the chloroplasts from the cortical gel is injurious to the cell. The speed of rotation should therefore be controlled so that the chloroplasts (at the centrifugal end) are detached from only 10–30% of the whole cell area. In *Nitella flexilis* this is obtained in 3 min at 600–1,000*g* (Kamitsubo 1972a).

On strong centrifugation, the motile fibrils are removed with the chloroplasts to which they are attached, and cytoplasmic streaming is no longer observed. Fig. 15–7 shows a schematic view of a centrifugal end of the *Nitella* cell. The endoplasm (B) is accumulated above the layer of chloroplasts (A). The moving fibrils are rarely found in B, since the optical conditions are probably poor there. However, the moving fibrils are often found in D, where only a thin layer of the

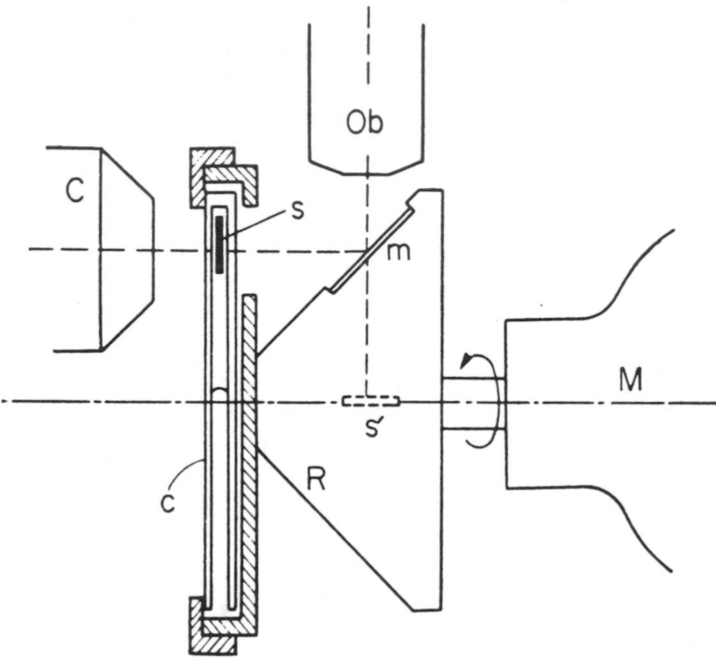

Fig. 15–6. Schematic view of Brown-type centrifuge microscope (1940). Condenser (C), vessel (c) containing specimen(s), rotor (R), mirror (m), objective of the microscope (Ob), motor (M), virtual image of s(s′). (From Tazawa et al. 1969.)

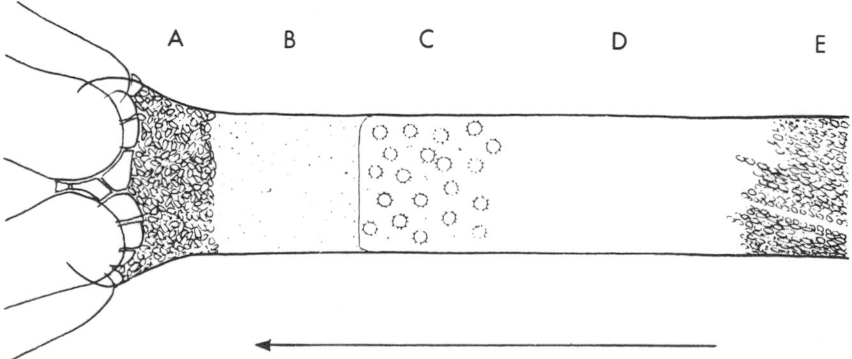

Fig. 15–7. Part of a *Nitella* internode after centrifugation in the direction of the arrow showing a mass of chloroplasts (A), endoplasmic mass (B) in which nucleic are contained, vacuolar inclusions (C), clear zone from where chloroplasts were removed (D), and a zone where chloroplasts remain attached to the cortex (E). (From Kamitsubo 1972a.)

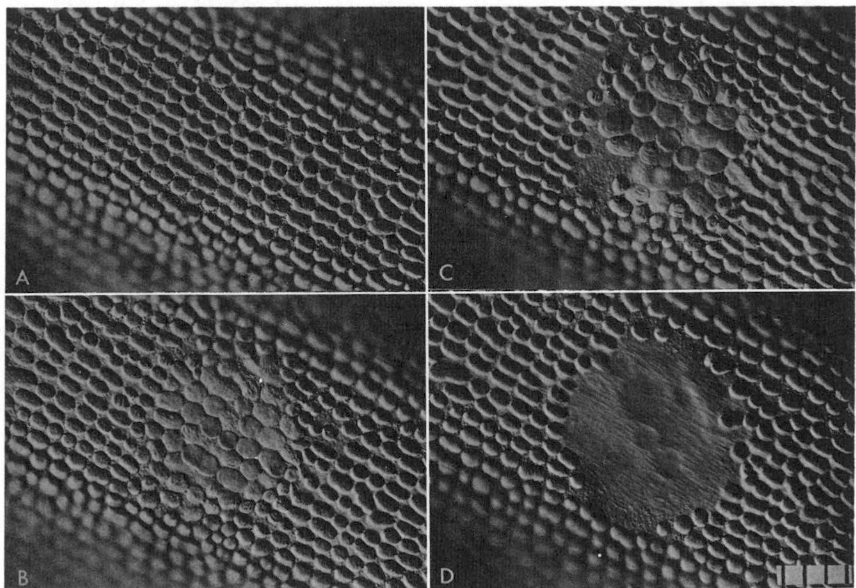

Fig. 15–8. Surface view of a part of a *Nitella* internode. A: before microbeam irradiation. B: 1 h after microbeam irradiation; chloroplasts are bleached and swollen. C: 2 h after irradiation; some chloroplasts are dislodged. D: 1 day after irradiation; at the window lacking chloroplasts, the cytoplasmic streaming is nearly normal. (From Kamitsubo 1972b.)

endoplasm remains after centrifugation. The moving fibrils, detached from the cortical gel, form polygonal loops of various shapes and rotate very rapidly (Kamitsubo 1972a). These loops are structures homologous to the loops found in isolated endoplasm (Jarosch 1956; Kuroda 1964). However, the polygonal loops are more easily observed because they are coated with particles.

D. Window technique

Microbeam irradiation experiments can be performed on *Nitella* internodes (Fig. 15–8A–D). Strong, white light bleaches the chloroplasts of the irradiated area, with the bleaching being proportional to the light intensity. When the chloroplasts are thin, as in *Nitella axillaris* or *Nitella axilliformis,* 3–5 min irradiation with an Osram Mercury Arc HBO 200 W/4 is sufficient. In an irradiated area cytoplasmic streaming is significantly slowed down, indicating that the site generating the motive force is destroyed. The damaged chloroplasts, which swell (Fig. 15–8B), can be removed by centrifugation (ca. 200*g*) 45–60 min after irradiation. A few days after irradiation, fibrils are regenerated (Fig. 15–8D) and cytoplasmic streaming recovers completely (Kamitsubo 1972b). The window technique makes it possible to study cytoplasmic structure and motile processes in the cell at the highest magnifications offered by the light microscope. The cortical fibrils are easily observed in this way with suitable optical systems such as phase contrast or differential interfence microscopy.

IV. Acknowledgment

The author is indebted to Dr. T. Shimmen for his help in preparing the manuscript.

V. References

Brown, R. H. J. 1940. The protoplasmic viscosity of *Paramecium. J. Exp. Biol.* 17, 317–24.

Dainty, J., and Hope, A. B. 1959. The water permeability of cells of *Chara australis* R. BR. *Aust. J. Biol. Sci.* 12, 136–45.

Fensom, D. S., and Dainty, J. 1963. Electro-osmosis in *Nitella. Can. J. Bot.* 41, 685–91.

Hayashi, T. 1957. Some dynamic properties of the protoplasmic streaming in *Chara. Bot. Mag.* (Tokyo) 70, 168–74.

Hope, A. B., and Walker, N. A. 1975. *The Physiology of Giant Algal Cells.* Cambridge University Press, Cambridge. 201 pp.

Jarosch, R. 1956. Plasmaströmung und Chloroplastenrotation bei Characeen. *Phyton* (Buenos Aires) 6, 87–108.

Kamitsubo, E. 1972a. Motile protoplasmic fibrils in cells of the Characeae. *Protoplasma* 74, 53–70.

Kamitsubo, E. 1972b. A 'window technique' for detailed observation of characean cytoplasmic streaming. *Exp. Cell Res.* 74, 613–16.

Kamiya, N. 1959. Protoplasmic Streaming. In Heilbrunn, L. V. and Weber, F. (eds.), *Protoplasmatologia,* vol. VIII 1a. Springer-Verlag, Vienna.

Kamiya, N., and Kuroda, K. 1956. Artificial modification of the osmotic pressure of the plant cell. *Protoplasma* 46, 423–36.

Kamiya, N., and Kuroda, K. 1958. Measurement of the motive force of the protoplasmic rotation in *Nitella. Protoplasma* 50, 144–8.

Kamiya, N., and Kuroda, K. 1973. Dynamics of cytoplasmic streaming in a plant cell. *Biorheology* 10, 179–87.

Kamiya, N., and Tazawa, M. 1956. Studies on water permeability of a single plant cell by means of transcellular osmosis. *Protoplasma* 46, 394–422.

Kikuyama, M., and Tazawa, M. 1976a. Tonoplast action potential in *Nitella* in relation to vacuolar chloride concentration. *J Membr.Biol.* 29, 95–110.

Kikuyama, M., and Tazawa, M. 1976b. Characteristics of the vacuolar membrane of *Nitella J. Membr. Biol.* 30, 225–47.

Kiyosawa, K., and M. Tazawa. 1973. Rectification characteristics of *Nitella* membranes in respect to water permeability. *Protoplasma* 78, 203–14.

Kuroda, K. 1964. Behaviour of naked cytoplasmic drops isolated from plant cells. In Allen, R. D. and Kamiya, N. (eds.), *Primitive Motile Systems in Cell Biology,* pp. 31–41. Academic Press, New York.

Kuroda, K., and Kamiya, N. 1975. Active movement of *Nitella* chloroplasts in vitro. *Proc. Japan Acad.* 51, 774–7.

Nagai, R., and Rebhun, L. I. 1966. Cytoplasmic microfilaments in streaming *Nitella* cells. *J. Ultrastruc. Res.* 14, 571–89.

Nakagawa, S., Kataoka, H., and Tazawa, M. 1974. Osmotic and ionic regulation in *Nitella. Plant and Cell Physiol.* 15, 457–68.

Shimmen, T. 1978. Dependency of cytoplasmic streaming on intracellular ATP and Mg^{++} concentration. *Cell Structure and Function* 3, 113–21.

Shimmen, T., and Tazawa, M. 1977. Control of membrane potential and excitability of *Chara* cells with ATP and Mg^{++}. *J. Membr. Biol.* 37, 167–92.

Shimmen, T., Kikuyama, M., and Tazawa, M. 1976. Demonstration of two stable potential states of plasmalemma of *Chara* without tonoplast. *J. Membr. Biol.* 30, 249–70.

Tazawa, M. 1964. Studies on *Nitella* having artificial cell sap. I. Replacement of the cell sap with artificial solutions. *Plant and Cell Physiol.* 5, 33–43.

Tazawa, M. 1968. Motive force of the cytoplasmic streaming in *Nitella. Protoplasma* 65, 207–22.

Tazawa, M., and Kamiya, N. 1966. Water permeability of a characean internodal cell with special reference to its polarity. *Aust. J. Biol. Sci.* 19, 399–419.

Tazawa, M., and Kishimoto, U. 1964. Studies on *Nitella* having artificial cell sap. II. Rate of cyclosis and electrical potential. *Plant and Cell Physiol.* 5, 45–59.

Tazawa, M., and Nishizaki, Y. 1956. Simultaneous measurement of transcellular osmosis and the accompanying potential difference. *Jap. J. Bot.* 15, 227–238.

Tazawa, M., Kuroda, K., and Kamiya, N. 1969. Centrifugation of living cells. In Ebashi, S. and Kamiya, N. (eds.), *Research Methods in Biophysics of Cells,* Vol. 1, pp. 199–214. Yoshioka, Kyoto.

Tazawa, M., Kishimoto, U., and Kikuyama, M. 1974. Potassium, sodium, and chloride in the protoplasma of Characeae. *Plant and Cell Physiol.* 15, 103–10.

Tazawa, M., Kikuyama, M., and Nakagawa, S. 1975. Open-vacuole method for measuring membrane potential and membrane resistance of Characeae cells. *Plant and Cell Physiol.* 16, 611–22.

Tazawa, M., Kikuyama, M., and Shimmen, T. 1976. Electric characteristics and cytoplasmic streaming of Characeae cells lacking tonoplast. *Cell Structure and Function* 1, 165–76.

Williamson, R. E. 1975. Cytoplasmic streaming in *Chara*: a cell model activated by ATP and inhibited by cytochalasin B. *J. Cell Sci.* 17, 655–68.

16: Microbeam irradiation in Mougeotia

WOLFGANG HAUPT

University of Erlangen-Nürnberg,
D-8520 Erlangen, Federal Republic of Germany

CONTENTS

I. Introduction

Phytochrome is a photoreceptor pigment which is responsible for many photomorphogenetic processes in plants. It seems to be universally present in algae and in higher plants. Localization of phytochrome in cells is important for understanding the mechanism of its action; however, this localization is difficult to recognize because phytochrome occurs in very small quantities. Thus, localized physiological responses induced by localized phytochrome photoconversions should be analyzed. Microbeam irradiation of portions of a single cell has been used successfully in localizing phytochrome in the green filamentous alga *Mougeotia.* This alga has been found very suitable for these studies, because a focused microbeam of light can cause a fast phytochrome-controlled response (Haupt 1973), which is expressed by the rotation of the single chloroplast.

Mougeotia is characterized by having one flat chloroplast, the length and width of which correspond, normally, to the length and diameter of the cylindrical cell (ca. 200 μm and 20 μm, respectively). Thus the cell with its vacuole is divided by the chloroplast, into two half-cylinders (Fig. 16–1b). The chloroplast is able to turn in the cell so as to expose its face (full surface) or its profile (side view) to unidirectional light. Under natural conditions face position is found in low-intensity light, and profile position in high-intensity light. The "low-intensity movement" to face position is obtained optimally in red light, and this red effect is reversible by subsequent far-red light. Thus, phytochrome is the photoreceptor pigment (Haupt and Schönbohm 1970), and it is the aim of this chapter to deal with the low-intensity response and its application to phytochrome localization in *Mougeotia.*

In addition, the chloroplast reorientation depends on the orientation of the electrical vector of linearly polarized light (= action dichroism): the low-intensity movement requires that at least part of the light vibrates perpendicularly to the long axis of the cell (Fig. 16–1a). Such observations indicate a specific orientation pattern of dichroic phytochrome molecules, and this, in turn, points to well-defined sites of localization of phytochrome within the cell (Haupt 1973).

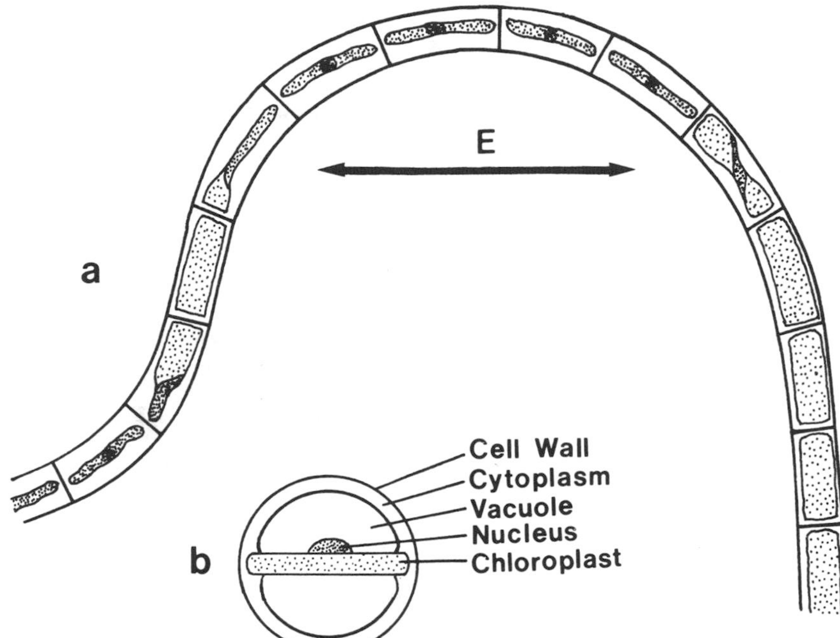

Fig. 16–1. *Mougeotia* after irradiation with polarized red light with the electrical vector vibrating along the double arrow (a). At the beginning of the experiment, the chloroplast was in profile position in all cells. Schematic cross section through a *Mougeotia* cell (b). (After Haupt 1970b; and Haupt and Schönbohm 1970.)

II. Materials and methods

A. Experimental organism

Mougeotia sp. cells, originally obtained from a natural habitat in South Germany and since then cultivated in several laboratories, can be grown in pond water under artificial illumination (Haupt 1959). In our laboratory, they are grown in fluorescent light (100–250 lux; 10–25 ft-c) with a LD cycle of 14:10 h at 15–20°C. We normally use a soil-water extract medium such as that described by Weisenseel (1968).

B. Equipment for microbeam irradiation and experimental conditions

Partial illumination of microscopic fields was used by Buder (1915), who adapted an objective system in place of the condenser system on a light microscope. In this way he was able to obtain an image of a slit in the object plane.

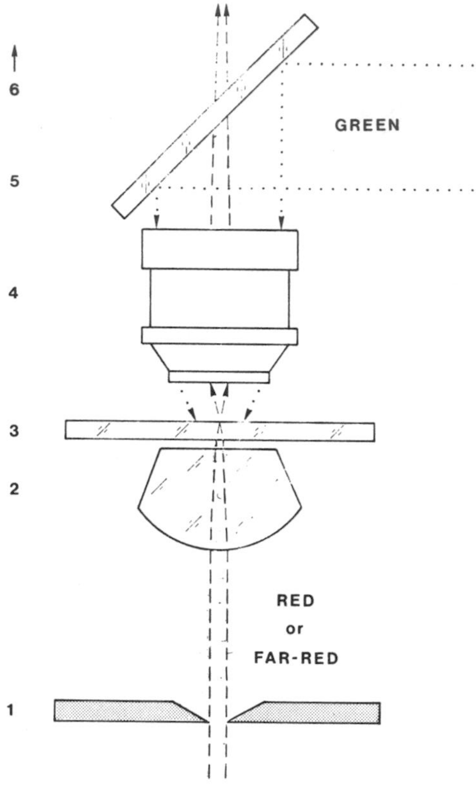

Fig. 16–2. Schematic drawing of the microbeam equipment. The small diaphragm (1) is imaged by the condenser (2) on the cell at the surface of the slide (3). The cell is observed by the microscope with its objective (4) and eyepiece (6). Total irradiation of the microscopic field by green light, reflected by the semitransparent or dichroic mirror (5).

1. Microspectrophotometer. We have used a commercial microspectrophotometer by Zeiss (Fig. 16–2) (Bock and Haupt 1961). With this instrument, a circular or square beam can be imaged on the cell, and by selecting the proper diaphragm, we obtain microbeams of 3, 6, 12, or 24 μm diameter as compared to the cell diameter of ca. 20 μm.

2. Conditions required

a. Choice of filters. To obtain monochromatic light, interference filters can be inserted between the light source and the diaphragm. As always in photobiology, one has to find the proper compromise between the intensity needed and the best possible spectral purity. In our experiments, either red (660 nm) or far-red light (720 nm) are used in order to phototransform phytochrome P_r to P_{fr} or vice versa. Under these conditions, interference band filters (Schott, Mainz) with a half-width of 20–25 nm are suitable. Contamination of red by far-red and vice versa is negligible, and the intensity allows for successful induction (chloroplast rotation) in the range of one minute. In addition, since light of the long wavelength end is used, no special requirements for the light source have to be met. We mostly use a tungsten point source.

b. Specimen mounting. For microbeam experiments, a small sample (if possible, only one filament) is mounted on a slide and coverslip in growth medium. One should be able to define and recognize individual cells in the filament. By irradiation of the whole preparation with red light either from above or from the side, the chloroplasts will be oriented to the face or profile position, respectively, as the desired starting point.

c. Regulation of light dose. By irradiating populations of whole cells with red light, we established that the percentage of cells induced (chloroplast turned from profile to face position) increases either with increasing intensity (I) or with increasing duration (t) of the inducing light. Reciprocity holds in *Mougeotia* as long as irradiation times are in the range of less than 5 min (Haupt 1959). Hence, in order to vary the energy of the inducing light, it is sufficient to vary its duration and keep the intensity constant.

d. Localized irradiation. To position the microbeam at a desired area in the cell without transiently touching a "wrong" area, the total microscopic field has to be illuminated by light that does not cause a response. As a safelight, we use green light, which is obtained by a separate lamp by insertion of either an interference filter (e.g., 537 nm) or a combination of glass filters (e.g., Schott BG 18, 4 mm thick, supplemented by GG 14, 2 mm thick). The green light is reflected to the microscopic field by a semitransparent, or a dichroic mirror, which is inserted either in the light path of the microbeam below the microscopic field or in the light path between the objective and eyepiece (thus illuminating the cells from above, Fig. 16–2). Green light at sufficient energy can photoconvert some phytochrome (Haupt 1959); however, with the time and intensity needed to select, localize, and orient the proper cell, the green light does not cause any measurable response. The eyepiece of the microscope is provided with a marker (e.g., spider lines) to facilitate the proper orientation and localization of cell and microbeam.

If the chloroplast in profile position is irradiated with the red microbeam at one end of the cell, only half of the chloroplast or less responds by turning to the face position, that is, the chloroplast twists (Fig. 16–3b). We make use of this autonomy in all our experiments. The microbeam is always placed at one end of the cell, the response of the chloroplast at this end is observed, and the other end is taken as a

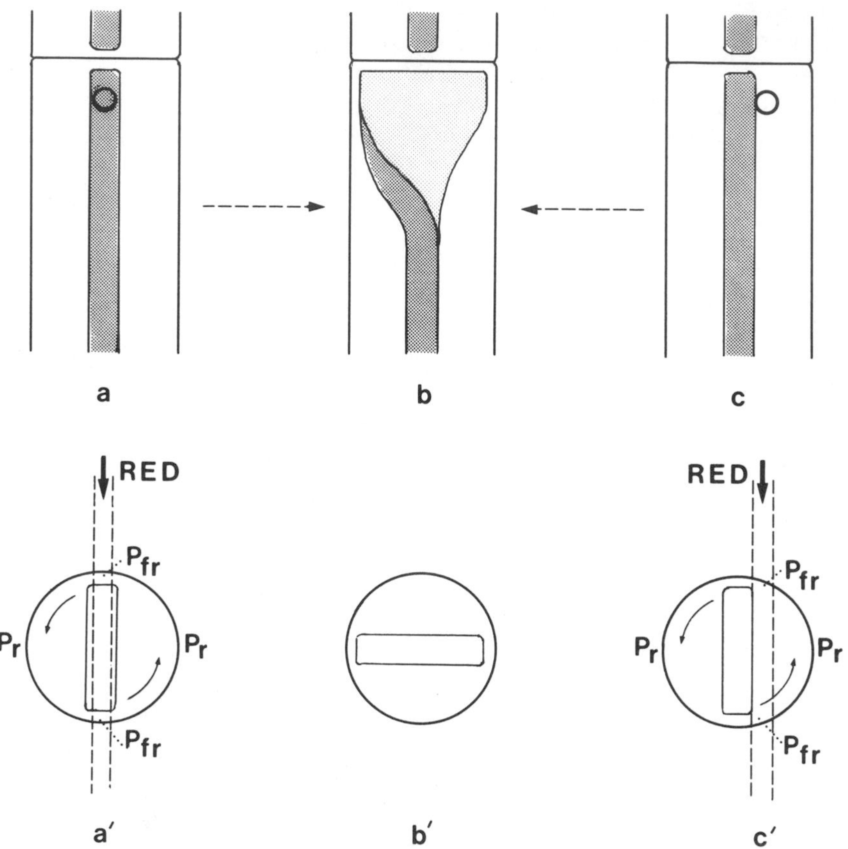

Fig. 16–3. Schematic drawing of *Mougeotia* cells in surface view (above) and in cross section (below) with the positioning of the microbeam at a, a′ or beneath c, c′ the chloroplast. The resulting P_{fr} gradient is symmetric (a′) or asymmetric (c′); the response of the chloroplast is shown in b, b′. (After Bock and Haupt 1961.)

blank (that is, only those cells are scored in which reorientation has not occurred at that other end).

Next, the smallest microbeam is placed either on the profile of the chloroplast (Fig. 16–3a), or as close to it (Fig. 16–3c) as possible. A response occurs in both cases, that is, whether or not the chloroplast is irradiated in addition to the cytoplasm. To be sure that this result is not due to scattered or diffracted light reaching the chloroplast, dose-response curves for the "chloroplast position" and the "cytoplasm position" have to be compared. The effects of both these positions differ by a factor of 3 to 4, whereas with scattered and diffracted light the effect of cytoplasm position should not exceed a few percent of that of chloroplast position (Bock and Haupt 1961).

Table 16–1. *Phytochrome as photoreceptor pigment for chloroplast orientation in Mougeotia.*

Experimental condition[a]	Chloroplast position[b]
Before irradiation	Profile
R	Turns into face
R–FR	Remains in profile
R–FR–R	Turns into face

[a] R = red, FR = far-red, 1 min each.
[b] Response within 30 min of darkness after the short irradiations.
Source: After Haupt (1959; 1970b).

Photoreversion experiments can be done using the irradiation pattern of Fig. 16–3 by following the red with a microbeam of far-red (same location and beam size). (See also II.B.2.a.)

III. Results and conculsions

Physiological effects of low-energy red light are suggestive of a phytochrome effect. In *Mougeotia* the induction (by red light) and the response are separated from each other in time, and therefore, reversion experiments can be tried. If the red induction is followed immediately by a brief far-red irradiation, the effect can be cancelled in the classical way (Haupt 1959) (Table 16–1). This red/far-red antagonism is definite proof of phytochrome's being the photoreceptor pigment in the chloroplast movement of *Mougeotia,* and this movement, therefore, can be used as a criterion of phytochrome action.

More detailed experiments can also be performed with slightly different starting conditions (Haupt 1970a). If the chloroplast is in the face position (phytochrome being mainly in the P_r form), and the red microbeam is placed at the edge of the cell, phytochrome in that region becomes transformed to the far-red form (P_{fr}). A gradient becomes established, causing movement of the chloroplast. If, however, the same site of the cell is postirradiated with far-red, no response is observed. This far-red reversibility is strictly localized; if the far-red microbeam is placed at a site slightly different from the inducing red microbeam, no adverse effect of far-red is elicited. Far-red light can act only where P_{fr} has been established by the red microbeam irradiation. This is another clear demonstration that phytochrome is the photoreceptor pigment for the response in question.

The chloroplast of *Mougeotia* orients in an intracellular P_{fr} gradient and hence can be used as an indicator for a gradient of P_{fr}; that is,

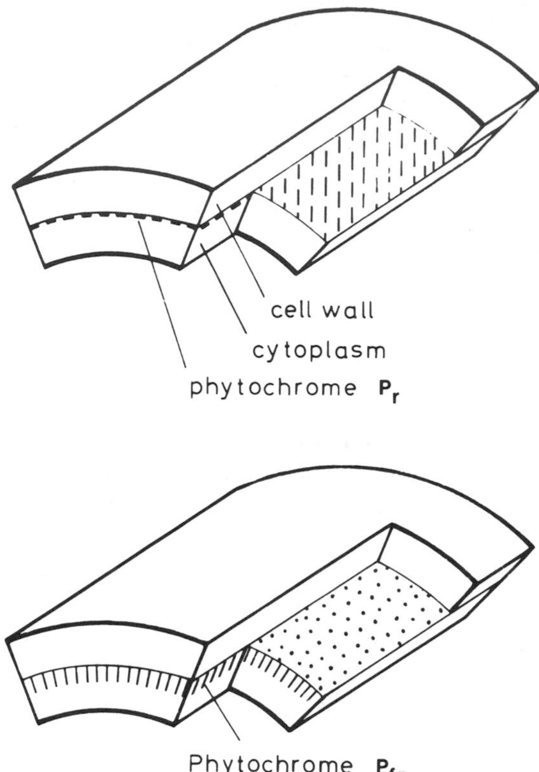

Fig. 16–4. Schematic drawing of the outer part of a *Mougeotia* cell with the dichroic orientation of the phytochrome molecules (dashes) in the P_r and P_{fr} form, respectively. (After Haupt 1970b.)

ideally the distribution of P_r and P_{fr} in the cytoplasm after different types of irradiation can be derived from the movement. This has been used for rather complicated experiments with polarized red and far-red microbeams to find out the pattern of dichroic orientation of the phytochrome molecules (Haupt 1970a). The important results are demonstrated in Fig. 16–4. The photoreceptor molecules P_r are arranged mainly parallel to the cell surface, an orientation common to other photoreceptor substances in other organisms. The P_{fr} molecules, however, are always oriented normal to the cell surface. From these stable and well-defined orientations it has been inferred that they are located in the outer cytoplasmic membrane (plasmalemma). This has also been suggested by experiments on phytochrome extractions from higher plants (e.g., Boisard et al. 1974; Schäfer 1974; Marmé et al. 1976); but it is only by microbeam experiments that this association can be proven in vivo for phytochrome that triggers a physiological effect.

The close association of phytochrome with the membrane suggests the hypothesis that phytochrome is primarily a membrane affector. Expecially in *Mougeotia,* we assume that P_{fr} controls membrane properties such as ion fluxes, which in turn can be a controlling factor for the motor apparatus of the intracellular movement (Weisenseel and Smeibidl 1973; Haupt and Weisenseel 1976).

IV. Problem areas

In the method described, there are two main sources of error: diffraction effects and light scattering. The more the microbeam size is reduced, approaching the wavelength of the light, the stronger are the diffraction rings. We found that a diameter of 3 μm is the minimum allowable. The diffraction rings of a 1 μm microbeam are too strong and too far-reaching. Even with a 3 μm microbeam, one has to perform control experiments to ensure that a wrong area is not hit by diffracted light or that the effect of this diffracted light is negligible because of its reduced intensity.

Light scattering occurs at the cell wall and at the chloroplast. Though scattering can be easily seen in the microscope, it is impossible to have exact information about the quantity of scattered light that hits neighboring cell areas. In addition, partial depolarization occurs by scattering. This might be a source of error when one is using polarized microbeams. Taking this into account, final conclusions can be drawn from microbeam experiments only if strong differences are found with different treatments.

V. Final remarks

By microbeam experiments we could demonstrate unambiguously that functional phytochrome in *Mougeotia* is associated with cytoplasmic structures and that it undergoes a drastic change in dichroic orientation (and thus, probably, in conformation of its protein moiety) upon photoconversion $P_r \rightleftharpoons P_{fr}$. Hopefully, this could be a first step toward understanding the mechanism of phytochrome action. Furthermore, in this alga the light direction is measured by the cytoplasm, and the chloroplast orients itself in a cytoplasmic gradient. It should be added that the same is true in other organisms that make use of a yellow dichroic pigment instead of phytochrome for photoperception. This too has been found with the aid of microbeam irradiation (Fischer-Arnold 1963; Mayer 1964).

VI. References

Boisard, J., Marmé, D., and Briggs, W. 1974. In vivo properties of membrane-bound phytochrome. *Plant Physiol.* 54, 272–6.

Bock, G., and Haupt, W. 1961. Die Chloroplastendrehung bei *Mougeotia.* III. Die Frage der Lokalisierung des Hellrot-Dunkelrot-Pigmentsystems in der Zelle. *Plants* 57, 518–30.

Buder, J. 1915. Zur Kenntnis des *Thiospirillum jenense* und seiner Reaktion auf Lichtreize. *Jahrb. Wiss. Bot.* 56, 529–84.

Fischer-Arnold, G. 1963. Untersuchungen über die Chloroplastenbewegung bie *Vaucheria sessilis. Protoplasma* 56, 495–520.

Haupt, W. 1959. Die Chloroplastendrehung bei *Mougeotia.* I. Uber den quantitativen und qualitativen Lichtbedarf der Schwachlichtbewegung. *Planta* 53, 484–501.

Haupt, W. 1970a. Uber den Dichroismus von Phytochrom$_{660}$ und Phytochrom$_{730}$ bei *Mougeotia. Z. Pflanzenphysiol.* 62, 287–98.

Haupt, W. 1970b. Localization of phytochrome in the cell. *Physiol. Vég.* 8, 551–63.

Haupt, W. 1973. Role of light in chloroplast movement. *BioScience* 23, 289–96.

Haupt, W., and Schönbohm, E. 1970. Light-oriented chloroplast movements. In Halldal, P. (ed.), *Photobiology of Microorganisms,* pp. 283–307. Wiley, London.

Haupt, W., and Weisenseel, M. H. 1976. Physiological evidence and some thoughts on localized responses, intracellular localization, and action of phytochrome. In Smith, H. (ed.), *Light and Plant Development,* pp. 63–74. Butterworth, London.

Marmé, D., Bianco, J., and Gross, J. 1976. Evidence for phytochrome binding to plasma membrane and endoplasmic reticulum. In Smith, H. (ed.), *Light and Plant Development,* pp. 95–110. Butterworth, London.

Mayer, F. 1964. Lichtorientierte Chloroplasten-Verlagerungen bei *Selaginella martensii. Z. Bot.* 52, 346–81.

Schäfer, E. 1974. Evidence for binding of phytochrome to membranes. In Zimmerman, U., and Dainty, J. (eds.), *Membrane Transport in Plants,* pp. 435–40. Springer-Verlag, Heidelberg.

Weisenseel, M. 1968. Vergleichende Untersuchungen zum Einfluss der Temperatur auf lichtinduzierte Chloroplastenverlagerungen. Die Wirkung verschiedener Lichtintensitäten auf die Chloroplasten-Ordnung und ihre Abhängigkeit von der Temperatur. *Z. Pflanzenphysiol.* 59, 56–69.

Weisenseel, M., and Smeibidl, E. 1973. Phytochrome controls the water permeability in *Mougeotia. Z. Pflanzenphysiol.* 70, 420–31.

17: Phototropism: determination of an action spectrum in a tip-growing cell

HIRONAO KATAOKA

Institute for Agricultural Research,
Tohoku University, Sendai 980, Japan.

CONTENTS

I. Introduction

This chapter deals with the method for determining an action spectrum for phototropism of *Vaucheria geminata.* The half-side illumination and the null-point method are used for this purpose. The xanthophycean coenocytic alga *Vaucheria* has several advantages for studying phototropism. It is a coenocytic alga, which has a convenient size (40–80 μm diameter) for irradiation. The tip grows at a rate of ca. 100–200 μm/h.

The half-side illumination method is based on the technique developed by Buder (1920) and used extensively by Gettkandt (1954) in her study on negative phototropism of some parasitic fungi. With improvement in the method, it is possible to simultaneously irradiate separate halves of an organism (5–300 μm in size), placed on the microscope stage, with two monochromatic light beams closely adjacent to each other. In the half-side illumination method the direction of phototropic bending is almost perpendicular to that of the light beam. Therefore, fixing the alga with a small cover slip has little effect on bending because the alga is immersed in culture solution and placed parallel to the surface of the microscope stage. Another advantage of this method is that a great difference in light intensity and quality (wavelength) between the two sides can be attained. Thus, the possibility of a lens effect, as observed in unilaterally illuminated *Phycomyces* sporangiophore (Blaaw 1914), can be eliminated.

The tip of *Vaucheria* shows positive bending in the half-side illumination as well as in unilateral illumination. However, in some plants phototropism observed with the half-side illumination is different from that with unilateral illumination. For example, a *Puccinia menthae* germling shows negative bending when it is unilaterally illuminated, but it bends towards the lighted side in half-side illumination (Gettkandt 1954). A sporangiophore of *Phycomyces* shows positive phototropism by unilateral illumination (Blaaw 1914; Castle 1933; Thimann and Curry 1960; Bergman et al. 1969), but it will bend away from the lighted side with half-side illumination. These different responses are due to different mechanisms of phototropism (Kataoka

1975b). Although this will not be discussed further in this chapter, it is first necessary to determine whether the material shows positive or negative phototropism with unilateral illumination before analyzing phototropic bending using the half-side illumination method.

A clear, general treatment on action spectra determination was recently provided by Shropshire (1972).

II. Equipment

A. Microscope and accessories

1. An ordinary tube-sliding type microscope with a fixed stage (Fig. 17–1a) (Nikon K-type microscope, Nihon Kogaku Kogyo Co.) is suitable.
2. Two illuminators: Olympus LSD (Olympus Kogaku Kogyo Co.) or Nikon Standard Illuminators can be set up at right angles (Fig. 17–1a) and should be attached to DC voltage regulators (10 V, 18 A).
3. A suitable mirror, half of the front surface covered, can be made in a vacuum evaporator.

B. Spot intensity measuring apparatus

1. A silicon photocell (SBC-510) (Hayakawa Electrics Co.) or (SPC-780-18BK) (Hamamatsu TV Co.) can be mounted on a stage micrometer equipped with a special holder.
2. A thermopile radiometer (Kipp and Zonen Model E-20) and a microvoltmeter (TOA PM-17A, Toa Electronics Ltd.) can be used to record the intensity at the specimen level.
3. An electric recorder or electronic voltmeter are suitable for measuring the photocell output voltage.

C. Interference filters

Interference filters (Toshiba KL-series) are mounted as shown in Figs. 17–1a and 17–1b. Glass filters (Toshiba AT-series) can be used for blocking secondary transmission in the primary filters.

III. Test organisms

Vaucheria geminata (Vauch.) De Condolle var. *geminata,* (UTEX LB 1035) is more sensitive to light effects (Kataoka 1975a) than *Vaucheria sessilis* (LB 745-lb), from the Göttingen collection. Chlorophycean tip-growing algae such as *Cladophora, Bryopsis,* and *Caulerpa* are also suitable organisms for application of the techniques described in this chapter.

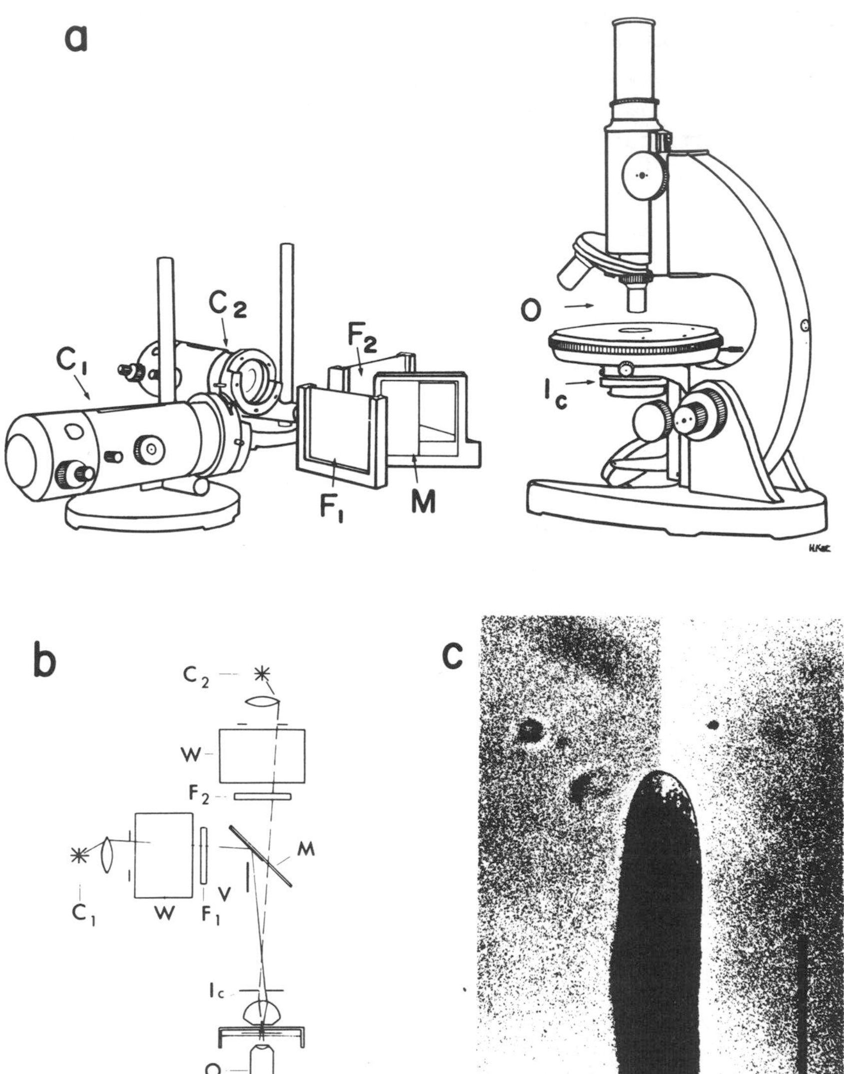

Fig. 17–1. Optics employed for the half-side illumination method. (a) Arrangement optics: objective lens (O); iris diaphragm (I_c); and light source (C_1, C_2); filters (F_1, F_2); and mirror (M). (b) Schematic diagram of the optics including water filters (W) and vane (V). (c) Photograph of an algal tip with half-side illumination. (Bar indicates 100 μm.)

IV. Methods

A. Culturing

1. Culture medium. The culture solution is a combination of Uspenski's (Hustede 1957) and Darden's (1966) media with certain modifications. The following ingredients are dissolved in glass-distilled water to give a final volume of 1 liter: 25 mg KNO_3, 25 mg $MgSO_4{\cdot}7H_2O$, 100 mg $Ca(NO_3)_2{\cdot}4H_2O$, 13.6 mg KH_2PO_4, 582 μg $FeCl_3{\cdot}6H_2O$, 246 μg $MnCl_2{\cdot}4H_2O$, 30 μg $ZnCl_2$, 12 μg $CoCl_2{\cdot}6H_2O$, 24 μg Na_2MoO_4, 4.5 mg Na_2-EDTA, 0.1 μg Biotin, and 0.1 μg Vitamin B_{12}. The medium is buffered with 0.1 m*M* TRIS–HC1 (pH 7.2) or, alternatively, with 0.1 m*M* HEPES (pH 7.0) and autoclaved at 120°C for 20 min. This medium supersedes that mistakenly listed previously by Kataoka (1975a).

2. Unialgal culture. To eliminate contaminating orgamisms, the alga is shaken gently in a culture medium containing sterilized activated charcoal powder (Norit A) and is then pulled several times over an 0.8% agar-plate, before being transferred to a new agar-plate (Ohiwa, personal communication). Although it is possible to attain axenic cultures by this method, unialgal cultures are satisfactory for phototropism studies.

3. Culture conditions. For optimum growth, both *Vaucheria* species are kept under white fluorescent light (1,000–1,500 lux; 95–140 ft-c) on 12 : 12 h LD cycle. The optimum temperature for growth of *V. geminata* is 20°C, and that of *V. sessilis,* 23°C.

B. Optics and half-side illumination method

Experiments are carried out on a microscope stage in a darkroom at 20°C. Figs. 17–1a and b show the diagram of the optical arrangement, and the local illumination of a small region of the alga with two beams of monochromatic light. C_1 and C_2 are light sources. One half of a glass plate (M) (Kodak slide cover glass) is made into a surface mirror by evaporation of aluminum in a vacuum evaporator; the other half is left transparent. To make a sharp edge on the mirrored half of a glass plate, a thin cover slip (50 × 25 mm) is placed on the plate and set just below the basket of the evaporator. The optics are set up as follows.

1. Two lamps and a microscope are placed as shown in Fig. 17–1a.
2. A flat surface mirror (30 × 30 mm, made from the cover slip) is glued onto the microscope mirror for elimination of double images.
3. The glass plate (M) is placed at a 45° angle against the optical axis of the microscope. With this arrangement, the light beam from C_1 reflects at the mirror half of the plate, and the light from C_2 passes

through the transparent half so that the two beams reach the right and left field of the microscope, respectively.

4. A black vane (V) is placed in front of M to eliminate reflected light at the plain glass surface.

5. Glass cells containing water (W) and heat absorbing filters (IRQ-80, Toshiba) are placed in front of each lamp.

6. Then the desired interference filters (F_1 and F_2) are inserted between the heat absorbing filters and M.

7. The image of the edge of M is focused on the specimen plane.

A light spot can thus be obtained, one half of which is composed of monochromatic light from F_1 and the other half from F_2. This, with good Köhler illumination, produces two light beams illuminating the sample as shown in Fig. 17–1c. The condenser aperture must be as small as possible, usually 2–3 mm in diameter. The axis of both beams should coincide exactly with that of the microscope.

C. *Determination of small spot intensity*

1. Principle. The energy of the light spot is determined on the microscope stage. Since the area of the spot is smaller than the smallest area measurable with the thermopile, the spot intensity must first be measured with a silicon photocell. Simulation of a larger illuminated area with an intensity equal to the measured voltage is made, and the larger area is measured with a thermopile.

2. Method.

a. The silicon photocell is glued onto the shutter plate of the thermopile box, using the light source with all required filters, the output voltage of the silicon photocell is calibrated with a thermopile radiometer at selected wavelengths. Fig. 17–2 demonstrates the spectral sensitivity of the silicon photocell (SBC-510). In this figure the sensitivity factor f_s (that is, the ratio of mV value at each wavelength to V_{479} at a constant intensity) is also shown. Figure 17–3 shows the relationship between the intensity at 479 nm (I_{479}) and the output voltage of the SBC-150 photocell (V_{479}). The intensity at a given wavelength I_λ can be obtained by f_s and from Fig. 17–3:

$$V_{479} = V_\lambda \times f_s \qquad (1)$$

where V_λ is the output voltage at a certain wavelength.

b. The area of the light spots is measured, and the right and left areas are expressed as S_{sr} and S_{sl} respectively.

c. The same silicon photocell (SBC-510), mounted to a holder, is placed over the microscope stage and adjusted so that the surface of the photocell is parallel with the microscope stage.

d. The lamp is switched off. The photocell is shifted upward until

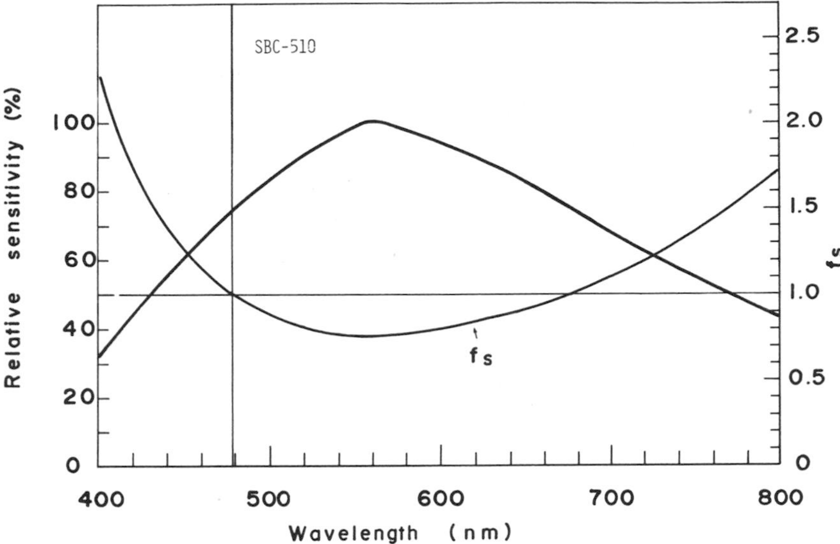

Fig. 17–2. Spectral sensitivity of SBC-510 and the sensitivity factor f_s.

the light beam covers its square surface. In this condition, the margin of the light beam is faint because the light spot is out of focus. Ignoring a small margin of error, it can be considered that the whole surface is illuminated with all photons composing the light beam. The output voltage of the photocell is equal to $S_{sr}/25$ or $S_{sl}/25$ of the light intensity at the specimen level for the right or left spot, respectively.

e. The relationship between the size of the condenser aperture and the intensity at the specimen level is determined. At a constant intensity and wavelength, the ratios of these intensity values is expressed as f_a. For irradiation, the size of the aperture is generally 2 or 3 mm in diameter; for the intensity calibration it is 10 mm.

f. Finally, the spot intensity at the specimen level $I_{\lambda s}$ is obtained by f_a and I_λ (eq. 1; Fig. 17–3). Namely,

$$I_{\lambda s} = I_\lambda \times f_a \tag{2}$$

D. Dose–response relationship

For determining an action spectrum, a good way to minimize internal and external factors causing sensitivity fluctuation and amplification factors that might be involved in the dark processes is to use a null-point method. In the half-side illumination, when both halves of a tip are stimulated equally and the tip bends neither right nor left, the null point is reached. The advantages of this method were clearly described by Shropshire (1972).

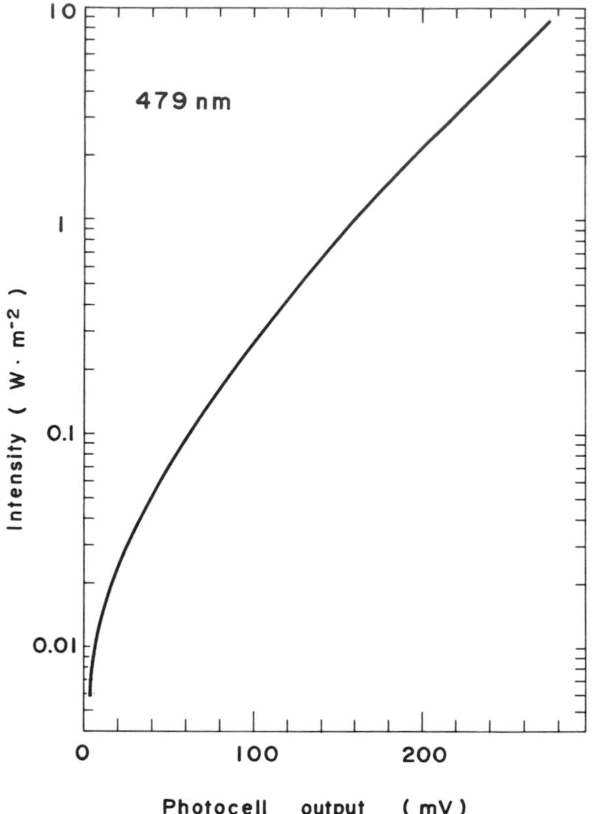

Fig. 17–3. Relationship between the light intensity at 479 nm(I_{479}) and the output voltage of the silicon SBC-510 photocell V_{479}.

However, to choose an appropriate energy and a wavelength of the reference light source, determination of dose–response relationship is necessary. Furthermore, the null-point method cannot provide absolute values of quantum efficiency. If one once obtains a dose–response curve at a given wavelength, one can easily calculate absolute quantum efficiencies. Thus, we first deal with the method of determination of dose–response relationship and then with that of an action spectrum.

1. Selection of the material, adaptation period, and experimental period. A specimen that has been cultivated for 3–4 days in a petri dish is covered with a coverslip and preincubated in a dark room at 20°C for 1–4 h. Actively growing tips (100–200 μm/h) should be selected. After the preincubation, one longitudinal half of the alga is illuminated with actinic blue light (e.g., 450 nm). The other half is illumi-

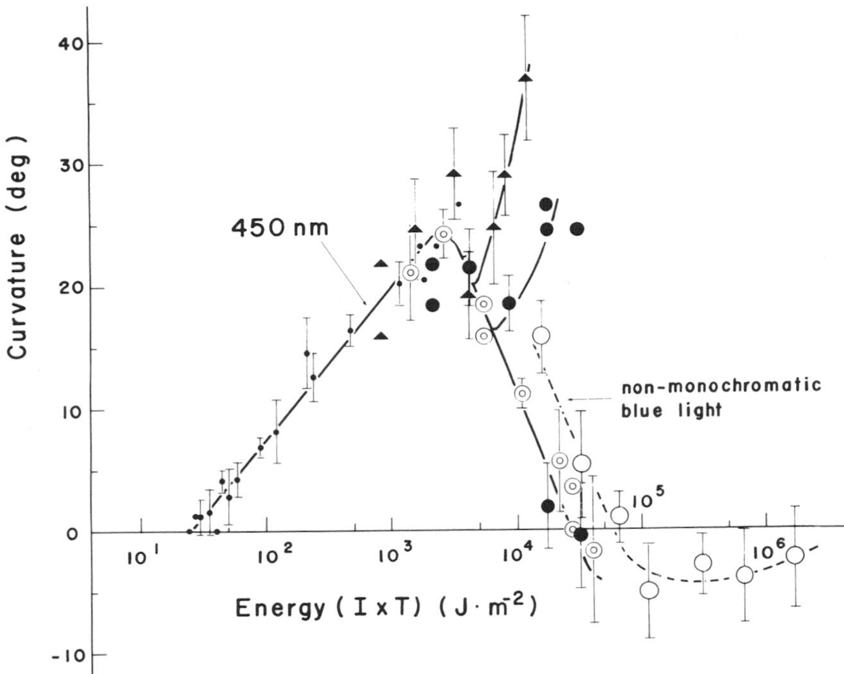

Fig. 17–4. Dose–response curve for phototropism of *V. geminata*. The algal tips were half-side illuminated with monochromatic light at 450 nm (solid lines) and with non-monochromatic light (broken line). The intensities of 450-nm light were below 5 W/m² (•), at 13.7 W/m² (▲), 36.1 W/m² (●), and 46.0 W/m² (◎). Data obtained with nonmonochromatic light at 3 intensities (505, 1110, and 1860 W/m²) (light from halogen lamp, through a 4% $CuSO_4$ liquid filter) are superimposed (○). (After Kataoka 1977a.)

nated with a dim safelight (e.g., 660 nm, 0.1 W/m²). The curvature is measured microscopically 15 min after the onset of irradiation. As described previously (Kataoka 1975a), the experimental period is set at 15 min, taking into consideration the time required for the completion of bending and the growth rate. Measurements should not be repeated on the same tip until a 2 h "resting period" in the dark has been completed.

2. *Reciprocity law.* The curvatures produced by the half-side illumination are measured at varying intensities I and times t to determine the reciprocity range. In *V. geminata* the reciprocity law holds in the range between 5 W/m² × 6 sec and 0.03 W/m² × 900 sec for the onset curvature (0°) at 450 nm (Kataoka 1975a). The curvatures are plotted against the logarithm of the stimulus ($I \times t$) at a given wavelength; this results in a straight line within a certain range of the ($I \times t$) value (Fig. 17–4). For the null-point method, it is sufficient to know the

values of $I \times t$ of the usable range for several important wavelengths (e.g., 450, 480, 660, 730 nm).

E. Determination of an action spectrum

One longitudinal half is irradiated with a reference light of constant intensity, at 479 nm (e.g., 1.05 W/m² or 0.11 W/m² if longer than 500 nm), while the other half is irradiated with monochromatic light of various intensities and wavelengths of the same duration. A null-point intensity is determined for every wavelength. This is expressed as the quantum flux density $nE/\text{m}^2/\text{sec}$ and can be calculated by the following equation:

$$nE/\text{m}^2/\text{sec} = 8.36_\lambda \text{ (nm)} \times I \text{ (W/m}^2\text{)} \qquad (3)$$

where λ is wavelength and I, intensity. The action spectrum (Fig. 17–5) is determined by obtaining the ratios of nE-value of the reference wavelength to that of the balancing test wavelength (quantum flux density reference/quantum flux density test light at null point) at each wavelength. Since the irradiance times are the same for both lights, comparing the quantum flux densities is sufficient. If necessary, quantum density nE/m^2 can be obtained by multiplying the time t with the respective quantum flux densities.

V. Some applied methods

A. Strong light

To obtain light stronger than 5 W/m², the actinic light source (C_2 in Fig. 17–1) can be replaced with a halogen–tungsten lamp (Nikon Halogen lamp: 12 V,100 W). Figure 17–4 was obtained with a halogen lamp.

B. UV region

To survey the action spectrum in the UV region, quartz optics must be used, that is, UV-transparent dishes (e.g., Falcon dish), UV condenser, UV-transparent glass plate, and the mercury lamp. Although it is possible to construct the optics, a difficulty may arise owing to the vibration of the mercury arc; this makes it difficult to obtain a stable microbeam. *Caution* must be observed to protect eyes and extremities from UV damage. Protective eye covers, glasses, or goggles should be worn.

C. Phototropic hunting and microbeam irradiation

If the mirror (M in Fig. 17–1, IV.B) is changed to a plain glass plate and a small window is made with a razor blade, or a piece of photogra-

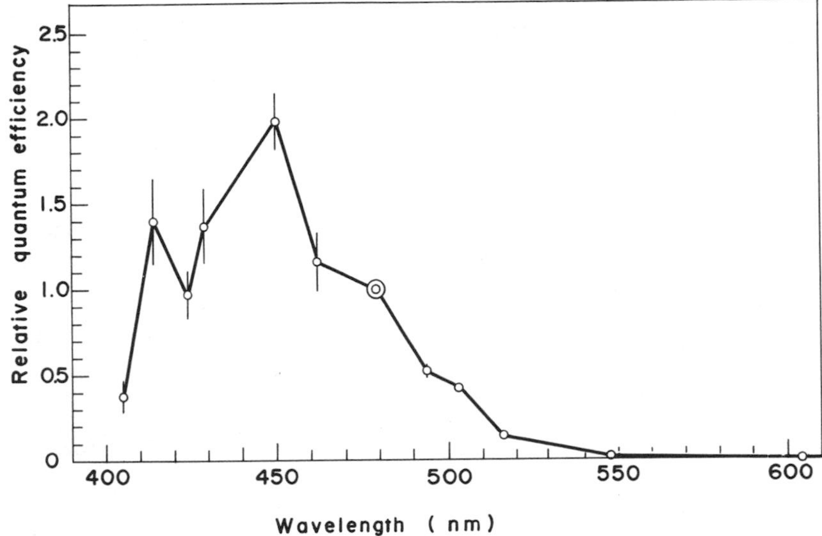

Fig. 17–5. Action spectrum for first positive phototropism of *V. geminata.* Each plot represents average values from 4 to 8 determinations. Bars indicate standard errors. (Redrawn from Kataoka, 1975a.)

phic film with a spiral pattern is inserted between the glass plate and F_2, one can obtain a microbeam or a spiral light stripe. In this condition, light from C_1 reflects at the plain glass surface and the microbeam is composed of two monochromatic lights. When part of a growing tip of *Vaucheria* is irradiated thus with a microbeam (blue plus dim red), the alga bends towards the light spot. When the alga is moved so that the light beam always irradiates the same position relative to the tip, the alga continues to bend (Fig. 17–6a). When the alga is brought to the spiral light stripe, it continues its growth following the stripe (Fig. 17–6b). Page and Curry (1962), using a similar microbeam irradiation system, described how the young sporangiophore of *Pilobolus* bent away from the light spot.

VI. Possible problem areas

A. Usable range of growth rate

Taking into consideration the bending velocity (ca. 2°/min) and the spot size (ca. 500 μm in diameter), the usable range of the growth rate will lie between 50 and 500 μm/h. This method is therefore not applicable to slow-growing algae. One drawback of this method is that only one specimen can be tested at a time.

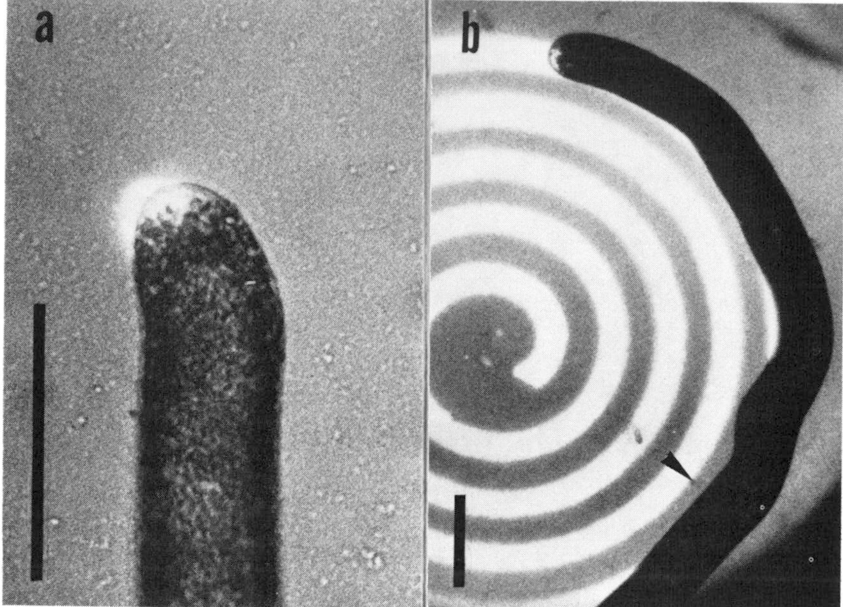

Fig. 17–6. Phototropic hunting. (a) The alga was trapped in the moving light spot, and when (b) the alga came into the light stripe (arrow) it continued to bend. (Bar indicates 100 μm.)

B. Size of the material

Taking into consideration the angle between the two monochromatic beams, the size of the material should be smaller than 300 μm.

C. Size of growth region

The tip of *Vaucheria* has a large hyaline cap. The hyaline cap coincides with the active growth zone and acts as a steering apparatus (Kataoka 1975b; 1977b). This transparent hyaline cap is of great advantage to the half-side illumination method. Cells whose hyaline caps are small, such as fronds of *Bryopsis* and *Acetabularia,* rhizomes of *Caulerpa,* and rhizoids of *Boergesenia,* are rather unsuitable materials for this method.

VII. Special remarks

Bending in *Vaucheria* occurs through formation of a new growth center ("bulging," Green et al. 1970) when the flank of the apical dome is illuminated with the blue light. The apical plasmalemma is considered to play an essential role, because the photoreceptor seems to occur in the outermost layer of the cytoplasm of the hyaline cap (Kataoka 1975a; b).

Recently, I found that 3′, 5′-cyclic AMP attenuated the phototropic sensitivity leaving the growth unaffected. This indicates that the phototropism can be experimentally separated from growth itself. The concentration of endogenous cyclic AMP in the cytoplast of *V. geminata* is kept at a constant level around 3×10^{-6} *M*. If the level is increased by exogenous cyclic AMP, the phototropic sensitivity becomes less, but if the level is lowered by imadazole, the sensitivity increases (Kataoka 1977b).

VIII. Acknowledgments

This study was partly supported by a postdoctoral fellowship from the Japan Society for the Promotion of Science. I wish to express my thanks to Prof. N. Kamiya of the Department of Biology, Faculty of Science, Osaka University, for his valuable discussions and kind advice. Thanks are also due to Drs. C. and L. Schilde of Tübingen Universität for their kind donation of *V. sessilis.*

IX. References

Bergman, K., Burke, P. V., Cerdá-Olmedo, E., David, C. N., Delbrück, M., Foster, K. W., Goodell, E. W., Heisenberg, M., Meissner, G., Zalokar, M., Dennison, D. S., and Shropshire, W., Jr. 1969. Phycomyces. *Bacteriol. Rev.* 33, 99–157.

Blaaw, A. H. 1914. Licht und Wachstum I. *Z. Bot.* 6, 641–703.

Buder, J. 1920. Neue phototropische Fundamentalversuche. *Ber. Dtsch. Bot. Ges.* 38, 10–20.

Castle, E. S. 1933. The physical basis of the phototropism of *Phycomyces. J. Gen. Physiol.* 17, 49–62.

Darden, W. H., Jr. 1966. Sexual differentiation in *Volvox aureus. J. Protozool.* 13, 239–55.

Gettkandt, G. 1954. Zur Kenntnis des Phototropismus der Keimmycelien einiger parasitischer Pilze. *Wiss. Z. Univ. Halle. Math.-Nat. Jahrg.* 3, 691–701.

Green, P. B., Erickson, R. O., and Richmond, P. A. 1970. On the physical basis of cell morphogenesis. *Ann. N.Y. Acad. Sci.* 175, 712–31.

Hustede, H. 1957. Untersuchungen über die stoffliche Beeinflussung der Entwicklung von *Stigeoclonium falkandicum* und *Vaucheria sessilis* durch Tryptophanabkömmlinge. *Biol. Zentralbl.* 76, 555–95.

Kataoka, H. 1975a. Phototropism in *Vaucheria geminata.* I. The action spectrum. *Plant and Cell Physiol.* 16, 427–37.

Kataoka, H. 1975b. Phototropism in *Vaucheria geminata.* II. The mechanism of bending and branching. *Plant and Cell Physiol.* 16, 439–48.

Kataoka, H. 1977a. Second positive and negative phototropism in *Vaucheria geminata. Plant and Cell Physiol.* 18, 473–6.

Kataoka, H. 1977b. Phototropic sensitivity in *Vaucheria geminata* regulated by 3′, 5′-cyclic AMP. *Plant and Cell Physiol.* 18, 431–40.

Page, R. M., and Curry, G. M. 1962. Studies on phototropism of young sporangiophore of *Pilobolus kleinii. Photochem. Photobiol.* 5, 31–40.

Shropshire, W., Jr. 1972. Action spectroscopy. In Mitrakos, K. and Shropshire, W., Jr. (eds.), *Phytochrome,* pp. 162–81. Academic Press, New York.

Thimann, K. V., and Curry, G. M. 1960. Phototropism and phototaxis. *Compar. Biochem.* 1, 243–309.

18: Chromatic adaptation in Fremyella diplosiphon and morphogenetic changes

JOHN F. HAURY*

Biological Laboratories, Harvard University, Cambridge, Massachusetts 02137

CONTENTS

* Present address: Plant Growth Laboratory, University of California, Davis, California 95616

I. Introduction

A growing interest in the biology of chromatic adaptation can be seen in recent work on its action spectra (Diakoff and Scheibe 1973; Haury and Bogorad 1977) as well as in a review of the subject by Bogorad (1975), and from a survey made by Tandeau de Marsac (1977). This chapter describes an experimental approach to the analysis of chromatic adaptation and photomorphogenesis in the filamentous blue-green alga, *Fremyella diplosiphon.* Chromatic adaptation occurs with changes in the content of the three photosynthetic accessory pigments, allophycocyanin (APC), phycocyanin (PC), and phycoerythrin (PE) according to variations in the wavelength of incident light. In red-light-grown cells PE disappears, but the PC content is enhanced. On the other hand, PE production is enhanced by green light. These pigments as well as chlorophyll *a* and the carotenoids can be routinely and accurately measured as a proportion of the total cell mass.

Morphogenetic changes have been observed by Bennett and Bogorad (1973), where growth under red incandescent light results in short, PC-rich filaments, and growth under fluorescent light (rich in the green wavelengths) results in long PE-rich filaments. Individual cell morphology changes as well; red-grown cells are more spherical and fluorescent-grown cells are longer and more cyclindrical in shape.

Action spectroscopy has been used in an attempt to identify the chromatic adaptation photoreceptor(s) by its absorption properties in *F. diplosiphon.* For more extensive treatments on the theory of action spectroscopy and guidelines on technique, Setlow (1957), Jagger (1967), and Shropshire (1972) should be consulted. The illumination sources and filters listed were those available to the author; it should be noted that other appropriate equipment can be substituted for determinations of action spectra.

II. Materials and methods

A. Culture conditions

Fremyella diplosiphon (UTEX 481) is grown in medium C (Kratz and Myers 1955) with ferric diethylenetriamine penta-acetate (Sequestrene, Ciba-Geigy) substituted for ferric sulfate and without sodium citrate. Axenic stock cultures are maintained under fluorescent lights on 1% agar slants. Liquid cultures are grown at 35°C in 1-liter lots in 2,800-ml Fernbach flasks on a rotary shaker (100 rpm) enclosed in a growth chamber under one of two light conditions. A red-light chamber is illuminated with six 60 W red incandescent bulbs (General Electric 60A 21/R) with a total output of ca. 10 $\mu W/cm^2/nm$ between 380 and 725 nm. A white-fluorescent-light chamber is illuminated with fluorescent tubes (Sylvania F30 T12-W-RS) with an output of 38 $\mu W/cm^2/nm$ between 380 and 725 nm. Under these two conditions, growth occurs with a cell dry weight increase of 1.5–3.0 $\mu g/ml/h$ (this increase is greater with fluorescent illumination) and yields two extremes of chromatically adapted cells: PE-rich and PC-rich (Haury and Bogorad, 1977).

B. Experimental illumination conditions

1. Initial characterization. Preliminary characterization of the responses are carried out using broadband filters (50–100 nm half band width). The results from these indicate the active spectral regions and the required intensities. Colored transparent Plexiglas sheets (blue, green, and red, Nos. 2424, 2092 and 2423 respectively, Rohm and Haas) are used both within the growth chambers for illumination of flask cultures and for illumination of test tube samples by slide projectors. The Plexiglas is used within the chamber with incandescent lighting, the green Plexiglas is used with white fluorescent light bulbs, and the blue Plexiglas is used with 20 W blue fluorescent lamps (Westinghouse F20 T12/B). The output of these light conditions is demonstrated in Fig. 18–1.

2. Monochromatic illumination. Eighteen interference filters, ranging from 365 to 730 nm at wavelength of maximum transmission (10 nm half band width; Baird Atomic and Jena Glaswerk) in combination with slide projectors (General Electric model DAK, 500 W lamp) were used for illumination. It is necessary to examine all filters to ascertain if sidebands exist (A split-beam spectrophotometer is most useful, because its reference beam can be attenuated with a neutral density filter to enhance detection of minor transmission peaks). If secondary bands exist in the monochromatic filters, they must be blocked out or taken into account in the interpretations of the action spectra.

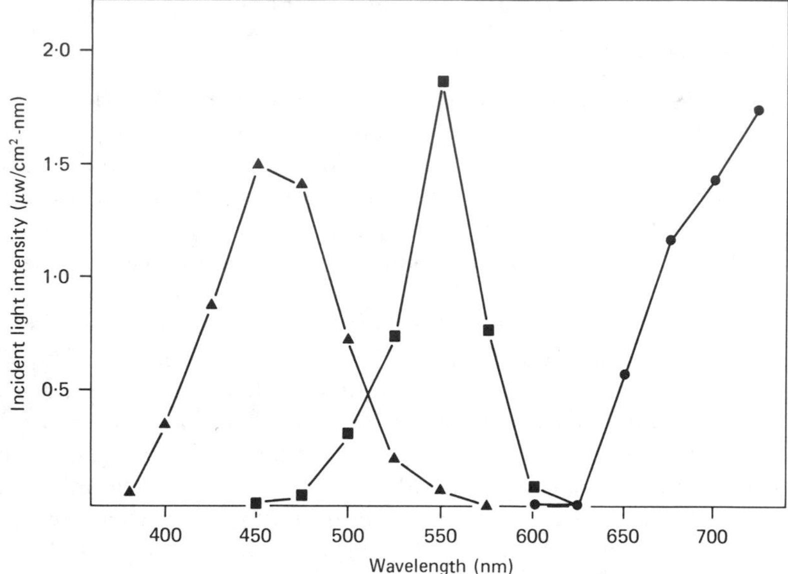

Fig. 18–1. Incident light intensity from fluorescent and incandescent lamps filtered by transparent Plexiglas filters as used within the growth chambers. These illumination conditions are used to initially characterize complementary chromatic adaptation and photomorphogenesis. The total incident energy summed for each 25-nm interval from 380 nm to 725 nm is as follows ($\mu W/cm^2/nm$): red, 5.0 (● – ●); green, 3.8 (■ – ■); and blue, 5.3 (▲ – ▲).

3. Light intensity measurements and regulation. Light intensity measurements are made using the direct incidence head detector of a spectroradiometer (ISCO model SR); for wavelengths less than 365 nm these are made with an Eppley thermopile. All measurements are made with the experimental conditions duplicated as nearly as possible. Wavelength-independent values of light intensity (e.g., $\mu W/cm^2/nm$) yield the quantum-flux data necessary for calculations. The intensity is regulated with a variable powerstat and by variation of the distance from light source to sample.

C. Cell and pigment assays

1. The cell dry weight is determined by turbidity at 730 nm under conditions described by Bennett and Bogorad (1973).

2. Phycobiliprotein contents are determined from the supernatant of broken cells. Cells are cavitated (sonifier model W185D, Heat Systems – Ultrasonics, Inc.) for 40 sec per ml in 0.01 *M* sodium phosphate buffered saline 0.15 *M*, pH 7.0. Cell fragments are removed by centrifugation (81,000*g*, 30 min), and absorption spectra are made of

the supernatant. The relative amounts of the phycobiliproteins are calculated by the simultaneous equations of Bennett and Bogorad (1973).

$$\mathrm{PC} = \frac{A_{615} - 0.474\ (A_{652})}{5.34} \tag{1}$$

$$\mathrm{APC} = \frac{A_{652} - 0.208\ (A_{615})}{5.09} \tag{2}$$

$$\mathrm{PE} = \frac{A_{562} - 2.41\ (\mathrm{PC}) - 0.849\ (\mathrm{APC})}{9.62} \tag{3}$$

III. Chromatic adaptation

A. Conditions critical to light treatment

1. Cultures should be fully adapted, having undergone at least two weekly transfers in the chosen growth-light condition.

2. Cells should be transferred using asceptic techniques and handled at all times under the proper safelights. An excellent safelight is the growth light for the cells being handled.

3. Excessive self-shading is avoided by continuous agitation of the cells, and by keeping the cell density low (between 0.05–0.15 mg dry weight/ml in the system described).

4. A constant time period of experimental illumination is suggested (6 h in these experiments); thus a possible failure of reciprocity between time and dose rate is avoided. If sufficient light-dose variations cannot be achieved by regulation of the light intensity, then variations in the exposure time become necessary.

B. Experimental procedure

1. To eliminate the necessity of nearly 20 h of monochromatic illumination to yield the responses measured here, a light pretreatment procedure can be adopted. Cultures grown in the red incandescent light chamber are transferred to the fluorescent chamber (and vice versa) for 12–15 h immediately prior to the monochromatic illuminations. The pigment changes of chromatic adaptation become significant only after 6–12 h; the broadband colored light treatment is given during this lag period (Haury and Bogorad 1977).

2. From cultures pretreated in the desired light conditions, 20-ml samples are dispensed into test tubes (25 × 150-cm). Each test tube contains an internal stainless steel U-tube connected to a water bath for temperature regulation (36–39°C), and each has an air-inlet tube (100 μl capillary) for aeration and agitation. Humidification and sterile filtration of the air are suggested.

3. The remainder of the original culture (500 ml or more) is placed back in the culture chamber for an additional 6 h (the time during which the experimental samples receive monochromatic light). The net pigment synthesis in this sample serves as a control value for culture activity (see III.C.1).

4. Each experimental tube is illuminated at the chosen wavelength(s) for 6 h at doses ranging from 0.5 to 30,000 ergs/cm^2/sec.

5. Six hours of continuous light treatment is ended by immediate dark storage at 4°C, and the necessary pigment and dry-weight assays are carried out within 48 h.

6. The pigment content is expressed as micrograms of individual phycobiliprotein (PC, PE, or APC) per mg cell dry weight. Subtracting the pigment content of the cells prior to monochromatic illumination from that found after illumination yields the net pigment change in the 6-h experiments. This response is expressed, for example, for PE, as net microgram PE-content change per milligram cell dry weight per hour. Net pigment decreases per cell dry weight occur when the growth rate is greater than the pigment-synthesis rate. Such negative responses are not used in the subsequent calculations.

C. Data analysis

1. Dose response curves. The net accumulation of specific pigment per cell mass per hour (the chromatic adaptation response) for the experimentally treated samples is normalized as a ratio to the chromatic adaptation response of the remainder of the original culture placed back into the pretreatment light (section III.B.3). For each narrow wavelength band the normalized response is plotted as the ordinate versus the log of the dose of incident quanta (Fig. 18–2).

One expects an increased response with increased light dose over a range of intensities where light is the limiting factor. At a high dose of a wavelength absorbed by the photoreceptor(s), the response is expected to saturate and then to remain high and constant with further increases in the dose. In such cases, the dose of light needed to yield a constant, a small fraction of the maximum response is used as a measure of the wavelength's effectiveness expressed as an action spectrum. For active wavelengths, the pigment accumulation response in *F. diplosiphon* becomes significant at a dose rate in excess of 10^{12} quanta/cm^2/sec given for 6 h; less active wavelengths need tenfold higher quantum-flux densities for a response.

2. Action spectrum. Contrary to expectation, the dose–response curves (e.g., Fig. 18–2) for *F. diplosiphon* show that the maximum response is followed by a decreased response at higher doses. This is a complicated dose–response behavior, and it may be due to bleaching, or it

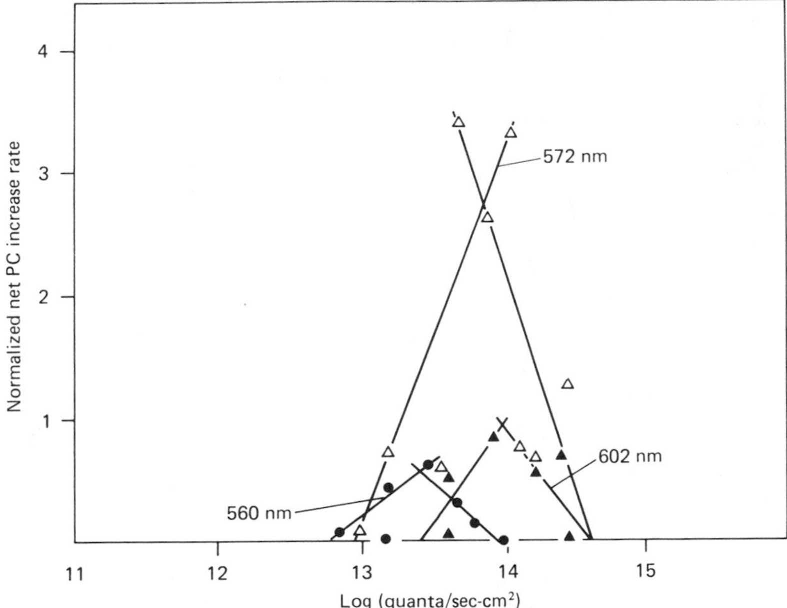

Fig. 18–2. Typical dose–response curves for phycocyanin accumulation. Least-squares analysis was applied to the data on either side of the maximum response to yield one ascending and one descending straight line. The decreased response at high doses is discussed in the text (III.C.2). (After Haury and Bogorad 1977.)

may represent the destruction of a factor (perhaps the photoreceptor itself) involved in phycobiliprotein synthesis. Under these circumstances, to minimize dark amplification steps (Shropshire 1972), a low absolute-response value is used for comparison of the dose–response plots and calculation of the action spectra (Fig. 18–3). A standard response is chosen well within the linear ascending portion of all the dose–response curves, and the ordinate values for the action spectra are determined as the inverse of the dose necessary to attain the standard response. Nonparallel dose–response curves will yield different action spectra, depending upon the standard response chosen for comparison. In *F. diplosiphon,* for example, standard responses of ca. 0.27, 0.55, and 0.83 (that is less than the response of the controls, but nevertheless a positive response), yield comparable action spectra differing only in the relative heights of the major peaks. The most active wavelengths are in the blue (463 nm) and red (641 nm) regions for PC accumulation and in the blue (387 nm) and green (550 nm) regions for PE accumulation (Haury and Bogorad 1977).

Interpretations of action spectra must take possible shading pigments into account. For example, *F. diplosiphon* has significant absorp-

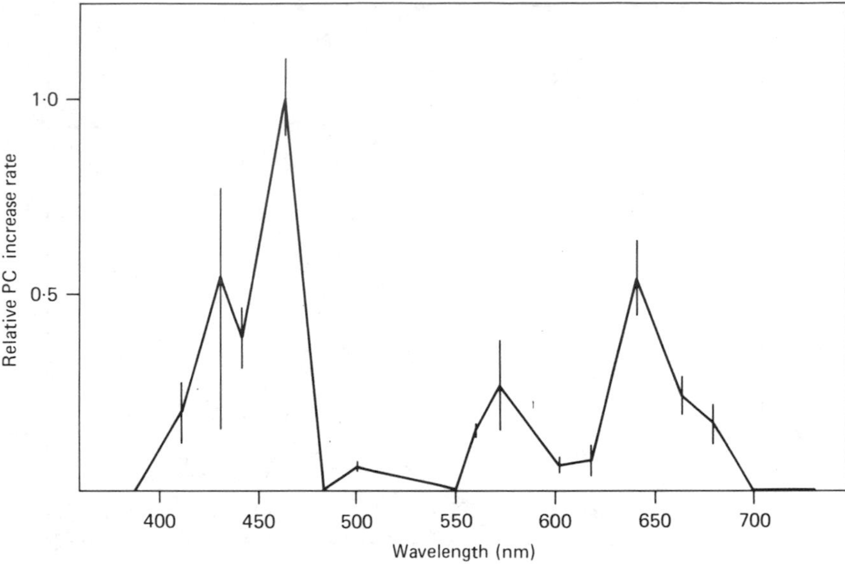

Fig. 18–3. Action spectrum for increase of phycocyanin during chromatic adaptation. The ordinate values are the inverse of the dose rate at which the assending dose–response curve for the pigment (Fig. 18–2) reaches a chosen rate [normalized relative to the original culture (III.C.1)]. In this case, a standard response of 0.555 was used, a rate just over one half the response of the controls. The response at 463 nm is arbitrarily set at a value of 1.0. (After Haury and Bogorad 1977.)

tion in the spectral region below 500 nm, due to chlorophyll and carotenoids (Bennett and Bogorad 1973). Longer wavelength shading is to be expected by the biliprotein pigments themselves. The action spectra show the blue activities to be as large or larger than the red and green activities, yet blue-region shading is consistently larger and therefore would be expected to diminish the actual activity seen in the blue. The blue light absorption by the photoreceptor(s) may therefore be relatively stronger than the action spectra show.

IV. Morphogenetic changes

The spectral response of photomorphogenesis, as with chromatic adaptation, is analyzed after one has allowed growth and consequent adaptation to occur under light of defined bands of the visible spectrum. Knowledge of the spectral response is a good foundation for action spectral analysis. For the growth-chamber illumination conditions described here, red and blue light induce or maintain the short filament habit (less than 100 μm average filament length), whereas green and white fluorescent light induce or maintain a larger average filament length (longer than 300 μm) and a broader frequency distribution.

A. Sample treatment

1. Growth of 100-ml cultures in 250-ml flasks proceeds for up to 16 days under appropriate Plexiglas broadband filters, with incubation as described previously in a growth chamber (II.A and II.B). Filaments are allowed to achieve a steady-state length, which is attained when consecutive samplings show insignificant differences.

2. Filament lengths are determined by photographing filaments through a light microscope (ca. 400×) from slides prepared by applying the sample with a transfer loop in a drop of medium, avoiding use of a cover slip (this minimizes artifactual filament breakage). The filament image is enlarged by projecting the negative, and the filament lengths are determined with a map measurer, a stage micrometer image at the same magnification serving as a standard.

B. Application

The morphological response to light can be given as average filament lengths or as frequency distributions. Light control of filament shortening may be measured by the appearance of necridia (dying cells across which the filaments breaks) (Lamont 1969). The quantitation of necridia formation with light dose would allow an action spectral analysis of filament breakage and an analysis of inhibition of light-induced necridia formation (e.g., an antagonistic effect of long wavelengths versus others) would yield an action spectrum of filament elongation. Filament morphology analysis in combination with pigment synthesis data is important to an understanding of their possible interactions.

The short filament habit and PC predominance occur in red-light-grown cultures; long filaments and PE predominance occur in green-light-grown cultures. In contrast, short filaments and PE abundance result from growth in the blue-light conditions (Haury 1975). This inconsistency and the existence of active wavelengths in the blue for both PE and PC synthesis (Haury and Bogorad 1977) suggest that there is photomorphogenetic spectral-response detail in the blue region of the spectrum that has yet to be resolved by action spectral analysis.

V. Alternative techniques

Action spectral analyses of chromatic adaptation using another filamentous blue-green alga, *Tolypothrix tenuis,* have used a pigment-bleaching step in a nitrogen-free medium (greater than 20 h at ca.

2,000 ft-c or ca. 20,000 lux of light from fluorescent bulbs) to greatly diminish the phycobiliprotein content and decrease shading in the green and red regions of the spectrum (Fujita and Hattori 1962; Diakoff and Scheibe 1973). Chlorophyll and carotenoid contents remained practically unchanged by this procedure (Hattori and Fujita 1959). After bleaching, chromatic lights induced pigment synthesis in a medium supplemented with nitrate.

Photomorphogenesis has also been examined using *Nostoc muscorum* A., where homogenized trichomes were plated onto an agar-solidified medium and allowed to develop in darkness. The trichomes were arrested with this treatment in the aseriate stage. The cells were then exposed to measured quantities of light at different wavelengths and examined 4 days later for developing microcolonies and aseriate microcolonies. Red light, peaking at 650 nm, was found to be the most effective in inducing development of filaments. This development was most effectively reversed by light in the wavelength range from 500 to 600 nm (Lazaroff and Schiff 1962; Lazaroff 1973).

VI. Acknowledgments

The research upon which this chapter is based has been presented as a thesis in partial fulfillment of the requirements for the Ph.D. degree from the Department of Biology of Harvard University. This work was in part supported by a grant to Professor L. Bogorad from the National Science Foundation. The author was a predoctoral trainee on USPHS grant TO1-GM-00036.

VII. References

Bennett, A., and Bogorad, L. 1973. Complementary chromatic adaptation in a filamentous blue-green alga. *J. Cell Biol.* 58, 419–35.

Bogorad, L. 1975. Phycobiliproteins and complementary chromatic adaptation. *Ann. Rev. Plant Physiol.* 26, 369–401.

Diakoff, S., and Scheibe, J. 1973. Action spectra for chromatic adaptation in *Tolypothrix tenuis. Plant Physiol.* 51, 382–5.

Fujita, Y., and Hattori, A. 1962. Photochemical interconversion between precursors of phycobilin chromoproteids in *Tolypothrix tenuis. Plant Cell Physiol.* (Tokyo) 3, 209–220.

Hattori, A., and Fujita, Y. 1959. Effect of preillumination on the formation of phycobilin pigments in a blue-green alga, *Tolypothrix tenuis. J. Biochem.* (Tokyo) 46, 1259–61.

Haury, J. F. 1975. "Complementary chromatic adaptation in blue-green algae: photoresponse and ecological aspects." Ph.D. thesis, Harvard University, Cambridge, Mass.

Haury, J. F., and L. Bogorad. 1977. Action spectra for phycobiliprotein synthesis in a chromatically adapting cyanophyte, *Fremyella diplosiphon. Plant Physiol.* 60, 835–9.

Jagger, J. 1967. *Introduction to Research in Ultraviolet Photobiology.* Prentice-Hall, Englewood Cliffs, N.J. 164 pp.

Kratz, W. A. and Myers, J. 1955. Nutrition and growth of several blue-green algae. *Amer. J. Bot.* 42, 282–7.

Lamont, H. C. 1969. Sacrificial cell death and trichome breakage in an oscillatoriacean blue-green alga: the role of murein. *Arch. Mikrobiol.* 69, 237–59.

Lazaroff, N. 1973. Photomorphogenesis and nostocacean development. In Carr, N. G., and Whitton, B. A. (eds.), *The Biology of Blue-green Algae.* pp. 279–319. University of California Press, Berkeley, Calif.

Lazaroff, N., and Schiff, J. 1962. Action spectrum for developmental photoinduction of the blue-green alga *Nostoc muscorum. Science* 137, 603–4.

Setlow, R. 1957. Action Spectroscopy. In Lawrence, J. H., and Tobias, C. A. (eds.), *Advances in Biology and Medical Physics,* vol. 5, pp. 37–74. Academic Press, New York.

Shropshire, W., Jr. 1972. Action Spectroscopy. In Mitrakos, K., and Shropshire, W., Jr. (eds.), *Phytochrome,* pp. 161–81. Academic Press, New York.

Tandeau de Marsac, N. 1977. Occurrence and nature of chromatic adaptation in cyanobacteria. *J. Bacteriol.* 130, 82–91.

19: How to detect the presence of a circadian rhythm

BEATRICE M. SWEENEY

Department of Biological Sciences, University of California, Santa Barbara, California 93106

CONTENTS

I. Introduction

Circadian rhythms are already known in many different kinds of organisms, including a number of algae. Why should one want to find more? There are two quite different motivations for attempting to detect a circadian rhythm in an organism. First, a new circadian rhythm might offer just the particular set of characteristics that could provide insight into the mechanism of the circadian oscillator, one of the perplexing problems in cell biology today. The second is a more practical reason. If a physiological or morphological process is being studied in an organism that, unbeknownst to the investigator, is rhythmic, the data points, taken at different times of day and averaged, contain a large systematic error because the values are changing with time.

There is a good probability of finding a rhythm in any alga chosen, because circadian rhythms in algae are by no means uncommon. They have been examined with especial thoroughness in the unicellular alga *Euglena* (Bruce and Pittendrigh 1956; Brinkmann 1966; Edmunds et al. 1976); *Chlamydomonas* (Bruce 1972); *Gonyaulax* (Sweeney and Hastings 1960; Sweeney 1969; Christianson and Sweeney 1973); and *Acetabularia* (Sweeney and Haxo 1961; Mergenhagen and Schweiger 1973). The study of each organism offers its own advantages in understanding circadian systems. *Euglena* phototaxis can be measured automatically (Bruce and Pittendrigh 1956; Brinkmann 1966). In *Chlamydomonas,* genetic analysis is feasible, and mutants with different free-running periods in the rhythm of phototaxis have been characterized (Bruce 1972). In *Gonyaulax,* a number of different physiological processes, including bioluminescence, cell division, and photosynthesis, are controlled by the circadian oscillator (Sweeney 1969). *Acetabularia* can be enucleated and grown for a long time in this condition, still showing its rhythm in photosynthesis (Mergenhagen and Schweiger 1973). Circadian rhythms in macroscopic algae have been less frequently studied. However, a rhythm in chloroplast movement has been discovered in *Ulva* (Britz and Briggs 1976), and a convenient automatic monitoring system for studying this rhythm has been devised (Britz et al. 1976). These are only a few examples of algae in which circadian rhythms have been demonstrated and have proven interesting.

Definition of terms

Before beginning experiments to detect whether or not the alga of interest has a circadian rhythm, it is important to know what a circadian rhythm really is. The term is used in two quite different senses in the literature today, a difference of opinion nicely set forth in two recent letters to *Science*(Kanciruk 1976). One meaning of circadian rhythm, the more correct meaning to my way of thinking, is "an endogenous oscillation with a period of about 24 h in constant conditions of irradiance and temperature." However, the term has come to be used by some investigators, particularly in medicine, for any function that varies over the course of 24 hours when the organism is in a light–dark environment, and this without any test under constant conditions. Unless oscillations can be shown to continue under constant conditions, the term circadian rhythm should not be applied. In a light–dark cycle, one cannot be sure that the changes are not the direct result of the cyclic environment.

A few of the terms by which the characteristic features of circadian rhythms are described are included here. For a more complete glossary, see references by Halberg and Katinas (1973), Palmer (1976), and Sweeney (1969). Period, amplitude, and phase have the same meaning as in the physics of oscillatory phenomena. The *period* is the length of time needed to complete one cycle of the rhythm and can be measured from any clearly defined point of time in one cycle to the corresponding point in the next cycle, for example, the beginning of activity or the maximum bioluminescence. The *amplitude* is the excursion from the mean value, both positive and negative, and is a measure of the magnitude of the rhythm. The *phase* is the state of the rhythm at any particular time (maximum, minimum, decreasing, or increasing). A *phase shift* is a change in the relationship between a particular part of the rhythm and external time, as for example the shift in the maximum luminescence from midnight to noon that occurs when a culture of *Gonyaulax* is transferred from a light–dark environment where the period corresponds to night in the outside world to one where darkness occurs during normal daylight. These features of circadian rhythms are all measured by examining the processes controlled by the circadian system. It has not so far been possible to measure the parameters in the circadian system directly, since we do not yet know what those parameters are.

In studying circadian rhythms, it is convenient to have a measure of intracellular time or phase. This is called *circadian time* (c.t.). The usual convention is to begin the cycle at the phase corresponding to the beginning of the light period in natural light or a light–dark cycle, calling this 00 c.t. There are 24 circadian hours in one cycle of the rhythm. Time is given according to the Navy system of 0 to 24 h, midnight thus being 18 c.t. This prevents confusion between A.M. and P.M. It is sometimes convenient to think of a circadian cycle as a circle divided into 360 degrees. For example, one can speak of one rhythm being 90° (that is, 6 hours) out of phase with another.

Circadian rhythms have been shown to be *entrained* (set) by either light or temperature cycles in the environment, if these external cycles do not vary too greatly from a 24-hour period. When entrained, the period of the rhythm is exactly that of the environmental cycle. In the presence of entrainment (absence of the setting stimulus), the rhythm is said to be *free running,* and now shows its *natural* or *free-running period,* a characteristic of the endogenous oscillator.

II. Criteria that must be satisfied to establish the presence of a circadian rhythm

A. Rhythmicity in constant conditions

The first and most important criterion for establishing the presence of a circadian rhythm is rhythmicity. This cannot be demonstrated as long as there is a cyclic light or temperature environment such as day and night. The process to be measured must be examined under constant conditions, that is, constant temperature and continuous darkness. For plants, constant light is preferable because they starve without photosynthesis. It is safest to choose a relatively low irradiance level, since many organisms are known to loose rhythmicity in bright, continuous light (Harris and Wilkins 1976). One should follow the process for at least two complete cycles in constant conditions, preferably much longer, since the aftereffects of a light–dark (LD) cycle sometimes persist for more than 24 h. In addition, it is desirable to know the length of the period of a rhythm one is studying under different conditions. The inherent variability of many types of physiological measurements demands the measurement of a number of cycles for an accurate determination of the period.

In practice, the organism to be tested for rhythmicity is first grown in a known light–dark environment, usually one in which the alternating light and darkness are each 12 h long (12 : 12 LD cycle). This schedule should be continued for at least a week prior to transfer of the organism to constant conditions, so that the rhythm is entrained and the phase is known. It is advisable to begin measurements during this time, for if a rhythm is not evident in a LD cycle, one is probably not present, and the experiment can be discontinued. The transfer to constant conditions should be made at a time when the phase will not be disturbed by the change in conditions, at the beginning or the end of the environmental light exposure.

B. Temperature compensation

Once rhythmicity has been established (under constant conditions), the variability of the period with temperature should be examined. Circadian rhythms typically show little change in the period with temperature; much less than a factor of 2 with a 10° change in temperature. The period may become longer, not shorter, as the ambient temperature is raised, as in *Gonyaulax* (Sweeney and Hastings 1960). For this reason, it is thought that circadian rhythms are temperature-compensated, and this is considered to be one of their characteristic properties. Experiments to test for this property are carried out under constant conditions but at different temperatures. It is advisable to allow the alga to adjust to the new temperature during the previous light–dark entrainment period before beginning to make measurements. The degree of temperature compensation varies rather widely from organism to organism, from a Q_{10} of about 0.8 to one of 1.3 (Sweeney and Hastings 1960). Few rhythms are completely insensitive to ambient temperature.

C. Phase shifts

The third important criterion of a circadian rhythm is a change in phase in response to single or repeated exposures to light or to darkness. All rhythms that have so been examined show a very similar response to single, short exposures to light, usually plotted as a *phase–response curve* (Fig. 19–1). Such curves show in a concise form that the circadian oscillator responds much more to light during the part of the cycle corresponding to night than during that associated with day. Furthermore, if the light during the pulse is bright, the phase–response curve shows a sudden change in the *direction* of the phase shift in the middle of the night phase. Light pulses in the early part of the night delay the next cycle of the rhythm but advance the phase later. This shift in direction is typical of circadian rhythms.

To test for this characteristic response to short light exposures, the organism must be kept in otherwise constant conditions, preferably darkness, since phase shifts are then of larger magnitude than if the organism were kept in dim, continuous light (Christianson and Sweeney 1973). Different cell samples are irradiated at different times, and the rhythm is followed after they are returned to continuous darkness. The subsequent time of maximum activity or other

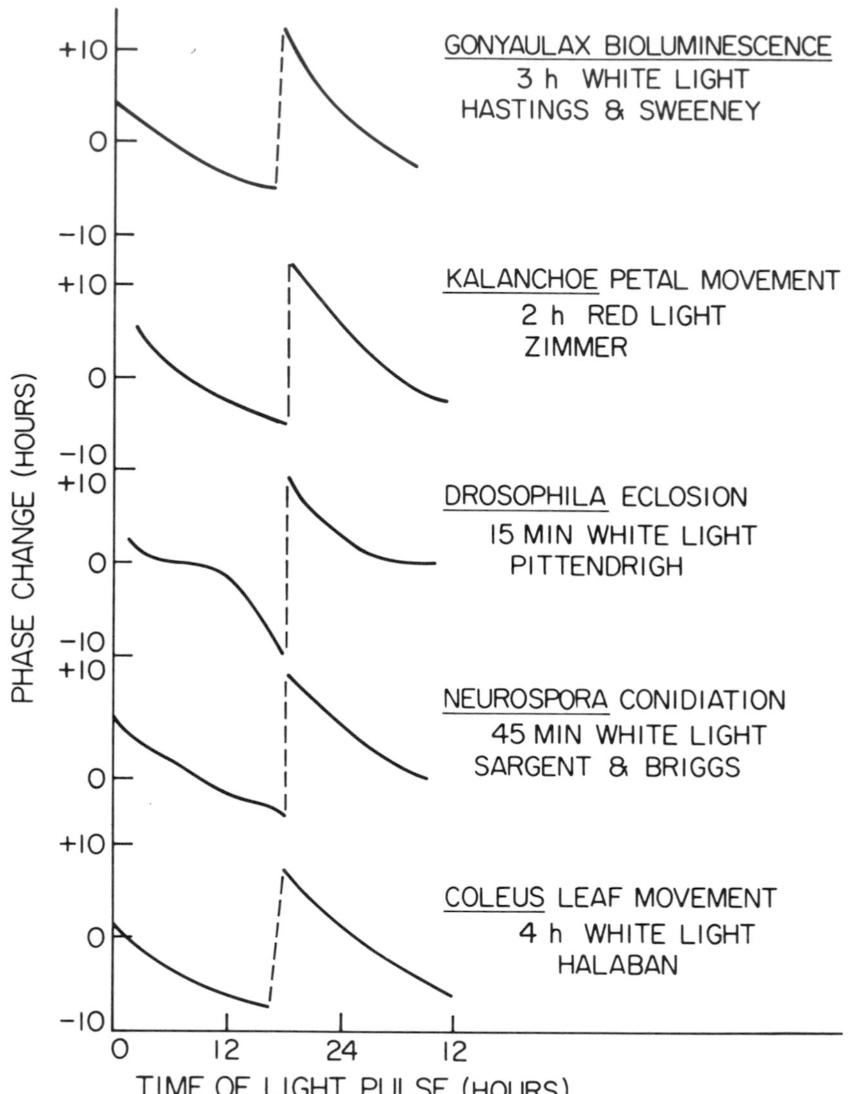

Fig. 19–1. A comparison of some visible-light phase–response curves for circadian rhythms in five different organisms from several laboratories is given on the right of each curve, along with the light used, the organisms, and the investigators who obtained the data. If, following a light exposure, the next cycle occurs earlier than in the control, the phase shift is considered to be positive; if later, negative.

phase point is then compared with that of an unirradiated control cell sample. The effective intensity and duration of the light exposure must be determined for each organism. Various rhythmic systems vary greatly in this regard, *Drosophila* responding to 10 $\mu W/cm^2$ blue light of only a few seconds' duration (Winfree 1971), while *Gonyaulax* must be irradiated in the visible wavelength region for 3 h at 2000 $\mu W/cm^2$ to produce maximal phase shift (Sweeney 1969).

III. Other considerations regarding method

Once the presence of a circadian rhythm has been established in the organism being studied, it can be investigated further, particularly if it shows interesting and promising features. There are obvious advantages in being able to take data by some method of automatic recording. This may be worthwhile if a long series of measurements are being planned. The students in my laboratory and I find it difficult to make measurements without automation at such hours as 2:00 and 4:00 in the morning. In fact, in a recent review (Hillman 1976), Hillman referred to the "Pittendrigh principle": "If you can't automate a rhythm, don't study it," attributed originally to Colin Pittendrigh, who has invented many clever systems for automation. However, one should not become committed too soon to a particular automation system.

It is important to be aware that the circadian rhythms in some organisms are capricious, sometimes disappearing without apparent cause. Furthermore, rhythmicity may not be expressed under all conditions. For example, the rhythm in bioluminescence and cell division in *Gonyaulax* gradually damp out in continuous darkness because of starvation. In wild-type *Neurospora,* CO_2 inhibits the rhythm in conidiation, which resumes if the mycelium is flushed with air to remove the excess CO_2 (Sargent and Kaltenborn 1972). The rhythm of cell division in *Euglena* sometimes disappears spontaneously but can be reestablished by the presence of thiol- and sulfur-containing substances (Edmunds et al. 1976). These effects are probably concerned with the processes being measured as indicators rather than the oscillator itself. The moral is: Try different conditions before concluding that your organism lacks circadian rhythm.

IV. Acknowledgment

The author gratefully acknowledges the support of the National Science Foundation (Grant No. BMS72-02221A03).

V. References

Brinkmann, K. 1966. Temperatureinflüsse auf die circadiene Rhythmik von *Euglena gracilis* bei Mixotrophie und Autotrophie. *Planta* 70, 344–89.

Britz, S. J., and Briggs, W. R. 1976. Circadian rhythms of chloroplast orientation and photosynthetic capacity in *Ulva. Plant Physiol.* 58, 22–7.

Britz, S. J., Pfau, J., Nultsch, W., and Briggs, W. R. 1976. Automatic monitoring of a circadian rhythm of change in light transmittance in *Ulva. Plant Physiol.* 58, 17–21.

Bruce, V. G. 1972. Mutants of the biological clock in *Chlamydomonas reinhardi. Genetics* 70, 537–48.

Bruce, V. G., and Pittendrigh, C. S. 1956. Temperature independence in a unicellular "clock". *Proc. Nat. Acad. Sci. U.S.* 42, 676–82.

Christianson, R., and Sweeney, B. M. 1973. The dependence of the phase response curve for the luminescence rhythm in *Gonyaulax* on the irradiance in constant conditions. *Intern. J. Chronobiol.* 1, 95–100.

Edmunds, L. N. Jr., Jay, M. E., Kohlmann, A., Liu, S. C., Merriam, V. H., and Sternberg, H. 1976. The coupling effects of some thiol and other sulfur-containing compounds on the circadian rhythm of cell division in photosynthetic mutants of *Euglena. Arch. Mikrobiol.* 108, 1–8.

Halberg, F., and Katinas, G. S. 1973. Chronobiologic glossary. *Intern. J. Chronobiol.* 1, 31–63.

Harris, P. J. C., and Wilkins, M. B. 1976. Light-induced changes in the period of the circadian rhythm of carbon dioxide output in *Bryophyllum* leaves. *Planta* 129, 253–8.

Hillman, W. S. 1976. Biological rhythms and physiological timing. *Annu. Rev. Plant Physiol.* 27, 159–79.

Kanciruk, P. 1976. Circadian rhythms. *Science* 193, 8 (with reply by J. W. Lang).

Mergenhagen, D., and Schweiger, H. G. 1973. Recording the oxygen production of a single *Acetabularia* cell for a prolonged period. *Exp. Cell Res.* 81, 360–4.

Palmer, J. D. 1976. *An Introduction to Biological Rhythms.* Academic Press, New York. 375 pp.

Sargent, M. L., and Kaltenborn, S. H. 1972. The effects of medium composition and carbon dioxide on circadian conidiation in *Neurospora. Plant Physiol.* 50, 171–5.

Sweeney, B. M. 1969. *Rhythmic Phenomena in Plants.* Academic Press, London. 147 pp.

Sweeney, B. M., and Hastings, J. W. 1960. Effects of temperature upon diurnal rhythms. *Cold Spring Harbor Symp. Quant. Biol.* 25, 87–104.

Sweeney, B. M., and Haxo, F. T. 1961. Persistence of a photosynthetic rhythm in enucleated *Acetabularia. Science* 134, 1361–3.

Winfree, A. T. 1971. Corkscrews and singularities in fruitflies: resetting behavior of the circadian eclosion rhythm. In Menaker, M. (ed.), *Biochronometry,* pp. 81–109. National Academy of Sciences, Washington, D.C.

Section II

Light and electron microscopy: preparative methods

20: Photomicrography and special microscopic techniques

J. ROBERT WAALAND

Department of Botany, University of Washington, Seattle, Washington 98195

CONTENTS

I. Objective

This chapter is designed to acquaint the experimentalist with certain aspects of photomicrography. It discusses photomicrographic equipment, characteristics of light sources and specimens, filters and films, and mentions techniques for special situations. The references at the end of the chapter cover many of these topics in greater detail. Where specific products are mentioned, it is because the author has used them; usually others can be substituted.

II. Equipment considerations

A. Microscope

A high-quality, stable microscope is essential for photomicrography. A rotating stage is an asset in orienting objects for the rectangular format of most films. For joining attachment cameras to the microscope, a trinocular or monocular body tube should be used. The substage should accommodate the different condensers needed for bright-field, dark-field, phase-contrast, Nomarski interference-contrast, and other optical and analytical equipment.

Flat-field apochromatic lenses (planapochromats) are capable of the highest resolving power and give the truest color rendition; they are the first choice for photomicrographic objectives. Next in quality and cost are the semi-apochromats (fluorite objectives), which may have special proprietary names (e.g., Neofluar, Zeiss). Fluorite objectives are found on many research microscopes. Flat-field achromats (planachromats) make good photomicrographic objectives. The ordinary achromats are of the lowest quality and are generally less desirable.

With apochromatic, fluorite, and planachromatic lenses, only compensating eyepieces should be used; they compensate for the fact that blue and red light do not come to the same focus after passing through an apochromatic or fluorite objective. In combination with high quality objectives and eyepieces, one must also employ a high quality substage condenser and illuminating system, or the excellence of the other optical components will be wasted. An achromatic (sometimes called achromatic-aplanatic) condenser is best for photomicrography, because it is corrected for spherical and chromatic aberration. The numerical aperture (N.A.) of the condenser must match the objective for the full resolving power to be realized. To realize the full numerical aperture of a high-N.A. condenser (N.A. > 0.95) immersion oil (Type B) must be used between it and the slide. The numerical aperture of a condenser is regulated by its iris diaphragm, so that its aperture can be matched with that of the objective (Hartley 1964; Chamot and Mason 1958).

B. Light sources

1. Filament lamps. Tungsten-filament (incandescent) lamps are the most common, but tungsten-iodide lamps are now coming into wide use. These offer several practical advantages: they do not blacken with use as does a standard tungsten lamp; they are compact and thus fit into small lamp housings and integral illuminators; they have a higher color temperature output (discussed in IV.A.2); and they are very bright and can even be used for fluorescence microscopy.

2. Arc-discharge lamps. Xenon arc lamps, in which an arc is struck in xenon gas in a quartz envelope, are often used. They have an intense and a broad spectral output range (from ultraviolet through infrared). This makes them useful for fluorescence microscopy and for color photography. Mercury arc lamps, with proper filtration, can be used to provide monochromatic blue, green, or yellow light, or ultraviolet light (**Caution:** protective filters must be used to protect eyes, skin, and optics from UV and IR with both types of arc-discharge lamps).

3. Electronic flash. This is, essentially, a xenon lamp that releases an intense flash (1/1000 sec or less) of light, which is advantageous in photomicrography with moving specimens or shaky microscopes.

C. Cameras

Almost any camera can be adapted for photomicrography. Detailed directions for using an ordinary camera with its lens for photomicrography can be found in the Kodak Publication P-2 (Kodak 1974); however, best results are obtained with cameras designed or adapted for photomicrography.

1. Attachment cameras. In these, a 35 mm (or other film size) camera body or film holder is attached to an intermediate piece that contains a beam splitter, focussing eyepiece, and leaf shutter (these produce the least vibration); recent models also usually contain a miniature exposure meter. Such cameras are attached to a monocular or trinocular tube on the microscope. Some attachment cameras incorporate automatic exposure-control devices and flash synchronization.

2. *Large format cameras.* With these units, large sheet film (4 × 5 in.) can be used, and only the microscope lenses are used in order to produce the highest quality images. They must be supported on a rigid stand. A clear screen in the ground glass and a magnifier aid focusing. Sensitive exposure meters can be used at the film plane to estimate the exposure. Polaroid copy cameras (models MP-3, and MP-4) also accept 4 × 5 in. sheet film holders, and a light-shielding tube plus lensless leaf shutter are available.

3. *Integral cameras.* The most sophisticated research microscopes have built-in cameras (usually 35 mm) with automatic exposure controls, color temperature meters, and automatic flash. Since they have so many lenses and beam splitters, integral cameras do not necessarily take better pictures than simple cameras; however, they do offer convenience.

III. Equipment alignment and specimen photography

A. Köhler illumination

For general observation, photomicrography, and techniques such as phase-contrast microscopy, it is essential to align the optical system precisely. The Köhler illumination method that follows was developed especially for photomicrography.

1. The lamp is centered with respect to the lamp housing and the lamp condenser lens.
2. The lamp is positioned with respect to substage mirror (not necessary with in-base illuminators).
3. The lamp filament image is focused on the condenser iris (if necessary, remove the microscope lens and use a tissue paper screen) and centered in the plane of the condenser iris.
4. A specimen is placed on the stage and brought into focus using a low power (10×) objective; then the lamp iris is closed, and the substage condenser is focused until the edge of the lamp iris is visible in the field of the specimen.
5. With the centering drives, the substage condenser is centered until the illuminated area is surrounded by the image of the field-limiting lamp iris. With separate lamps, it may be necessary to move the mirror slightly to center the illuminated field.
6. The lamp iris is opened to just fill the field of view; for thorough illumination of the specimen, the lamp filament image must be of sufficient size to completely illuminate the substage iris opening.

7. For phase-contrast microscopy, the lamp iris must be fully opened. Depending on the equipment, some fine adjustment of the lamp position and focus may be necessary to achieve Köhler illumination. Intensity adjustment is made with a rheostat or neutral-density filters, but *not* by closing the substage condenser iris or by lowering the condenser.

B. Camera attachment and alignment

Most attachment cameras are self-centering. The bellows of large-format cameras are normally extended so that the eyepiece-to-film distance is 25 cm; the lensless shutter of such cameras should have a large opening and be located at or very near the eyepoint of the ocular. A light shield will prevent stray light from entering the camera during exposure.

Vibration during exposure can be avoided by using a sturdy support for the microscope and camera and by avoiding shutter vibration. Focal-plane shutters (usually in 35-mm cameras) are subject to the greatest vibration problems. In such a case, it may be best to hold the camera's shutter open, using the bulb or time setting, and to expose the film by turning the microscope lamp on and off or by using a dark card as shutter.

C. Selecting optics

In photomicrography, it is essential to select that combination of optics, camera, and film that will result in the best picture of a particular specimen. Careful consideration of the interaction of magnification, resolving power, depth of field, and the numerical aperture of the objective, the resolving power of the film, and the final enlargement of the photograph is essential in achieving that result. Kodak Publication P-2 (1974), Allen (1958), Barron (1965), Gander (1969), Lawson (1972), Loveland (1970), McClung-Jones (1950), Needham (1958), and Slayter (1970) provide further details on these topics.

The most precise way of determining the magnification of a subject photographed with a particular set of optics is to photograph a stage micrometer with the same optics and to enlarge this negative to the same degree as the print of the specimen.

D. Properties of specimens

1. Specimen thickness. Thin specimens (one cell or less) are easiest to photograph because of the limited depth of field required and because they are subject to minimal stray light. Details in transparent specimens can be emphasized by the use of phase-contrast or Nomarski interference-contrast optics. If thick material must be photographed, often only lower magnification can be used effectively. Epi-illumination equipment may sometimes be used to photograph surface detail of a thick specimen, but scanning electron microscopy should be considered for such cases.

2. *Motile specimens.* Motile specimens, which may be difficult to photograph, can be slowed down by chemicals that increase the viscosity of the mounting solution (Methyl Cellosolve) or can be immobilized by fixatives (1–2% formaldehyde or 1% osmium tetroxide), or by narcotizing agents (dilute IKI, $NiSO_4$). Small, motile specimens such as spores, gametes, or phytoplankton can be immobilized by mounting on slides that have been precoated with a very thin layer of 1% gelatin or agar. If none of these techniques can be used, an electronic flash can "freeze" the motion of rapidly moving algae. The major limitation of electronic flash is that it frequently requires a high-speed film, which, typically, is grainy.

3. *Colored specimens.* To record the best possible image of a colored specimen, proper film and filters (see III.E and IV.A) are important to enhance contrast. Many biological specimens are quite transparent and require special optics such as phase contrast and Nomarski interference contrast, to produce a contrasty image. In bright-field microscopy, control of contrast can be achieved not only by the use of filters and special optics, but it can also be achieved by judicious manipulation of the substage condenser iris. To maximize contrast in the field of view, the lamp iris should be adjusted so that only the field under observation is illuminated. The contrast of the image can be regulated further by adjusting the substage condenser iris so that only light that enters the objective is supplied by the condenser. Usually, with bright-field microscopy a compromise between maximum resolution and adequate specimen contrast must be reached. In such cases, one typically adjusts the substage condenser iris so that when the back of the object is viewed with the ocular removed, about two-thirds of the back of the objective appears brightly illuminated by a circle of light. With the ocular in place, this setting can be monitored by closing down the lamp iris until it is just visible at the edge of the field; then, the condenser iris is closed until glare visible in the darkened portions of the specimen is minimized.

E. *Selection and use of filters*

1. *Filters for black-and-white photography.* Filters are used to regulate contrast and improve resolution of specimen details. A filter that absorbs the same color as a pigmented portion of a specimen will make that part of the specimen appear darker in the resulting photograph. Thus, a green filter will make the green absorbing chloroplasts of red algae appear dark; a blue filter will make the chloroplasts of green

Table 20–1. *Filters for contrast enhancement with black-and-white film*

Specimen color	Contrast filter color	Suggested Kodak Wratten filter No.[a]
Blue, blue-green, or green	Red	25
Red, magenta, or purple	Green	58
Yellow or brown	Blue	47
Violet	Yellow	15

[a] These are available as gelatin filters from Kodak distributors; many other filter manufacturers use Kodak reference numbers for their glass-mounted filters.

algae appear darker. For improving resolution, one usually chooses shorter wavelength filters (blue) or even interference filters (which produce very pure colors), see Table 20–1 for some specific suggestions.

2. *Special filters.* For the protection of specimen, optics, and observer, a heat-absorbing (or reflecting) filter is used to remove IR radiation produced by high wattage or arc lamps. For protection against UV radiation, a Kodak Wratten 2B filter or equivalent should be used. Neutral-density filters, which absorb evenly across the visible spectrum, are used to reduce the intensity of bright-light sources and are essential for color photography. Color-balancing and compensating filters needed for color photomicrography are discussed with color films (IV A.2).

IV. Films

A. Types

1. *Black-and-white film.* The major factors to consider in selecting a film are its wavelength sensitivity, resolving power, contrast, and speed. A fine-grain panchromatic (sensitive to all visible wavelengths) film such as Kodak Panatomic-X is suitable for general purposes. Lower-speed films have a finer grain pattern and hence greater resolving power. Contrast can be controlled with filters and by selection of high- or low-contrast films and/or developers. For electronic flash photography and for recording of dim fluorescent specimens, higher-speed films such as Kodak Tri-X may be required.

2. *Color films and filters.* Negative film or positive transparency film are the major color film types; the latter is more widely used because of its convenience and economy. Some color emulsions can be processed by

Table 20–2. *Filters for use with color films*

Light source and approximate color temperature	Filters for type A film[a] (3400°K)	Filters for daylight film[a] (5500°K)
6-Volt ribbon filament tungsten (3000°K)	82C	80A and 82A
6-Volt coil filament (3100°K)	82B	80 A and 82
12-Volt tungsten iodide (3200°K)	82A	80A
Xenon arc	85	None

[a] All are Kodak Wratten filters; filters with the same characteristics are available in glass mounts from other sources.

the user, but most are handled by commerical processing labs. Except for special applications, the lower-speed positive transparency films (e.g., Kodachrome 25 or Kodachrome 64) are more widely used. A major consideration in using color films is matching the color responsiveness of the film to the color output (color temperature) of the light source. A light source with low color temperature (expressed in °Kelvin) produces much red light, and a lamp with a high color temperature produces much blue light. Daylight is about 5500°K, and daylight color film is balanced for this source. Table 20–2 lists the color temperature of some common microscope lamps as well as filters to use with such sources and certain color films. Filters are most often used to match the color output of the light source with the film's color response. Blue color-converting filters are used with tungsten light for daylight color films, and reddish-yellow filters with daylight illumination are used with tungsten-balance film. With very long exposures, reciprocity effect can alter the color-recording capabilities of color film, and color-compensating filters may be necessary. The rendition of some biological stains, such as eosin and fuchsin, on color film can be enhanced by use of a didymium filter. Detailed information on these and other color filters can be found in Kodak Publications P-2 (Kodak 1974) and B-3 (Kodak 1970).

B. *Exposure determination*

With automated or semi-automated photomicrographic cameras, the investigator should not hesitate to experiment with film speed settings and/or shutter speeds above and below the manufacturer's recommendations. If the photomicrographic unit lacks a built-in light meter, a sensitive photographic exposure meter with a cadmium sulfide or silicon "blue" cell sensor can be used for most applications.

The small size and high sensitivity of these sensors make it possible to determine the exposure by measuring light at the eyepiece. When using such meters, one has to make a series of trail exposures to estimate the effective speed, or *f*-ratio, of the microscope–camera combination; then it is easy to use films of different speeds with the same meter. Trial exposures should be made with specimens that fill the same portion of the field that typical specimens occupy. With dark-field or fluorescence microscopy, the trial exposures are made separately because such subjects usually require a different meter setting.

V. Special methods

Most of the following methods require specialized, expensive equipment; more detailed explanations of these methods can be found in the references at the end of the chapter (Oster and Pollister 1955; 1956; Pollister 1966; 1969; Oster 1971).

A. Phase contrast microscopy

In phase contrast microscopy, transparent specimens are made visible by visualization of differences in optical-path length in the specimen and between the specimen and background. With phase-contrast optics, diffracted light is manipulated so that optical-path differences result in contrasty images, which the observer can detect. Phase microscopy is particularly useful for observing and recording small structures such as spines and flagellae on live specimens. Detailed discussions of phase contrast microscopy are found in Bennett et al. (1951) and Ross (1967).

A second method of creating contrast differences from optical-path-length differences is by means of Nomarski interference-contrast optics, which are discussed in the following chapter (Chap. 21). For some specimens, polarization microscopy or interference microscopy can be used to advantage to visualize particular details. Two additional contrast-enhancing types of optics are available: the "Anoptral" contrast method devised by Wilska (1954) and marketed by Reichert and American Optical Corp., and the Hoffman modulation contrast system (Modulation Contrast Optics, Inc.).

B. Dark-field and oblique illumination

In dark-field microscopy, a special condenser directly below or above the specimen illuminates it in such a way that light from the condenser does not directly enter the objective. Only light reflected or scattered by the specimen enters the objective; thus, the object appears self-luminous and the surrounding field is dark. With low to moderate numerical aperture objectives, such illumination can be obtained by a central stop in a high-N.A. condenser or with certain of the phase annuli of phase-contrast condensers. Dark-field illumination is particularly useful for revealing the outlines of specimens and for detecting very small objects which may be below the resolving power of light-microscope optics.

Oblique illumination is an illuminating technique that produces images that appear three dimensional, resembling the view produced by Nomarski optics; image interpretation may be difficult because spurious images can be produced by such illumination. Wessenberg and Reed (1971) discuss the technique in some detail.

C. *Fluorescence microscopy*

Many substances fluoresce if illuminated with the correct wavelengths of light. If such objects are illuminated and viewed so that the actinic light does not reach the observer but the fluorescence does, the object appears self-luminous. Some biological objects are themselves fluorescent (e.g., chloroplasts). Others must be labelled with fluorescent dyes such as Calcofluors (Chap. 9.III.A), rhodamine, and fluorescein. Fluorescent labels, especially antibodies conjugated to fluorescent dyes, are a very sensitive tool for detecting and locating certain compounds (see Chap. 27.IV). Fluorescent materials always fluoresce at longer wavelengths than they absorb. In the simplest setups, the exciting radiation is isolated by an exciter filter (between lamp and specimen), which passes only those wavelengths shorter than the wavelength at which fluorescence is expected. Fluorescence emanating from the excited specimen enters the objective and passes through a barrier filter, which blocks the light passed by the exciter filter but allows the fluorescent wavelengths to reach the observer. The exact choice of exciter and barrier filters depends on the wavelength of excitation and fluorescence emission. An IR filter is essential between lamp and microscope optics, and the observer should protect his or her eyes and skin from stray UV radiation from the lamp. Since fluorescence is usually dim, it is necessary to use high-speed film such as Tri-X (Kodak) and to use a developer such as Acufine (Acufine, Inc.). Fluorescence from the microscope optics ("fluorite" lenses), mounting media, and immersion oil should be noted and avoided if possible. In addition to transmitted light fluorescence, epi-illumination fluorescence microscopy is frequently employed because it permits simultaneous viewing of fluorescence patterns and bright-field or phase-contrast transmitted light images; it also permits fluorescence examination of thick or opaque specimens that the exciting light cannot penetrate or in which the fluorescence would be absorbed (Goldman, 1968; Kawamura 1969).

D. Time-lapse cinemicrography and perfusion chambers

Time-lapse cinemicrography is useful for recording changes that occur during growth and development and is probably the most frequently used cinemicrographic technique for biological specimens. Time-lapse cinemicrography requires a microscope, a motion-picture camera capable of making single-frame exposures, and an intervalometer for regulating the interval between exposures. Most such units are expensive. Some have a shutter or switch for illuminating the specimen only while it is being photographed. More detailed discussions of time-lapse cinematography can be found in many of the references in Kodak Publication N-2 (Kodak 1975).

One of the major problems with long time-lapse sequences is maintaining the live specimen in a healthy and active state (see also Chap. 1). For some optical setups, hanging-drop cultures or water-immersion caps are suitable, but phase-contrast and Nomarski optics impose strict limitations on the optical path. For such situations, a perfusion chamber may be necessary. It must be selected carefully because not all chambers work with high magnification, phase-contrast, or interference-contrast optics. A good comparative chart is available in Poyton and Branton (1970), and another multipurpose chamber is described by Dvorak and Stotler (1971). For many algae, long-term exposure on the microscope requires a cooling stage. A few are commercially available (e.g., from Leitz) that pump a refrigerated fluid through the stage and can accommodate perfusion chambers without seriously altering the geometry of the optical setup. Heat can also be removed with a Peltier-effect cooling stage (Cloney et al. 1970). To insure efficient heat transfer, the cold stage and slide or perfusion chamber may be joined with silicone grease or silicone grease plus zinc oxide. Temperature can be monitored by using a tiny thermoelectric or thermistor thermometer probe (Bailey Instruments) that fits under the cover glass or on the slide. Condensation on cold coverslips can be prevented by use of preparations such as Spray-Kleen (American Optical Corp.).

E. Other special microscopic techniques

Ultraviolet microscopy is one technique for increasing resolution. It requires special UV-transmitting lenses or reflecting optics and is used for some microspectrophotometric studies of nucleic acids (Chap. 25). Most photographic emulsions are sensitive to UV, and certain films have extra sensitivity at the shorter wavelengths (Kodak Spectroscopic Film Type 103–0).

Infrared wavelengths can be used for detecting some chlorophyll fluorescence and for examining specimens in nonvisible wavelengths.

Infrared-sensitive film or IR-sensitive image tubes (TV) are necessary to visualize IR images. Focussing for IR photography can be aided by empirical corrections as described in Kodak Booklet P-2 (Kodak 1974). Trial exposures are necessary, because most light meters are not sensitive to IR radiation.

Television and videotape should not be overlooked as a means of observing and recording certain phenomena. Time-lapse videotape units are now available, and many colleges and universities already have much of the necessary videotaping equipment and monitors in their audiovisual departments. Usually, a special microscope adapter or camera is needed. Rapid feedback, the ability to erase and rerecord, and the ability to operate at a variety of wavelengths as well as in color are attractive features of videotape units. Their resolution, however, is far less than that of typical photographic films.

VI. References

Allen, R. M. 1958. *Photomicrography,* 2nd ed. Van Nostrand. Princeton, N.J. 441 pp.

Barron, A. L. E. 1965. *Using The Microscope,* 3rd ed. Chapman and Hall, London. 257 pp.

Bennett, A. H., Jupnik, H., and Osterberg, H. 1951. *Phase Microscopy.* John Wiley and Sons, New York. 320 pp.

Chamot, E. M., and Mason, C. W. 1958. *Handbook of Chemical Microscopy,* 3rd ed., vol 1. John Wiley and Sons, New York. 502 pp.

Cloney, R. A., Schaadt, J., and Durden, J. W. 1970. Thermoelectric cooling stage for the compound microscope. *Acta Zool.* 51, 95–8.

Dvorak, J. A., and Stotler, W. F. 1971. A controlled environment culture system for high resolution light microscopy. *Exp. Cell Res.* 68, 144–8.

Gander, R. 1969. *Photomicrographic Techniques for Medical and Biological Scientists.* Hafner Publishing Co., New York. 106 pp.

Goldman, M. 1968. *Fluorescent Antibody Methods.* Academic Press, New York. 303 pp.

Hartley, W. G. 1964. *How To Use A Microscope.* American Museum of Science Books. Natural History Press, Garden City, N. Y. 255 pp.

Kawamura, A. (ed.). 1969. *Fluorescent Antibody Techniques and Their Applications.* University Park Press, Baltimore. 203 pp.

Kodak. 1970. *Kodak filters for scientific and technical uses.* 1st ed., Kodak Publication B-3. Eastman Kodak, Rochester, N. Y. 88 pp.

Kodak. 1974. *Photography through the microscope,* 6th ed. Kodak Publicatión P-2. Eastman Kodak, Rochester, N.Y. 76 pp.

Kodak. 1975. *Cinephotomicrography.* Kodak Publication N-2. Eastman Kodak, Rochester, N.Y. 40 pp.

Lawson, D. F. 1972. *Photomicrography.* Academic Press, New York. 494 pp.

Loveland, R. P. 1970. *Photomicrography, A Comprehensive Treatise,* vols. 1 and 2. John Wiley and Sons, New York.

McClung-Jones, R. (ed.). 1950. *McClung's Handbook of Microscopical Techniques,* 3rd ed. Hafner Publishing Co., New York. 790 pp.

Needham, G. H. 1958. *The Practical Use of the Microscope.* C. C. Thomas, Publisher, Springfield, Ill. 492 pp.

Oster, G. (ed.). 1971. *Physical Techniques in Biological Research.* vol. 1A, *Optical Techniques,* 2nd ed. Academic Press, New York. 429 pp.

Oster, G., and Pollister, A. W. (eds.). 1955. *Physical Techniques in Biological Research,* vol. 1. *Optical Techniques.* Academic Press, New York. 564 pp.

Oster, G., and Pollister, A. W. (eds.). 1956. *Physical Techniques in Biological Research,* vol. 3, *Cells and Tissues.* Academic Press, New York. 728 pp.

Pollister, A. W. (ed.). 1966. *Physical Techniques in Biological Research,* 2nd ed., vol. 4A, *Cells and Tissues.* Academic Press, New York. 408 pp.

Pollister, A. W. (ed.). 1969. *Physical Techniques in Biological Research,* 2nd ed., vol. 3C, *Cells and Tissues.* Academic Press, New York. 341 pp.

Poyton, R. O., and Branton, D. 1970. A multipurpose microperfusion chamber. *Exp. Cell Res.* 60, 109–14.

Ross, K. F. A. 1967. *Phase Contrast and Interference Microscopy for Cell Biologists.* St. Martin's, New York. 238 pp.

Slayter, E. M. 1970. *Optical Methods in Biology.* Wiley (Interscience), New York. 757 pp.

Wessenberg, H., and Reed, M. K. 1971. The use of oblique illumination in microscopic observations of living protozoa. *Trans. Am. Microsc. Soc.* 90, 449–57.

Wilska, A. 1954. Observations with the Anoptral microscope. *Mikroskopie* (Vienna) 9, 1–80.

21: Polarized light, interference, and differential interference (Nomarski) optics

PAUL B. GREEN

Department of Biological Sciences, Stanford University, Stanford, California 94305

CONTENTS

I. Purpose

A. Introduction

Techniques using polarized light and interference optics can produce strong visible contrast in cells, or parts of cells, where no differential absorption of light occurs, that is, in "colorless" objects. Polarized-light and interference microscopes are quantitative instruments that can measure parameters related respectively, to the molecular orientation and to the density of minute objects. When used qualitatively all three instruments to be discussed yield images that correlate directly with physical properties of the object; there are no effects, such as the halos seen in phase microscopy, which are only indirectly related to the properties of the specimen. All three instruments are nondestructive of the specimen and, hence, are ideal for the study of living material.

B. Nature of the information in the images

1. Polarized light

a. Without compensator. In this simplest type of setup, the background is dark and the specimen shows various degrees of brightness (Fig. 21–1). For biological specimens, increasing brightness means an increase in the product of the birefringence of the object (an intrinsic measure of alignment of certain bonds) and its thickness. This product is termed *retardation* and can be measured properly only when the molecular bond alignment is at the correct angle on a rotating stage. In Fig. 21–2 a *Nitella* cell wall has been torn so that three regions of increasing thickness are presented, that is, a "staircase." One notes that increasing thickness (*a* to *b* to *w*) results in increasing brightness. The brightness is not a simple, linear function of thickness but is qualitatively correlated with it when the object is roughly homogeneous throughout its thickness, as is the case in Fig. 21–2.

b. With compensator. The compensator is a thin homogeneous crystal, which occupies the whole field. Because the compensator is crystalline, its rotation generates brightness over the whole field, as in Fig. 21–2 (unlabeled area). The light beam passes through both the

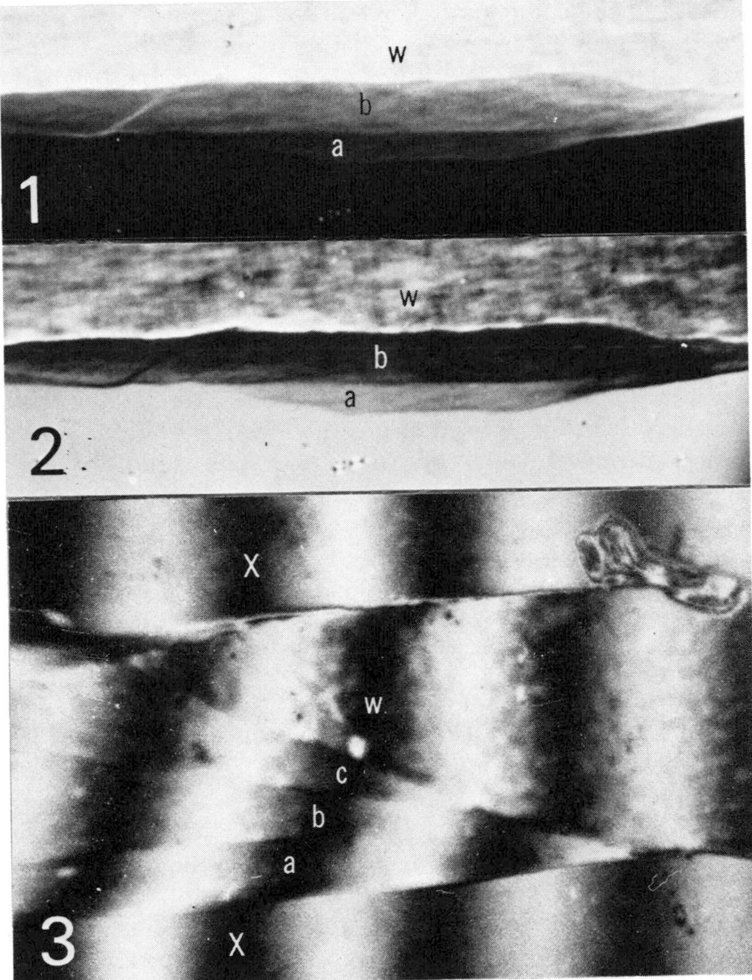

Fig. 21–1. Torn margin of a single *Nitella* cell wall, showing three increasing levels of thickness (*a*, *b*, *w*), the whole-wall thickness (*w*) being at the top. Increasing brightness correlates with increasing thickness. This was taken without a compensator.

Fig. 21–2. Same preparation as above. Lower half of the field is bright background because a compensator has been inserted and rotated to maximize contrast for the central; intermediate thickness of the wall.

Fig. 21–3. Interference image of a related specimen. The vertical dark bands (e.g., X – X) are deflected to the right to an extent directly proportional to the optical thickness (mass/area) of the specimen. Note that at left successive steps (*a*, *b*, *c*), lead to the full thickness of the wall (*w*).

specimen and the compensator. When the compensator is properly turned, the net result can be a cancellation of the specimen's action on the beam, darkening the specimen against a somewhat bright field. This condition is called *compensation* of the specimen.

The compensation condition is advantageous in two ways. First, it enhances contrast. Thus, the central step (*b*), seen in low contrast in Fig. 21–1, is seen more clearly in Fig. 21–2, where the compensator has been turned to compensate this step. Second, the compensated state is used for measurement of the retardation of the specimen. The compensator is first rotated to its null position, where it renders the field dark. As it is rotated further, the field gets brighter, and the specimen (properly oriented) gets dark. When maximal darkness of the specimen is obtained, the extent of rotation from the null position is noted, and this angle is used to measure the retardation. When retardation is divided by thickness (measured some other way) the result is birefringence, a measure of the degree of alignment of molecular bonds within the material. The direction of that alignment is also deducible from knowledge of the direction in which the compensator must be rotated to obtain compensation. There is no better way to obtain molecular alignment information in vivo.

2. *Interference*

a. Uniform field images. Figs. 21–1 and 2 could be images from an interference microscope. However, the reason for the contrast would be different. The increasing brightness in Fig. 21–2 would be due simply to increasing thickness (essentially independent of the degree of molecular bond alignment in the specimen). The thickness measured is, in fact, mass per unit area, rather than the geometrical distance between the two surfaces of the object. This optical thickness is actually more useful than geometrical thickness because it involves the substance of the object. Remarkably, the optical thickness involves only the *dry* mass of the object when it is immersed in water. This is because the water in the specimen and that surrounding it have equal actions on the light beam, giving no noticeable net action on the beam. Contrast arises from the refractility and thickness of the dry part of the specimen. The measurement is made by methods quite related to those for polarized light even though the construction of the instruments is different.

b. Fringed field method. This image (Fig. 21–3) has the advantage that it allows the direct measurement of optical thickness (optical path difference) in a single photograph. A vertical dark stripe pattern (X–X in Fig. 21–3) is generated by the insertion of a horizontal optical wedge elsewhere in the optical system. The specimen deflects the dark fringes, in this case to the right. Furthermore, it does so in direct linear proportion to the mass/area (thickness) of the region of the specimen concerned. Thus in Fig. 21–3 (left), a vertical fringe coming from the bottom of the field is deflected to the right four consecutive times and then finally shifts back to the left, to its original (undeflected) position. The specimen consists of a staircase of different thicknesses of *Nitella* cell wall. The differences in mass/area so readily visualized (and measured) are quite subtle, of the order of fractions of a milligram per square centimeter. Serial extractions of a specimen, each removing a given constituent, can yield the fractional chemical composition of a microscopic structure.

3. Differential interference microscopy (Nomarski). Differential interference (DI) microscopy gives very pleasing images that are ideal for the qualitative illustration of structure. Unlike the phase-microscope image, where contrast is enhanced by edge effects that give spurious halos around discontinuities, the images here correspond to a precise feature of the object. The image is the rate of change of the optical path through the object as one considers consecutive adjacent regions in the object. The optical path is the product of refractive index and distance. A light beam traversing a biological preparation almost always encounters gradients in optical path as one compares regions, (Fig. 21–4). In a regular interference microscope (uniform field) these differences in optical path would be directly related to intensity. In a DI image, the brightness seen is a function of the *change* in optical path. For example, if the optical path increases across a region, then that region appears relatively dark. Hence, the image gives a shadowed impression. Chromosomes in vivo look like "logs at sunset." This does not reflect any obliquity of illumination, but rather a directional bias built into the microscope. This bias converts any increase in optical path to brightness, any decrease to darkness. Constancy of optical path across a region gives this area the same intermediate brightness (gray) of the surround.

The DI instruments are justifiably popular. One special feature is a shallow depth of focus, which means that one can optically section small objects.

II. Equipment

Polarized-light microscopes are made by Leitz, Zeiss, Nikon, American Optical, and most other major optical firms. For biological, in contrast to geological, purposes it is essential to have the microscope equipped with a slot for a variable compensator. The action of most biological material on polarized light is relatively slight; a compensa-

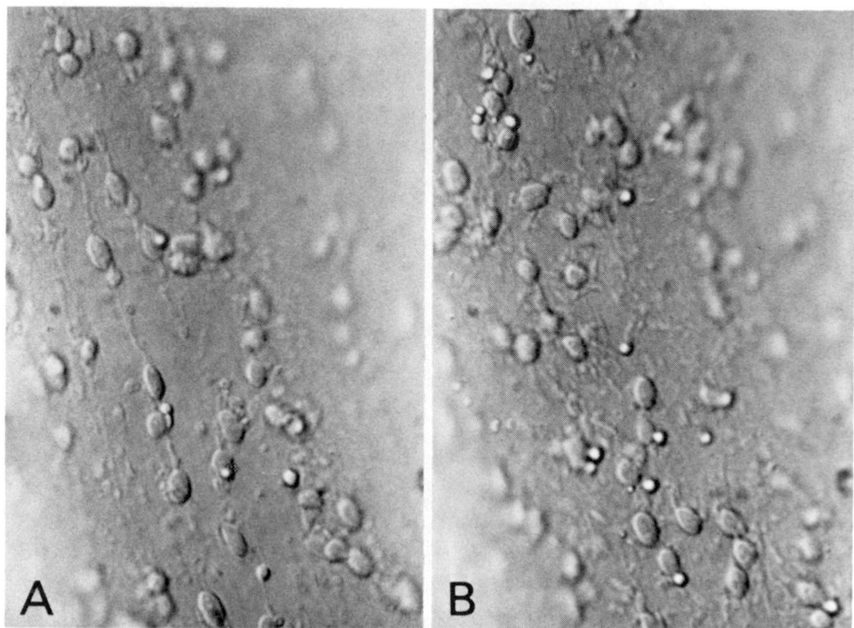

Fig. 21–4. Peripheral cytoplasm from *Vaucheria* cell taken with Nomarski optics. Along a NW–SE line, increasing optical path is brighter than background, decreasing is darker. The effect is most visible in the few small oil droplets (resembling white beads), somewhat less visible in the larger chloroplasts, and barely detectable in the subtle cytoplasmic strands in A. In B, blue light has disorganized the cytoplasmic strands (×455). (Courtesy of M. Blatt.)

tor greatly enhances contrast. For higher-power studies, the "rectified" optical system of Nikon is recommended.

Interference microscopes are made by Leitz and Zeiss. The Leitz instrument is quite expensive in that it consists of two matched microscope systems. It can deal with large specimens and is versatile in that it can be used in either the uniform-field mode, to give pleasing images, or in the parallel-fringe mode, where making measurements is particularly convenient (see Fig. 21–3). The Zeiss instrument uses a single-lens system, and while measurements are quite possible, they are not as convenient to make on a single image, as with the fringe-displacement method.

Differential interference microscopes are manufactured by the above-mentioned companies, as well as by Reichert and others. They all use a polarized-light system, so anyone interested in differential interference contrast can usually gain a good, workable, polarized-light microscope simply by ensuring that the microscope is equipped with a slot for a compensator. A suitable compensator costs less than $1000.

III. Methodology and requirements of the specimen

For in vivo work, the main requirement is that the specimen be thin and the particular region of interest be relatively isolated (that is, the nucleus in *Spirogyra* is best seen through gaps in the peripheral chloroplasts). The light beam is influenced by everything it encounters, in focus or not.

Details of methodology are best learned from manufacturers' promotional and instructional booklets. General features are described in the references.

An example of the combined use of polarized light and interference microscopy can be found in Green (1960). *Nitella* cell walls were analyzed for their internal arrangement of cellulose microfibrils; a gradient in orientation was demonstrated.

IV. Comments on general references

The general text by Slayter (1970) has a very good discussion of the theory of optics for biologists. A short treatment of optical theory is presented in the book by Lawson (1972). Bennett (1950) gives a highly readable and well-illustrated account of polarized light. Old but well-written articles on interference microscopy appear in Mellors (1959) and Davies (1958). An excellent account of differential interference microscopy, complete with outstanding color illustrations of the technique, is given in the paper by Allen et al (1969). The Zeiss information bulletins on the same subject are also good and are listed under Lang (1968; 1969).

V. General references

Allen, R. D., David G. B., and Nomarski, G. 1969. The Zeiss–Nomarski differential interference equipment for transmitted-light microscopy. *Z. wiss. Mikrosk.* 69, 193–221.

Bennett, H. S. 1950. The microscopical investigation of biological material with polarized light. In Jones, R. M. (ed.), *McClung's Handbook of Microscopic Technique,* 3rd ed., pp. 591–677. Macmillan (Hafner), New York.

Davies, H. G. 1958. The determination of mass and concentration by microscope interferometry. In Danielli, J. F. (ed.), *General Cytochemical Methods,* vol. 1, pp. 57–161. Academic Press, New York.

Green, P. B. 1960. Multinet growth in the cell wall of *Nitella. J. Biophys. Biochem. Cytol.* 7, 289–96.

Lang, W. 1968. Nomarski differential interference contrast microscopy. *Zeiss Information Bull.* 70, 114–20.

Lang, W. 1969. Nomarski differential interference contrast microscopy. *Zeiss Information Bull.* 71, 12–16.
Lawson, D. 1972. *Photomicrography.* Academic Press, New York. 494 pp.
Mellors, R. C. 1959. *Analytical Cytology.* McGraw-Hill, New York. 534 pp.
Slayter, E. M. 1970. *Optical Methods in Biology.* Wiley (Interscience), New York. 757 pp.

22: Preparation of algae for light microscopy

MARGARET E. McCULLY

Department of Biology, Carleton University, Ottawa, Canada

LYNDA J. GOFF

Center for Coastal Marine Studies, University of California, Santa Cruz, California 95064

PATRICIA C. ADSHEAD

Department of Biology, University of Ottawa, Ottawa, Canada

CONTENTS

I. Introduction

Investigators who prepare tissues for both optical and electron microscopy can easily lose sight of what they are hoping to elucidate, that is, the structure of living cells as they are in their natural environment. Anyone beginning a study of the algal structure should consult two books on cytological techniques by John Baker (1958; 1966). They outline two important principles for those who study the structure of cells and tissues. The first and most important of these principles is stated as follows (1966): "There is only one really reliable criterion by which we can determine whether the image that we see with the microscope is a good representation of what existed in life, and that criterion is comparison with the living cells." Baker (1958) states further: "Anyone who proposes to judge fixatives should equip himself for the work by prolonged experience in the study of living cells."

We, therefore, begin this chapter with a few hints on how to ensure that the specimens to be observed are in the best shape possible, and how to observe them while they are still alive. In the second part we examine many of the fixation, embedding and staining techniques which have been used successfully on algae.

II. Methods for living material

A. Collection and transport

1. Plankton. Regardless of how short the holding time before examination or fixation, it is essential to dilute samples immediately after collection to minimize adverse effects of concentration and storage. The following procedures should be followed:
All glassware should be clean (sterile if possible). Organisms should not be overcrowded in any given sample. Plankton tows should be greatly diluted (usually up to 1:10) with filtered lake or marine water immediately after collection depending on their density. In some cases (during blooms), it may be necessary to dilute unconcentrated samples with filtered water for greatest survival. Sample jars should be no more than two-thirds full, covered lightly with Parafilm, and stored in a cool, shady spot (covered with damp paper towels on hot

days) or on ice when samples are collected in cold weather or from under the ice. Sensitive species should be isolated into enriched lake water or a suitable medium in situ very soon after collection. More resistant species may be sorted later, and some samples may be diluted with a general-purpose medium [such as the medium of Chu as modified by Gerloff et al. 1950, supplemented by 1% peat extract as recommended by Guillard (personal communication)].

2. *Benthic forms.* Upon collection, benthic algae should be placed in separate plastic bags with a small amount of filtered local water to prevent crushing or desiccation. Large marine algae should be packed immediately in plastic (damp not wet), transported on ice, and kept in the dark.

B. *Observation of living algae*

1. *Whole mounts.* A wealth of structural detail can be seen in whole mounts of living unicellular and filamentous algae (see, for example, the superb photomicrographs of Leedale et al. 1965). Such preparations also allow observation over time so that developmental phenomena can be followed by still photomicrography or by cinephotomicrography (Brown et al. 1976; Chap. 1).

Phase-contrast and Nomarski differential interference contrast optics have been most useful for the observation of living cells, but the advantages of other optical systems, particularly polarizing microscopy and fluorescence microscopy (Chaps. 20 and 21), should not be overlooked. Fluorescence microscopy is useful, because in addition to chlorophyll, which autofluoresces, many algae contain strongly fluorescing secondary organic metabolites both in vacuoles and in walls and extracellular matrices, and these can show up structural detail at high resolution. Regardless of the optical system used, the microscope must be properly set up for Köhler illumination (Chap. 20.III.A). Optical brighteners will enhance walls and matrix material (Chap. 9.III.A.1) and nuclei can be vitally stained by the fluorochrome ethidium bromide (McCully 1976). Negative staining with India ink or nigrosin can be used particularly to observe flagella and extracellular mucilages.

2. *Hand sections.* Hand sections of the large algae are useful in studying the cell walls and extracellular matrices without the inevitable shrinkage and distortion introduced by dehydration. Successful hand-sectioning requires thin, flexible razor blades (e.g., surgical steel blades made by Ingram and Bell, Ltd.). Stainless steel blades are not satisfactory, being relatively dull and stiff.

The large brown algae are probably best cut by a slow, deliberate motion using a wet blade. Thinner, bladelike algae are most easily cut between two glass slides or between freshly split (longitudinally) carrot tissue. The material to be sectioned is placed on one flat surface and a second one is placed on top of the first with the material just under the edge of the top slide or tissue. By varying the angle of the razor blade against the top piece, a number of sections can be obtained. Sections should be floated on seawater or freshwater for about a minute and then mounted in water and examined directly. Some workers have found that the clinical chamber cryostats (International, Americal Optical, and others), and the Spencer table-top microtome with a CO_2 freezing attachment are also useful for cutting sections of some species.

3. Staining and mounting hand sections. Traditionally, the method of choice has been to stain sections in aniline blue as follows: The section is stained in 1% aqueous aniline blue for 2–3 min, washed in water for a few seconds, and then washed in 1% aqueous HCl solution until the desired definition is obtained; the section is washed briefly and mounted. Sections can also be stained in methylene blue (0.3% in 30% ethanol) and aniline blue-black (0.4% in 1% aqueous acetic acid). Newroth (1971) reports successful staining of red algae in glycerine jelly to which a small amount of crystal violet has been added. These methods give good morphological definition. If histochemical information is also required, toluidine blue (1% aqueous), Alcian Blue or Alcian Yellow (1% aqueous), or the PAS procedure (III.F.5), (but with shorter time in Schiff's reagent) should be used.

Coverslips, sealed down with nail polish, will preserve stained material for several days. Semipermanent slides can be made by mounting sections in glycerine, glycerine jelly, or glucose syrup (Karo) (Abbott 1971). A 50–60% aqueous Karo solution to which a small amount of phenol is added to prevent microbial growth is a convenient, inexpensive, and effective mounting medium.

In some cases it may be useful to fix (see methods below) material before it is hand sectioned, or sections can be fixed subsequently to allow the making of permanent preparations. In the case of the cartilaginous red algae, a modified Karpechenko solution (Abbott 1971) has been found useful to stabilize the matrix for cutting.

III. Methods for fixed algae

A. Fixation

Currently, most fixatives used in microscopy are the noncoagulative, protein-cross-linking fixatives (e.g., aldehydes and osmium tetroxide) which give much more faithful preservation of cell architecture than can be obtained with the best coagulative ones [see O'Brien et al. (1973) and references therein].

Evaluation of fixation procedures should always involve comparison of the fixed image with that of the living cell. In plants it is feasible to monitor the course of fixation in isolated cells by phase contrast microscopy (O'Brien et al. 1973; Mersey and McCully 1978). Such monitoring not only allows direct observation of gross distortion caused by shrinkage, swelling, or loss of extracellular mucilages but also allows evaluation of the preservation of cytoplasmic structure and can reveal procedure-induced changes in organization of the cytoplasm that would not otherwise be detectable (Mersey and McCully 1978).

Material fixed directly from the field gives the best results for the cytologist. Pickett-Heaps (1975) points out that material in culture frequently does not fix well and often has very dense cytoplasm not characteristic of field specimens. Culture conditions that simulate the natural environment as nearly as possible are essential (Lehman 1976), especially for species that are adapted to extreme conditions of temperature, salinity, and pressure (Hoham 1976).

1. Schedule I. Phosphate-buffered glutaraldehyde fixation is recommended for routine work.

a. Fixing Solution. 3% glutaraldehyde in 0.025 *M* (for freshwater forms, 0.1–0.2 *M* for marine forms) phosphate buffer (pH 6.8) at 4°C. The fixative should be freshly prepared; if storage is necessary it should be brief and at 4°C in the dark. Solutions with a precipitate, or stock solutions having a pH of 3.5 or less, should be avoided.

b. Procedure. Large algae are cut up in the cold fixative solution with a sharp razor blade in a fume hood. They should be cut into 2-mm^3 cubes or slices 1 mm thick. For techniques of handling single cells, see Glauert (1975) and Dawes (1971). Pieces are rapidly transferred with a spatula or a small brush to cold, fresh fixative (at least 50 × the volume of the tissue). Material is routinely fixed for 12–14 h, but 1 – 2 h may suffice for single cells. Difficult material may require up to 2 days, with a change of fixative.

c. Alternatives. Some algae respond better to unbuffered fixatives added to normal milieux such as sea water (Wetherbee and Wynne 1973) or the culture medium (Pickett-Heaps 1972). If buffering is thought necessary, HEPES (Buckley 1973) or PIPES (Salema and Brandão 1973) can be used. Cacodylate buffer may be toxic (Weakley 1977) to both algae and people. If plasmolysis or differential shrinkage occurs, adjustment of the tonicity of the fixative solution may be necessary (see Maser et al. 1967).

Mordanting of unfixed phenols to cell proteins and nonspecific staining may be prevented by fixation in formaldehyde–glutaraldehyde or postfixation with mercury-containing compounds (McCully 1966). For stabilization of surface polysaccharides, addition of cetylpyrdinium chloride to fixatives (McCully 1970) or postfixing in cyanuric chloride and/or *N*-methyl morpholine (Goland et al. 1967) can be tried. Success has been reported with addition of ruthenium red (Toth 1976), Alcian Blue (Behnke and Zeelander 1970), and Alcian Blue followed by lanthanum nitrate (Shea 1971).

Fixation at room temperature will speed fixative penetration, which may be aided by a mild pulling and release of a vacuum on specimens. Addition of dimethyl sulphoxide may also increase the penetration rate of fixative (Schwab et al 1970; Goff 1976) but may disrupt sensitive cells. The effects on penetration of adding small amounts of detergents to fixatives are not known; such additives may, however, be worth trying.

2. *Schedule II.* Unbuffered acrolein (McCully 1966; Feder and O'Brien 1968; Evans and Holligan 1972) may be useful for large algae, particularly for browns and reds.

a. Fixing Solution. A 10% acrolein solution is freshly prepared in tap water at 4°C. Stock acrolein should range from clear and colorless to pale yellow and should not form a precipitate in water. **Caution:** acrolein is very irritating to the eyes and respiratory system and is also capable of spontaneous explosion. It should be stored in dark bottles in a freezer. Commercial acrolein is stabilized by addition of hydroquinone, which should be added again after redistillation (see Feder and O'Brien 1968).

b. Procedure. Specimen pieces, which are cut up in cold fixative as in schedule I, can be somewhat larger because acrolein penetrates more rapidly. Regular fixation time is 12–24 h, or several days for difficult specimens. After fixation, material is placed in the dehydration series.

c. Alternatives. Mixtures of other fixatives with acrolein have been successful in some instances (e.g., Schwab et al. 1970). For other alternatives to be tried, standard manuals such as Glauert (1975), Dawes (1971, 1979), and Pearse (1968) should be consulted.

B. Dehydration

The most important consideration seems to be gentle dehydration, particularly, early in the schedule (initially, solvents are added dropwise to samples in excess of the fixative vehicle). It appears essential that dehydration be carried out in the cold. Damage suspected to occur during dehydration can be ascertained by microscopic examination at the different processing stages.

1. Standard schedule preceding glycol methacrylate embedding:
2-methoxyethanol (methyl Cellosolve) at 0°C, 2 changes in 1–24 h, depending on specimen permeability;
100% ethanol at 0°C, two changes as above;
n-propanol at 0°C, two changes as above;
n-butanol at 0°C, two changes as above.
Specimens can be stored in the freezer in the last two steps.

Feder and O'Brien (1968) found the above schedule superior to ethanol or acetone dehydration for a wide variety of plant tissues. It has been used successfully for the large brown algae (McCully 1966; Evans and Holligan 1972).

2. Procedure for epoxy resins. Dehydration in alcohol or acetone is carried out as described for electron microscopy (Chap. 23.III.D).

C. Embedding and sectioning: plastics

As embedding media, epoxy resins and methacrylates have greatly increased the resolution at the light-microscope level, mainly because much thinner sections (0.5–5.0 μm) can be prepared and can also be stained without removal of the plastic. Plastic embedding media in common use are of two types: those that yield sections which are freely premeable to dyes in aqueous solutions, and those that are impermeable. The first type include the hydrophilic esters of methacrylic acid: for example, glycol methacrylate (Feder and O'Brien 1968); JB-4 embedding medium (glycol methacrylate modified by butoxyethanol, Ruddell (1967); 2-hydroxypropyl methacrylate (Leduc and Holt 1965). The second type include the epoxy resins: for example, Spurr's (1969) resin, polydiallylphthalate (PDAP) (De Mets et al. 1973), Ladd's low-viscosity medium (Mascorro 1976) and Araldite and Epon resins (see Glauert 1975). The hydrophobic methyl and butyl methacrylates are also relatively impermeable to aqueous solutions. Because of the ease with which they can be thin-sectioned, epoxy resins are used when sections for electron microscopy and light microscopy are to be cut from the same blocks. The hydrophilic resins should be used if histochemistry is to be done.

1. Methacrylate resin embedding and sectioning. Glycol methacrylate (GMA) has become the most successfully used hydrophilic resin in light microscopy because of its many advantages. Its monomer is miscible with water and ethanol, and thus it easily permeates both soft and hard tissue even if the tissue is not completely dehydrated. The polymerized blocks can easily be sectioned, and routine staining techniques can be applied directly. *Note:* Care must be used in handling GMA because the reagents employed are highly sensitizing agents.

a. Embedding.

Monomer mixture:

purified glycol methacrylate (Hartung Associates)	95 cc
polyethylene glycol 200	5 cc
benzoyl peroxide	0.5–1 g
or 2,2′-azobisisobutyronitrile (Polyscience)	0.1–0.2 g

Note: The optimum proportion of accelerator must be determined for each batch of accelerator and monomer, and for different oven conditions. Too little accelerator will result in soft blocks, and bubbles may be produced if the concentration is not appropriate.

Upon dehydration, the tissue is transferred from *n*-butanol (room temperature) to the monomer mixture, where it is kept for 2 days in the dark, preferably with gentle rotation. Polymerization should be done in a nitrogen atmosphere at 60°C for 24–48 h. If a nitrogen atmosphere is not available tissue is embedded in No. 00 gelatin capules which are then filled to the top and capped, or tissue can be flat-embedded (2–3 mm resin depth) in stacked aluminum weighing dishes. Alternatively, polymerization can be effected by UV light (Cole and Sykes 1974). If this is done at low temperature, enzyme activity is preserved (Ashford et al. 1972).

If infiltration is poor, an initial step of butanol/monomer mixture 50:50 may be tried. Furthermore, a JB-4, a butoxyethanol-modified glycol methacrylate (Polysciences) may be used. It is completely miscible with water and polymerizes at room temperature. Successful results have been obtained with it on many marine algae (Goff, unpublished).

b. Sectioning. GMA- or JB-4-embedded materials are sectioned according to the following standard procedure:

i. Blocks are trimmed with a small file. Rectangular block faces (2 × 3 mm, or up to 2 × 7 mm) are easily sectioned and flattened. Specimens should be oriented so that poorly infiltrated walls or outer mucilage-coated surfaces are parallel to the long axis of the block face, and the leading block face should contain tissue along its entire length.

ii. Sections (1–2 μm thick) are cut with a dry, glass knife, and are picked up by a corner with fine forceps or a sharpened applicator stick. Concave side down, they are transferred to small drops of distilled water on slides (wiped clean with tissue).

iii. They are allowed to dry overnight in a dust-free environment, or on a warming plate (ca. 80°C).

With methacrylate resins, it is also possible to pick up and orient numerous sections in series individually and to obtain ribbons (Greany and Rubin 1971). Should smearing of badly fixed mucilages occur in such sections, it can be minimized by placing sections on aqueous $2 \times 10^{-3} M$ $CaH_4(PO_4)_2 \cdot H_2O$. Before staining excess calcium phosphate be should removed by water rinses (McCully 1966).

2. Epoxy-resin embedding and sectioning

a. Embedding. For embedment in epoxy resins, see Chap. 23.III.E.

b. Sectioning. Epoxy-resin-embedded algae are sectioned according to the following standard procedure:

i. Blocks are trimmed and sectioned as for electron microscopy.

ii. Gold or purple sections are picked up with a small platinum loop (cleaned with xylene to remove adhering sections) or a flat-tipped toothpick.

iii. Sections are floated in a small drop of water on a glass slide.

iv. To flatten sections, they can be exposed to xylene-soaked cotton wads in a covered dish. The expansion of the sections is observed through a dissecting microscope, and the dish is uncovered immediately upon completion of the expansion.

v. Sections are dried to the slide on a warming plate (not too hot to touch), and their position is marked with a diamond scribe on the reverse side. Recovery of serial sections of epoxy-embedded material is also possible (Roberts and Hutchinson 1975).

D. Embedding and sectioning: wax

The wax DGD (diethylene glycol distearate) (Taleporos 1974) introduces less distortion than paraffin, and because it is harder, thinner sections (0.5–2 μm) can be produced. It must be removed prior to staining, and this may cause some collapse of tissue structures; but it facilitates phase-contrast study of unstained low-contrast tissues. Enzyme and antigenic activity is retained during embedment (Taleporos 1974). This wax has been successfully used in cytological studies of many marine algae.

1. Embedding

a. Tissue is dehydrated by any standard dehydration method used for paraffin and cleared in two changes (30 min each) of xylene or benzene.

b. Tissue is infiltrated in xylene/DGD wax mixtures of 3:1, 1:1, and 1:3 (30 min each). This is followed with three 20-min changes of 100% DGD, the wax being kept at 60°C during the entire process.

c. The infiltrated tissue is then placed in a mold with fresh wax and this is hardened by cooling on ice. *Note:* DGD contracts considerably during hardening, and it may be necessary to pipette a few additional drops of melted wax into the central depression before hardening occurs around the tissue. Because DGD contracts and expands considerably upon heating and cooling, wax which is not in flake form should be stored in a polyethylene or polypropylene container.

2. *Sectioning.* DGD-embedded tissue can be sectioned according to the following schedule:

a. Blocks are trimmed with fresh razor blades and mounted to a microtome holder.

b. They are cut by an ultramicrotome with a glass knife or with a rotary microtome having a well-sharpened steel blade. In either case, a water boat is used to avoid section compression.

c. The DGD sections are transferred on a flat surface (e.g., pieces of bond paper) or a wire loop to a microscope slide into drops of water.

d. Excess water is withdrawn with a blotter from around the sections. (Heat should not be used to expand sections.)

e. To facilitate section adhesion, the slides are placed in a vacuum desiccator for at least 12 h.

f. Before staining, wax is removed by immersion in xylene (5–10 min), rinsing in acetone or absolute ethanol, and rehydration in water. Dewaxed sections can be stained by any histochemical and morphological staining procedures that can be used for paraffin-embedded sections (Jensen 1962).

E. Procedures for general stains

Unlike wax-embedded tissues, sections embedded in plastic are stained and studied without removal of the plastic, and thus the distortions introduced by its removal are avoided. Most staining methods developed for wax-embedded tissues can be modified and used for staining hydrophilic methacrylate resins, but not for epoxy sections. If section loss is a problem, slides can be precoated by dipping cleaned glass slides in a 0.25% gelatin solution (w/v) and heating them for 1 h at 60°C. For the following staining procedures unless otherwise noted, sections are mounted on slides.

1. *Azure II–methylene blue.* This is a nonspecific stain that works well on sections embedded in Araldite or Spurr's medium (see Chap. 23.II.F and III.E).

a. Staining solutions. 1% aqueous azure II and 1% methylene blue in 1% sodium borate. (Solutions are filtered just before being combined.)

b. Procedure and results

i. Sections are stained (1–5 min) in a fresh mixture of equal parts azure II and methylene blue solutions on a hot plate (not too hot to touch); one should take care not to let the stain dry.

ii. The stained sections are then rinsed with water, dried in air, rinsed in xylene, and mounted in a nonaqueous medium. If sections stain weakly, they can be pretreated with 1% periodic acid for 5 min and rinsed. Cytoplasmic components stain all shades of blue, lignified walls and tannins are usually green, while some walls stain metachromatically. However, because of the high pH, the histochemistry is not reliable.

2. *Paragon (multiple stain).* Paragon is a commercially prepared staining solution (Ladd Research Industries) containing toluidine blue O and basic fuchsin. It is a nonspecific stain suitable for epoxy or methacrylate embedded tissue.

a. Staining solution. Paragon multiple stain undiluted

b. Procedure and results. The entire procedure is carried out on a hot plate (not too hot to touch).

i. Sections on slides are prewarmed for 10 min, oxidized in 1% periodic acid for 5 min, washed in water, and dried.

ii. They are then stained in Paragon (5 min for epoxy, 2 min for methacrylate), washed, dried, and mounted. Cytoplasmic components stain all shades of blue; starch and some wall components stain magenta.

3. *Toluidine blue O for epoxy-embedded tissues*

a. Staining solution. 0.5% toluidine blue O in 1% sodium borate is filtered before use.

b. Procedure and results. Sections on slides are stained for 1–6 min on a hot plate (not too hot) and rinsed in water. When dry they are briefly rinsed in ethanol; this is followed by a brief xylene rinse and mounting under coverslips. Results are as for azure II–methylene blue, except that the colors are more purple than blue.

4. *Acid dyes for epoxy-embedded tissues (aniline blue-black, Coomassie brilliant blue)*

a. Staining solution. 1% aniline blue-black in 7% acetic acid, or 0.25% Coomassie brilliant blue in 7% acetic acid.

b. Procedure and results. Sections are stained in either dye solution for ca. 10 min on a hot plate, after which they are rinsed in 7% acetic acid, dried in air, and mounted. The cytoplasm and some wall components stain shades of blue. (Jensen and Fisher 1968.)

F. *Procedures for histochemical staining*

Cytochemical reactions can be used to localize proteins, carbohydrates, lipids, and nucleic acids in embedded tissues (for summary of results and suitable embedding materials see Table 22–1). Staining procedures for sections mounted on slides are given below. It should be noted that these histochemical reactions are not necessarily reliable for osmium tetroxide fixed tissue.

Table 22–1. *Specific stains for cytochemical studies*

Staining reaction	Compounds identified	Reaction color	Embedding media[a]	References
1. Bromphenol blue	Proteins	Blue	Methacrylates	Mazia et al. (1953); Ruthmann (1970); Chapman (1975)
2. Fast green	Proteins	Bright green	Methacrylates; Spurr's and Ladd's low-viscosity resins	Ruthmann (1970)
3. Ninhydrin–alloxan–Schiff's reaction	Proteins	Red-purple (brownish)	Methacrylates	Jensen (1962)
4. Acid fuchsin	Proteins	Red, pink	Methacrylates	McCully (1966); Robinow and Marak (1966)
5. Periodic acid Schiff's (PAS) reaction	Polysaccharides with vicinal hydroxyl groups	Red, magenta	Methacrylates, epoxy resins	McCully (1966); Feder and O'Brien (1968)
6. Toluidine blue O pH 4.4	Sulfated and carboxylated polysaccharides	Pink	Methacrylates	McCully (1966); Feder and Wolf (1965)
	RNA	Reddish purple		
	DNA	Green to purple		
	Phenolics	Green to bright blue		

7. Alcian Blue	Sulfates and carboxylated polysaccharides	Blue	Methacrylates	Quintarelli, et al. (1964)
8. Alcian Blue–Alcian Yellow	Sulfated and carboxylated polysaccharides	Blue/yellow	Methacrylates	Parker and Diboll (1966)
9. Oil red O	Lipids	Dark red	Methacrylates; Spurr's and Ladd's low-viscosity resins	Lillie (1965)
10. Sudan black B	Lipids	Black	Methacrylates	Ruthmann (1970)
	Lipoproteins	Black	Epoxy resins	Bronner (1975); Goff (1976)
11. Nile blue A	Fatty acids	Oxazine component: fatty acids – bright blue	Methacrylates; Spurr's and Ladd's low-viscosity resins	Ruthmann (1970); Tainter (1971)
	Neutral lipids	Oxazone component: neutral lipids – red		
12. Luxol fast blue MBS	Phospholipids	Gray-blue	Methacrylates; Spurr's and Ladd's low-viscosity resins	Pearse (1968); Goff (1976)
13. Feulgen reaction	Chromosomes/ chromatin	Red/purple	Methacrylates, epoxy resins	Pearse (1968; 1972)

[a] All reactions also work on fresh, frozen, or dewaxed sections.

1. Bromphenol blue

a. Staining solution. 1 g $HgCl_2$ and 0.05 g sodium bromphenol blue (Allied Chemical) are dissolved in 2% aqueous acetic acid.

b. Procedure

i. Sections are stained for 2–6 h at room temperature, or by gently warming over a bunsen burner.

ii. They are rinsed for 10 min each in 2 changes of 0.5% aqueous acetic acid.

iii. Sections are rapidly dried with a hair dryer.

iv. Immersion in xylene and transfer to 0.5% *n*-butylamine in xylene for a few seconds will produce a blue color. Tertiary butyl alcohol may be substituted for xylene.

v. This is followed by 2 changes (2–5 min each) in xylene and mounting.

2. Fast green

a. Staining solution. 0.1% fast green (Allied Chemical) is dissolved in distilled water at pH 2.0 (adjusted with 1 *N* HCl).

b. Procedure

i. Sections are stained by gently heating (over a bunsen burner) until stain begins to evaporate and edges appear metallic (30–60 sec).

ii. The stain is removed by gently rinsing with water before mounting.

3. Ninhydrin–alloxan–Schiff's reagent

a. Staining solution. 0.5% ninhydrin, or 1.0% alloxan, is dissolved in 100% ethanol.

b. Procedure

i. Sections are stained in ninhydrin, or alloxan, 20–24 h at 37°C.

ii. They are rinsed with 2 changes of 100% ethanol, and this is followed by 2 rinses in distilled water.

iii. Staining in Schiff's reagent (see III.F.5 below, and Chap. 25 IV.A) for 30–60 min, is followed by a water rinse.

iv. They are then incubated in 2% sodium bisulfite for 2 min, carefully rinsed in running tap water (10–20 min), dried in air, and mounted.

4. Acid fuchsin

a. Staining solution. 1% acid fuchsin (Baker Chemical) is dissolved in distilled water.

b. Procedure

i. Staining is carried out for 1–5 min at 40°C (on a hot plate if slow in staining). Care should be taken to avoid drying out the section.

ii. This is followed by rinsing in water until the surrounding plastic is free of stain, drying in air, and mounting.

iii. In some tissues more specificity of proteinaceous structures is obtained with the method of Robinow and Marak (1966), which uses a 0.005% solution of dye in 1% acetic acid, followed by a brief rinse in 1% acetic acid.

5. Periodic acid Schiff's (PAS) Stain

a. Staining solution. Commercially prepared Schiff's reagent is available (Fisher Scientific) or can be prepared as follows:

i. To 200 ml boiling distilled water is added 1 g basic fuchsin (G. T. Gurr, Ltd.) followed by shaking (5 min).

ii. Upon cooling to 50%C, the solution is filtered, 30 ml HCl and 3 g potassium metabisulfite are added, and the solution is stored in the dark for 24–48 h. During this time, the red color should disappear.

iii. To the decolorized solution is added 0.5 g vegetable charcoal with shaking for 1 min.

iv. The mixture is then filtered rapidly under vacuum and stored at 5°C in the dark.

b. Procedure

i. Tissue aldehydes are blocked in a saturated solution of DNPH (2,4-dinitrophenylhydrazine) for 30 min. (DNPH is prepared by adding 0.5 g DNPH to 100 ml 15% aqueous acetic acid, agitating for 1 h, and filtering.) Dimedon can substitute for DNPH.

ii. Slides are rinsed with running water (10 min) and dried in air.

iii. Sections are oxidized in 1% periodic acid for 5–10 min and then placed in Schiff's reagent for 30 min.

iv. They are then transferred quickly and directly to 3 successive baths of 0.5% sodium bisulfite (2 min each), rinsed in running water (5–10 min), dried and mounted.

6. Toluidine blue O

a. Staining solution. 0.05% in benzoate buffer (benzoic acid 0.25 g, sodium benzoate 0.29 g in 200 ml water) at pH 4.4 (McCully 1966; Feder and O'Brien 1968). For specific metachromasy (pink color) of sulfated polysaccharides, the solution is adjusted with 1 *N* HCl to pH 1.0–0.5 (McCully 1970).

b. Procedure and results

i. Sections are stained for 1–5 min and washed in running water for ca. one minute, until most of the stain has washed out of the plastic. If impure GMA is used, the plastic may not destain due to binding of basic dyes to methacrylic acid impurities.

ii. For mounting, one breathes on the section, and when the condensation has *just* disappeared nonaqueous mounting medium is applied. This procedure returns just enough water to tissues to insure proper metachromatic staining of carboxylated and sulfated polysaccharides. Carboxylated and sulfated polysaccharides stain pink to reddish purple; RNA, purple; DNA, blue or blue-green; polyphenols, turquoise, green, or blue-green; cellulose remains unstained.

7. Alcian Blue

a. Staining solution. 3 g Alcian Blue 8GX (Matheson Coleman and Bell) is dissolved in 100 ml 0.1 *N* sodium acetate buffer (pH 5.6). For best results, the dye should first be purified as follows: To 1 g dye, add 1 ml methanol, and dry in a rotary evaporator. Repeat 5 times with drying after each. The final product is stored as a 15% solution of Alcian Blue in acetate buffer.

b. Procedure

i. Sections are stained for 45 min at room temperature.

ii. They are then washed 5 min in tap water and dried. Results with higher plant tissues (McCully, unpublished) suggest that phenolic compounds are also stained.

8. Alcian Blue–Alcian Yellow

a. Staining solutions. Alcian Blue: 0.5 g Alcian Blue 8GX is dissolved in 100 ml acid water (pH 0.5, adjusted with 1 *N* HCl); Alcian yellow: 0.5 g Alcian Yellow (No. 836, ESBE Lab Supplies) is dissolved in acid water (pH 2.5, adjusted with 1 *N* HCl).

b. Procedure

i. Slides are stained for 30 min in filtered Alcian Blue.

ii. They are then washed for 10 sec in acid water (pH 0.5) and then with distilled water.

iii. Slides are stained for 30 min in filtered Alcian Yellow and washed in distilled water.

9. Oil red O

a. Staining solution. 0.5 g oil red O (Sigma Chemical Co.) is added to 100 ml 98% isopropanol. 6 ml of the stock solution is diluted with 4 ml water and after 30 min is filtered.

b. Procedure

i. Sections are rinsed in water; this is followed by rinsing in 60% isopropanol (freshly diluted), and staining (10 min) in freshly filtered oil red O solution.

ii. They are then allowed to differentiate briefly in 60% isopropanol (freshly diluted), washed in water, dried, and mounted.

10. Sudan black

a. Staining solution. A fresh solution is prepared by dissolving 0.4 g Sudan black B (Allied Chemical) in 70% ethanol. It is stored in a closed container for 12 h at 37°C, then filtered.

b. Procedure

i. Sections are pretreated in 70% ethanol (1–2 min), then stained in a freshly filtered solution of Sudan black B at 60°C for 1 h.

ii. They are rinsed in 70% ethanol for 1 min, washed in water, dried, and mounted.

11. Nile blue A

a. Staining solution. Nile blue A (NA 0686, Allied Chemical) is dissolved in distilled water (ca. 0.25 g/100 ml to make a saturated solution) to which 1.0 ml sulfuric acid is added; the solution is then boiled or refluxed for 2 h and filtered (2×). (If the solution is boiled, water must be added to maintain the original volume).

b. Procedure

i. Sections are stained in the above solution at 60°C for 90 min to 8 h. Alternatively, the stain can be dropped onto the slide, and the slide can be steamed over a burner until the puddle edge has a metallic sheen.

ii. Sections are rinsed in distilled water, differentiated in 5% acetic acid for 20–30 min, and rinsed again.

iii. They are further differentiated for 3 min in 0.5% HCl (without this step nuclei and all basophilic structures stain intense blue) and rinsed with distilled water.

12. Luxol fast blue MBS

a. Staining solution. 0.1 g Luxol fast blue MBS (or Luxol fast blue G) (Matheson Coleman and Bell) is dissolved in 100 ml 95% ethanol, and twice filtered.

b. Procedure

i. Sections are stained 6–18 h (56–60°C).

ii. They are rinsed in 70% EtOH (30–60 sec), then in water.

iii. They are differentiated in 0.05% aqueous lithium carbonate (1–2 min) and rinsed in water.

13. Feulgen reaction

a. Staining solution. See section 5 above for Schiff's reagent.

b. Procedure

i. Aldehydes in sections are blocked in a saturated aqueous solution of dimedone (5,5-dimethylcyclohexane-1,3-dione) for 16–24 h. (DNPH blockage is not satisfactory since it is reversed by HCl.)

ii. Slides are rinsed in running water (10 min) and dried to prevent section loss.

iii. They are hydrolyzed in 1 *N* HCl at 60°C for 10 min (optimum times vary with tissue), rinsed, and dried.

iv. Staining in Schiff's reagent is followed as for PAS reaction.

G. *General comments on staining of plastic-embedded sections for light microscopy*

Because sections are very thin and resolution is high, any contaminants that fall on the sections are distractingly visible. It is necessary to take special care to filter freshly all dye solutions. It is often convenient to use stain directly from a membrane-filter syringe. Membrane-filtered distilled water should be used for drying down sections, since bacteria often grow in water. The main limitation of plastic sections cut with glass knives is the small size of the block face. This, however, can yield perfectly adequate material for high-resolution work. One should not be discouraged by a few wrinkles: Plenty of information can be obtained between wrinkles!

Some stains fade after a year or so when sections are mounted in Permount. This problem seems to be overcome by mounting in immersion oil for viewing, then removing the oil with xylene and storing the slides dry, remounting as necessary. Most of the staining methods given above are also useful in enhancing phase contrast, particularly in very thin (gold) sections. Such sections can be mounted on slides or EM grids which can be viewed with the optical microscope.

IV. References

Abbott, I. A. 1971. On the species of *Iridaea* (Rhodophyta) from the Pacific Coast of North America. *Syesis* 4, 51–72.

Ashford, A. E., Allaway, W. G., and McCully, M. E. 1972. Low temperature embedding in glycol methacrylate for enzyme histochemistry in plant and animal tissues. *J. Histochem. Cytochem* 20, 986–90.

Baker, J. R. 1958. *Principles of Biological Microtechnique.* Methuen, London. 357 pp.

Baker, J. R. 1966. *Cytological Technique.* Barnes and Noble, New York. 149 pp.

Behnke, O., and Zeelander, T. 1970. Preservation of intercellular substances by the cationic dye Alcian blue in preparative procedures for electron microscopy. *J. Ultrastruc. Res.* 31, 424–38.

Bronner, R. 1975. Simultaneous demonstration of lipids and starch in plant tissues. *Stain Technol.* 50, 1–4.

Brown, D. L., Massalski, A., and Leppard, G. G. 1976. Fine structure of excystment of the quadriflagellate alga *Polytomella agilis. Protoplasma* 90, 155–71.

Buckley, I. K. 1973. Studies in fixation for electron microscopy using cultured cells. *Lab. Invest.* 29, 398–410.

Chapman, D. M. 1975. Dichromatism of bromphenol blue with an improvement in the mercuric bromophenol blue technic for protein. *Stain Technol.* 50, 25–30.

Cole, M. G., and Sykes, S. M. 1974. Glycol methacrylate in light microscopy: a routine method for embedding and sectioning animal tissues. *Stain Technol.* 49, 387–400.

Dawes, C. J. 1971. *Biological Techniques in Electron Microscopy.* Barnes and Noble, New York. 193 pp.

Dawes, C. J. 1979. *Biological Techniques for Transmission and Scanning Electron Microscopy.* Ladd Publication Co., Burlington, Vermont. 301 pp.

De Mets, M., Lagasse, A., and Goethals, E. J. 1973. Studies on embedding materials. I. Polydiallylphthalate (PDAP), a new embedding medium for biological tissues. *J. Ultrastruc. Res.* 42, 337–41.

Evans, L. V., and Holligan, M. S. 1972. Correlated light and electron microscope studies on brown algae. I. Localization of alginic acid and sulphated polysaccharides in *Dictyota. New Phytol.* 71, 1161–72.

Feder, N., and O'Brien, T. P. 1968. Plant microtechnique: some principles and new methods. *Amer. J. Bot.* 55, 123–42.

Feder, N., and Wolf, M. K. 1965. Studies on nucleic acid metachromasy II. Metachromatic and orthochromatic staining by toluidine blue of nucleic acids in tissue sections. *J. Cell Biol.* 27, 327–36.

Gerloff, G. C., Fitzgerald, G. P., and Skogg, F. 1950. The isolation, purification, and nutrient solution requirements of blue-green algae. In Brunel, J., Prescott, G. W., and Tiffany, L. H. (eds.), *The Culturing of Algae,* pp. 1–114. Antioch Press, Yellow Springs, Ohio.

Glauert, A. M. 1975. *Fixation, Dehydration and Embedding of Biological Specimens.* American Elsevier Publishing Co., New York. 207 pp.

Goff, L. J. 1976. The biology of *Harveyella mirabilis* (Cryptonemiales: Rhodophyceae). V. Host response to parasite infections. *Phycol.* 12, 313–28.

Goland, P., Grand, N. G., and Kotele, K. V. 1967. Cyanuric chloride and N-methyl morpholine in methanol as a fixative for polysaccharides. *Stain Technol.* 42, 41–51.

Greany, P. D., and Rubin, R. E. 1971. Serial sectioning of tissues in glycol methacrylate by supplementary embedding in a paraffin–plastic matrix. *Stain Technol.* 46, 216–17.

Hoham, R. W. 1975. Optimum temperatures and temperature ranges for growth of snow algae. *Arctic and Alpine Res.* 7, 13–24.

Jensen, W. A. 1962. *Botanical Histochemistry.* W. H. Freeman and Co., San Francisco. 408 pp.

Jensen, W. A., and Fisher, D. B. 1968. Cotton embryogenesis: the entrance and discharge of the pollen tube in the embryo sac. *Planta* 78, 158–83.

Leduc, E. H., and Holt, S. J. 1965. Hydroxypropyl methacrylate, a new water-miscible embedding medium for electron microscopy. *J. Cell Biol.* 26, 137–55.

Leedale, G. F., Meeuse, B. J. D., and Pringsheim, E. G. 1965. Structure and physiology of *Euglena spirogyra* I & II. *Archiv. für Mikrobiologie* 50, 68–102.

Lehman, J. T. 1976. Ecological and nutritional studies on *Dinobryon* Ehrenb.: seasonal periodicity and the phosphate toxicity problem. *Limnol. Oceanogr.* 21, 646–58.

Lillie, R. D. 1965. *Histophathological Technique and Practical Histochemistry.* McGraw-Hill, New York. 715 pp.

McCully, M. E. 1966. Histological studies on the genus *Fucus*. I. Light microscopy of the mature vegetative plant. *Protoplasma* 62, 287–305.

McCully, M. E. 1970. The histological localization of the structural polysaccharides of seaweeds. *Ann. N. Y. Acad. Sci.* 175, 702–11.

McCully, M. E. 1976. The use of fluorescence in the study of fresh plant tissues. *Proceedings Microscopical Society of Canada* 3, 172–3.

Mascorro, J. A. 1976. Vinyl cyclohexene dioxide [VCD/*n*-hexenyl succinic anhydride (HXDA)] an ultra-low viscosity embedding medium for electron microscopy. Ladd Scientific Products Bulletin, Burlington, Vermont.

Maser, M. D., Powell, T. E., and Philpott, C. W. 1967. Relationships among pH, osmolality, and concentration of fixative solutions. *Stain Technol.* 42, 175–82.

Mazia, D., Brewer, P. A., and Alfert, M. 1953. The cytochemical staining and measurement of protein with mercuric bromphenol blue. *Biol. Bull.* 104, 57–67.

Mersey, B., and McCully, M. E. 1978. Monitoring of the course of fixation of plant cells. *J. Microsc.* 114, 49–76.

Newroth, P. R. 1971. Redescriptions of five species of *Phyllophora* and an artificial key to the North Atlantic Phyllophoraceae. *Br. Phycol. J.* 6, 225–30.

O'Brien, T. P., Kuo, J., McCully, M. E., and Zee, S. Y. 1973. Coagulant and noncoagulant fixation of plant cells. *Aust. J. Biol. Sci.* 26, 1231–50.

Parker, B. C., and Diboll, A. G. 1966. Alcian stains for histochemical localization of acid and sulfated polysaccharides in algae. *Phycologia* 6, 37–46.

Pearse, A. G. E. 1968. *Histochemistry Theoretical and Applied,* vol. 1. Little, Brown, Boston. 759 pp.

Pearse, A. G. E. 1972. *Histochemistry Theoretical and Applied,* vol. 2. Little Brown, Boston. 759 pp.

Pickett-Heaps, J. D. 1972. Cell division in *Klebsormidium subtilissimum* (formally *Ulothrix subtillisima*) and its possible phylogenetic significance. *Cytobios* 6, 167–83.

Pickett-Heaps, J. D. 1975. *Green Algae.* Sinauer Assoc., Sunderland, Mass. 606 pp.

Quintarelli, G., Scott, J. E., and Dellovo, M. C. 1964. The chemical and histochemical properties of alcian blue. III. Chemical blocking and unblocking. *Histochemie* 4, 99–112.

Roberts, I. M., and Hutchinson, A. M. 1975. Handling and staining of epoxy resin sections for light microscopy. *J. Microsc.* 103, 121–6.

Robinow, C. F., and Marak, J. 1966. A fiber apparatus in the nucleus of the yeast cell. *J. Cell Biol.* 29, 129–51.

Ruddell, C. L. 1967. Embedding media for 1–2 micron sectioning 2. Hydroxyethyl methacrylate combined with 2-butoxyethanol. *Stain Technol.* 42, 253–5.

Ruthmann, A. 1970. *Methods in Cell Research.* Cornell University Press, Ithaca, N. Y. 368 pp.

Salema, R., and Brandão, I. 1973. The use of PIPES buffer in the fixation of plant cells for electron microscopy. *J. Submicr. Cytol.* 5, 79–96.

Schwab, D. W., Janney, A. H., Scala, J., and Levin, L. M. 1970. Preservation of fine structures in yeast by fixation in a dimethyl sulphoxide–acrolein–glutaraldehyde solution. *Stain Technol.* 45, 143–7.

Shea, S. M. 1971. Lanthanum staining of the surface coat of cells. *J. Cell Biol.* 51, 611–20.

Spurr, A. R. 1969. A low viscocity epoxy resin embedding medium for electron microscopy. *J. Ultrastruc. Res.* 26, 31–43.

Tainter, F. H. 1971. The ultrastructure of *Arceuthobium pusillum*. *Can. J. Bot.* 49, 1615–22.

Taleporos, P. 1974. Diethylene glycol distearate as an embedding medium for high resolution light microscopy. *J. Histochem. Cytochem.* 22, 29–34.

Toth, R. 1976. A mechanism of propagule release from unilocular reproductive structures in brown algae. *Protoplasma* 89, 263–78.

Weakley, B. S. 1977. How dangerous is sodium cacodylate? *J. Microsc.* 109, 249–51.

Wetherbee, R., and Wynne, M. J. 1973. The fine structure of the nucleus and nuclear associations of developing carposporangia in *Polysiphonia novae-Angliae* (Rhodophyta). *J. Phycol.* 9, 402–7.

23: Fixation, embedding, sectioning, and staining of algae for electron microscopy

BERNHARD E. F. REIMANN

Department of Biology, New Mexico State University, Las Cruces, New Mexico 88003

ELEANOR L. DUKE

Department of Biological Sciences, The University of Texas at El Paso, El Paso, Texas 79968

GARY L. FLOYD

Department of Botany, Ohio State University, Columbus, Ohio 43210

CONTENTS

I. Introduction

The technology for transmission electron microscopy has been developed from an esoteric art to a routine matter taught today even at the undergraduate level. It is impossible in one brief chapter to give the complete details of all the essential methods commonly used in electron microscopy. These are comprehensively presented in numerous excellent sources (Pease 1964; Kay 1965; Sjöstrand 1967; Hayat 1970; 1972; Meek 1976). It is assumed that the reader already has some preliminary knowledge of electron microscopy. This chapter presents most of the general fixation and embedding methods and the modifications required for the preparation of sections of freshwater and marine algae. For examination of whole cells and small, isolated components Chaps. 28–31 should be consulted. The discussion that follows offers many different choices in each category. We suggest that the investigator begin each project by preparing a complete outline of the chosen procedures and reagents to be used in the project.

II. Equipment and reagents

A. General remarks

The investigator should be *very cautious* in handling the chemicals (especially osmium tetroxide, aldehydes, collidine, propylene oxide, and embedding resins) because many are toxic and many are also strong oxidizing or reducing agents.

Common laboratory equipment will not be listed; it is assumed that the investigators will have this available. We will concentrate on items of particular importance for the preparation of cells and tissues existing in milieux of varying salinities.

In fixing marine algae, the salinity of the fixative should be nearly the same or slightly above that of the environment in which the organism grew. The salinity, usually measured in grams of dissolved salts in one liter of water, can range from 33‰ to 38‰. Normal seawater has the ionic strength of 1 *M* (= 0.5 *M* NaCl or 1 *M* sucrose solution) (McLachlan 1973).

Two instruments that can be used to measure salinity are the osmometer and the electrical conductivity bridge. Modern osmometers take advantage of the phenomenon that substances dissolved in water cause a lowering of the freezing point below that of pure water. Within limits, the degree of this freezing point depression is proportional to the concentration of the sum of dissolved substances (all ions as well as solutes that do not ionize). Osmometers are usually calibrated in mOsm/kg. In the absence of such relatively expensive instruments, the freezing point can be directly determined with a cryoscope. The function of a conductivity bridge depends on the movement of a certain amount of available ions, which correspond to the amount of dissolved salts in water. The ionic movement takes place between two electrodes of defined size and distance from each other using a defined electrical potential. Conductivity bridges are usually calibrated in Siemens (= Mhos = 1/Ω = 1/Ohms).

Substances that are not ionized are practically absent in seawater, and salinity values can be obtained through either osmolality or conductivity measurements. If nonionized molecules (aldehydes) are dissolved in water, the osmolality values are correspondingly higher than the conductivity measurements, and the difference between the two values can be used to distinguish between ions and nonions present in a solution.

B. *Equipment*

A standard electron microscopy laboratory should contain a transmission electron microscope, a vacuum evaporator, an ultramicrotome, diamond knives, and facilities to break glass knives. For specimen preparation a good light microscope, preferably with phase-contrast optics, is required, as is also a dissecting microscope with a magnification range from 5× to 100×. A clinical centrifuge with speeds up to 5,000 rpm and a rotor capacity of 4–6 tubes (10 ml each) is also necessary. Facilities for cold storage (4°C) of reagents (fixatives and embedding media) should also be available.

C. *Reagent quality*

Fixation agents, such as formaldehyde, paraformaldehyde, and glutaraldehyde must be of "electron microscope" (EM) grade. They can be obtained from several companies (see Appendix) that deal in electron microscopy supplies. Other reagents should be of "pro analysis" or so-called analyzed reagent grade.

D. *Buffers*

Seawater and defined media in which algae grow are usually buffered (Nichols 1973). However, in some cases, the buffer becomes inadequate upon addition of the fixative, or the buffer and fixative are incompatible (e.g., TRIS and aldehyde fixatives). Therefore, prior to fixation, media are often supplemented by addition of buffer, or cells are fixed in buffered fixatives directly. Formulae are given for two buffer systems: cacodylate buffer and phosphate buffer. Because buffers employing veronal acetate or TRIS react with aldehydes (Holt and Hicks 1961; Sabatini et al. 1963), they are not discussed here, although both have been applied either in conjunction with osmium tetroxide in electron microscopy or are widely used (especially TRIS) as buffers in growth media.

1. Cacodylate buffer. This is usually used in 0.05 *M* concentration between pH 6.0 and 7.4, as given in Table 23–1. The preparation of the final solution is best done by pipetting 100 ml of 0.1 *M* sodium cacodylate (21.4 g $Na(CH_3)_2AsO_2 \cdot 3H_2O$ dissolved in 1 liter of water is equivalent to 0.1 *M*) into a 200-ml volumetric flask. The addition of the amounts of 0.1 *N* HCl shown in the table and addition of the water to the 200-ml mark will yield the indicated pH.

2. Phosphate buffer. This buffer is used at concentrations between 0.1 and 0.05 *M*, in the same pH-range as cacodylate buffer. It is usually prepared from 0.2 *M* stock solutions of monosodium phosphate (27.58 g/liter $NaH_2PO_4 \cdot H_2O$) and disodium phosphate (28.38 g/liter Na_2HPO_4). The proportions of the two phosphates given in Table 23–2 will yield the indicated pH.

3. Other buffers. These are limited to special fixatives, such as osmium tetroxide or permanganates.

a. Bennett and Luft (1959) used s-collidine (2,4,6-trimethyl pyridine) to produce a buffer of pH 7.40–7.45. The buffer stock solution is prepared by dissolving 2.67 ml of pure s-collidine in about 50 ml of distilled water (= 0.2 *M*); then 9.0 ml 1 *N* HCl is added, and the mixture is diluted with distilled water to a total volume of 100 ml. The pH of this stock solution will be between 7.40 and 7.45. The fixative is prepared by adding 1 ml of the above buffer to 2 ml of 2% aqueous osmium tetroxide solution. The osmolality can be increased by addition of NaCl.

b. Dichromate–chromate buffers have been used in various formulas with osmium tetroxide (Dalton 1955; Wohlfahrth-Bottermann 1957; Schnepf 1960) and with potassium permanganate (Reimann and Volcani 1968). Dalton's chrome–osmium fixation solution contains 1% dichromate, but it is perfectly satisfactory to use a 0.005 *M* solution (0.15 g $K_2Cr_2O_7$ in 100 ml distilled water). The pH is adjusted by carefully adding small amounts of NaOH (pellets briefly washed in distilled water, dried on filter paper and crushed in a mortar to a powderlike consistency) until the pH is approximately 7.0 (pH 7.0–7.2).

Table 23–1. *pH Adjustment of cacodylate buffer*

ml 0.1 *N* HCl	pH
59.1	6.0
47.7	6.2
36.5	6.4
26.6	6.6
18.6	6.8
12.6	7.0
8.3	7.2
5.4	7.4

Table 23–2. *pH Adjustment of phosphate buffer*

ml 0.2 *M* NaH_2PO_4	ml 0.2 *M* Na_2HPO_4	pH
87.7	12.3	6.0
81.5	18.5	6.2
73.5	26.5	6.4
62.5	37.5	6.6
51.0	49.0	6.8
39.0	61.0	7.0
28.0	72.0	7.2
19.0	81.0	7.4

E. Fixatives

Three different types of fixatives have been generally used: aldehydes, osmium tetroxide, and permanganates. Aldehydes yield better cell preservation and permit location of certain enzymes and antibodies (Chaps. 24 and 27). Osmium tetroxide and permanganate have the advantage of providing electron contrast in addition to fixing certain cell components.

1. Formaldehyde. One of the most extensively used fixatives for light microscopy, formaldehyde (HCHO) has the disadvantage that it undergoes the Cannizzaro reaction producing formic acid and methanol. To avoid this, formaldehyde is prepared from its polymer paraformaldehyde. Paraformaldehyde (1.75 g) is dissolved in 50 ml of 0.1 *M* phosphate buffer (pH 7.4–7.6). This solution is heated to 70°C, stirred, and allowed to cool to room temperature. After filtration through filter paper, it is ready for use. This yields a 3.5% formaldehyde solution (pH 7.3–7.5). The osmolality is adjusted by adding NaCl crystals. In our laboratory we store all fixatives as 5-ml aliquots in rubber-stoppered test tubes (7-ml capacity) under vacuum (in Vacutainers, or Venojects) at 4°C. With these precautions, formaldehyde solutions are stable for at least 6 months.

2. Glutaraldehyde. 1,5-Glutar-di-aldehyde is available in EM grade. Ampules with various concentrations (25–70%), sealed under nitrogen, can be purchased and stored unopened for months. Before use, the glutaraldehyde is diluted to the desired concentration with cacodylate or phosphate buffer. Note that the ampule should be cleaned and handled with a clean cloth. With a file, the neck of the ampule can be scored and carefully opened (using lint-free sheets) by snapping off the top. For the fixation of fresh water algae, concentrations between 1 and 3% glutaraldehyde are recommended. For marine algae, we use a standard concentration of 3.5%.

3. Formaldehyde–glutaraldehyde. Karnovsky (1965) combined formaldehyde and glutaraldehyde, producing a fixative of particularly high osmolality. It is composed of 2 g paraformaldehyde plus 25 ml water, which is heated as described above (II.E.1). To this are added 1–3 drops of 1 *N* NaOH (stirring until the solution clears) and 5 ml of 50% glutaraldehyde. This solution is further diluted with 0.2 *M* buffer (phosphate or cacodylate, pH 7.4–7.6) to a total of 50 ml. This fixative should be used at pH 7.2.

4. Acrolein. Acrylic aldehyde is commercially available as a liquid of EM grade with 0.05% hydroquinone to inhibit autopolymerization. It is buffered with 0.1 *M* cacodylate or phosphate buffer (Luft 1959) and used for fixation at a concentration of 5–10% (v/v).

5. Osmium tetroxide. Now used almost exclusively as a contrast-yielding agent after aldehyde fixation, osmium tetroxide can also act as a direct fixative. It is available in highly purified (99.9%) crystalline form in ampules and is generally prepared as a 1% solution buffered with 0.1 *M* phosphate or cacodylate buffer (pH 7.2). **Caution:** Osmium tetroxide fumes are highly toxic and affect the eyes particularly. It should always be handled in a well-functioning fume hood. The ampule is opened as described in II.E.2. for glutaraldehyde, except that it is dropped into the buffer, and the crystals are allowed to dissolve overnight in a tightly stoppered bottle or flask. When kept under vacuum in sealed test tubes at refrigerator temperatures, the osmium tetroxide is stable for several weeks. Discoloration beyond straw-yellow indicates the expiration of usefulness.

6. Potassium permanganate. Among the fixatives, potassium permanganate occupies a somewhat isolated position. Introduced for use in electron microscopy by Luft (1961), it is still of interest and is valuable for producing enhanced images of biological membranes. It is applied at concentrations between 0.5 and 1.0%, unbuffered, or buffered with phosphate or dichromate buffers. An extremely strong oxidant, permanganate will react with any impurity. Consequently, permanganate solutions should be kept for at least one day and filtered before use. Cells grown in nutrients containing reductants such as glucose should be rinsed in isotonic inert solutions (e.g., buffer or filtered seawater) because a voluminous precipitate of manganese dioxide is usually the result of not following this rule.

F. Embedding resins

For conventional procedures, epoxide resins are used, which in their monomeric state are small enough to penetrate the objects but which become very hard upon polymerization. (Epon 812 had been widely used until 1978, when Shell Industries announced cessation of its production. A compound, LX-112, recently became available from Ladd Industries and is a suggested substitute for Epon 812.)

1. Precautions

a. A number of the components are anhydrides (e.g., dodecenyl succinic anhydride), which quite readily attract water from the air. As much as 5% water will cause the blocks to remain soft and exhibit poor cutting qualities. Resin combinations (Luft's Epon 812 mixtures A and B), stored in a refrigerator should, therefore, be brought to room temperature before being opened.

b. Epoxy resins react with their hardening components according to epoxide equivalents. These equivalents (W.P.E. = weight of epoxy resin containing one equivalent weight of epoxide) are imprinted by most distributors on the containers and should be considered. To ensure that the correct combination is achieved, graduated disposable syringes should be used, and the combinations should be thoroughly mixed before use.

c. The final dehydration agent must be compatible with the corresponding resin formula.

2. Resin mixtures

a. Araldite (Glauert et al. 1956; Glauert and Glauert 1958; Luft 1961; Mollenhauer 1964), developed by Ciba, is available in a number of variants. In the United States, two British products, Araldite Resin M and Resin CY-212, are available along with the domestically produced Araldites 502, 506, and 6005. The formula developed by Glauert et al. (1956) using a hardener and a plasticizer (dibutylphthalate) was modified by Luft (1961) to the more widely used formula:

Araldite 502	100 g
DDSA (dodecenyl succinic anhydride, hardener)	75 g
DMP-30 [2,4,6-tri(dimethylaminomethyl) phenol; accelerator] (added just before use)	2.5–3.5 g

Mollenhauer (1964) has included Epon 812 in a similar formula:

Epon 812	100 g
Araldite 502	55 g
DDSA	180 g
BDMA (benzyldimethylamine, accelerator)	10 g

b. Epon 812 (epoxy resin) is most widely applied in a formula proposed by Luft (1961):

Mixture A		*Mixture B*	
Epon 812	62 ml	Epon 812	100 ml
DDSA	100 ml	NMA (nadic methyl anhydride	89 ml

Mixtures A and B are used in various proportions. In our laboratories, 3 parts of A plus 7 parts of B yield a block consistency sufficiently good in its overall cutting qualities. Mixture A yields very soft blocks, mixture B very hard blocks. To a total of 10 ml of both mixtures, 0.15 ml DMP-30 accelerator are added before use.

c. Spurr's low-viscosity resin (1969) can be substituted for Araldite and Epon 812 if it is difficult to obtain a homogeneous and thorough penetration. The formula introduced by Spurr is as follows:

ERL-4206 (vinyl cyclohexene dioxide)	10 g
DER 736 (diglycidyl ether of polypropylene glycol)	6 g
NSA (nonenyl succinic anhydride)	26 g
DMAE (dimethylaminoethanol)	0.4 g

The medium is prepared by adding the components by weight in the order given. The low viscosity of the medium permits rapid mixing by shaking and swirling. This embedding medium is particularly suited for objects with thick cell walls and has, therefore, been successfully applied in many laboratories. The staining properties of material embedded in this resin can be effected by Paragon (a polychromate dye often used in orienting thick sections for thin sectioning, available from Ladd Industries. For staining procedure see Chap. 22.III.E.2).

d. The Maraglas 655 formula was introduced by Freeman and Spurlock (1962 a, b). Since then, it has been variously modified. In recent years, we have used Erlandson's (1964) modification:

Maraglas 655	72 ml
DER-732 (diepoxide flexibilizer)	16 ml
dibutyl phthalate (plasticizer)	10 ml
benzyldimethylamine (catalyst)	2 ml

This formula is particularly useful in providing good embedment of blue-green algal cell and siliceous cell walls, because the resin does not split between the surrounding organic structures and the surrounding medium. It permits the use of high-resolution electron micrscopy of relatively newly deposited silicic acid in diatoms after cell division. It has also been advantageous in the embedment of sediment samples from the ocean floor, as described by Bowles (1968) and Bowles et al. (1969).

III. Procedures

A. Specimen preparation

The ultrastructural appearance of an algal specimen depends upon its state of physiological health. To assess the conditions of the cells, colonies, or tissues and their development, inspection of the material should always be made by light microscopy (see also Chap. 22.II.B). The best way to pursue the chain of events of fixation, dehydration, and embedment is to inspect the material with the light microscope after each step. Suitable parts of the specimen are selected by using 1-μm-thick sections of the resin-embedded materials that can be stained in various ways or be inspected with a phase contrast microscope.

1. Procedures for microalgae. For the fixation, rinsing, and postfixation steps, small cells are concentrated by centrifugation. The speed of centrifugation should be fast enough to ensure that most of the cells will sediment but should not reach speeds that would permanently deform pliable cells or break brittle cell walls. After fixation, various ways of handling the cells or tissues can be followed:

a. The cells are embedded in agar. A 2–5% (w/v) aqueous agar solution is prepared by melting the agar in a bath of boiling water and keeping it in a liquid state at 60°C until use. Small drops (ca. 0.1 ml) of agar are placed on preheated (60°C) microscope slides, which are kept in petri dishes to prevent loss of water. The cells, concentrated during the preceding centrifugation and suspended in a small portion of liquid of the last rinsing step (buffer), are dropped on the agar and mixed with it. The agar-embedded specimens are then allowed to solidify either at room temperature or in a refrigerator. With a razor blade, (1 mm^3) blocks are cut from the solidified agar; these are handled subsequently like tissue blocks and are dehydrated or stained *en bloc.*

b. Cells large enough to be sedimented, even in the relatively viscous resin formulas, can be handled by centrifugation during fixation, dehydration, and penetration by the resin formula.

c. Very brittle objects or fragile colonies are prepared by a filtration method (this method was initially used in von Stosch's laboratory in Marburg/Lahn, Germany in 1958 and subsequently modified). A small Buchner-type funnel (ca. 2-ml capacity) with a frittered glass disc (coarse porosity) can be used. A short piece of rubber tubing (about 30 mm) with an adjustable hose clamp is attached to the stem of the funnel. The other end of the rubber tubing is slipped over a piece of glass rod mounted on a flat cork stand, or over the receptacle tube of a suction flask. The top of the funnel is sealed with a stopper or a piece of Parafilm. A piece of filter paper (Nuclepore) or a Millipore filter is fitted over the glass disc. Fragile cells are then collected, fixed, rinsed, dehydrated, and embedded on this filter always taking care to prevent drying out of cells. Supernatant liquids are removed after opening the hose clamp by simple gravity drainage, or if viscous resins are used, with the aid of suction. Various methods can be applied to remove the objects suspended in the final resin formula from the funnel: If the objects are very small and adhere to the filter, the filter is removed, cut in (1 mm^2) pieces and placed with the adhering resin in the molds for curing. Larger objects are removed with a Pasteur pipette. A third possibility is to pour the objects, suspended in the final resin formula, directly into molds for sedimenting during the penetration period (see E.2.c.below).

2. Dissection procedures for larger algae. Large objects are dissected while submerged in the fixative, preferably at the collection site. A cube, not excedding 5 mm on a side, is excised and placed in a small petri dish containing the fixing solution. A smaller slice is cut from this cube with a sharp razor blade. The slide is transferred to a piece of dental wax, where it is covered with a few drops of fixative and dissected still further into 1-mm^3 pieces. The cubes are then transferred on the point of a broken applicator stick into fresh fixative. This last transfer should be done without any physical stress to the object (do not use a tweezer). With the aid of a dissecting microscope, specific tissue parts can be selected at this step of the operation.

B. Fixation

1. Aldehydes. Aldehydes have a high rate of penetration, and fixation times can be kept short. For most purposes, 1 h at 0–4°C is sufficient.

Tissues or algae with thick cell walls may require a longer fixation time. It may also be necessary to increase the fixation temperature to 15–20°C. Often, microtubules are not preserved at low temperature, and a compromise may have to be reached, depending on the structures of interest. Tests are required in every instance to determine the procedure that yields the best results. This also applies to the choice of the aldehyde and pH.

2. *Osmium tetroxide.* This can be used in one of several ways, but the prerequisites for fixation with osmium tetroxide are the same as for the aldehydes.

a. Direct fixation of cells in 1% buffered osmium (1–2 h), may be useful for visualizing cell-wall and membrane layers, but is otherwise rarely used.

b. Postfixation in 1% buffered osmium (1–2 h) is now used routinely after aldehyde fixation, because the osmium imparts electron density. Phycologists have only recently begun to use the "semisimultaneous" fixation (Franke et al. 1969) pioneered by histopathologists. According to Pickett-Heaps (personal communication, 1977), good results with freshwater algae have been obtained by fixation in 1–3% glutaraldehyde for 15–16 min, followed by direct addition of osmium tetroxide to a final concentration of 1–2%. Total fixation time at room temperature is 1–2 h.

3. *Other fixation procedures*

a. For preservation of blue-green algae, the method developed by Kellenberger et al. (1958) for bacteria and modified by Pankratz and Bowen (1963) often yields superior results over aldehyde fixation.

b. The highly oxidative fixation with permanganate requires low temperatures (ice water or refrigerator temperatures). The exposure of cells and tissues to this fixative should be as short as possible. Single cells in suspension can be satisfactorily fixed in 15 min. Permanganate can also be used for prefixation, followed by osmium tetroxide fixation (Reimann and Volcani 1968).

C. *En bloc staining*

1. Uranyl acetate staining of nucleic acids was first described by Strugger (1956). This method is particularly useful for bacteria and cyanophytes (Kellenberger et al. 1958). After fixation, cells are rinsed and then incubated (2 h) in an acetic acid and uranyl acetate mixture (2.5 g uranyl acetate, 0.5 ml glacial acetic acid, 50 ml water; pH 3.5).

2. Another reagent for *en bloc* staining is phosphotungstic acid, which can be applied (2–12 h) in either aqueous or alcoholic solutions in a 1% concentration.

3. Ruthenium red (aqueous 0.15%) is applied for 10–15 min after fixation with aldehydes and postfixation with osmium tetroxide. It has some specificity for glucuronic acid-containing slimes in algae. (See Chap. 24.II.C.1, which provides additional details for this procedure.)

D. Dehydration

During dehydration, water in the cells or tissues is replaced with a solvent compatible with the embedding resin, such as ethyl alcohol or acetone; more recently, acidified 2,2-dimethoxypropane has been used (Muller and Jacks 1975). The dehydration is carried out by gradually increasing the dehydrating reagent (e.g., 30, 60, 90, and 100% acetone or ethanol) in successive steps. Many laboratories begin the dehydration at 60%.

The dehydration should be carried out as rapidly as possible. If unicellular algae with thin cell walls are used, a 4-min exposure to each step is sufficient. Algae with thicker cell walls and tissues require a longer exposure (15–30 min per step). At 100%, usually two steps of equal duration are used. The schedule of a conventional dehydration is as follows:

30% aqueous ethanol or acetone, 4–30 min (one change)
60% aqueous ethanol or acetone, 4–30 min (one change)
90% aqueous ethanol or acetone, 4–30 min (one change)
100% ethanol or acetone, 4–30 min (two or three changes)

Acetone is generally compatible with all epoxy resins. Some of the embedding resins are incompatible with ethanol; therefore, an additional step employing propylene oxide is required. Because propylene oxide is highly volatile (boiling point 35°C), a final step in this reagent is usually done even after acetone dehydration.

The most rapid dehydration is possible with acidified 2,2-dimethoxypropane (Muller and Jacks 1975). It is prepared by adding 1 drop (0.05 ml) of concentrated hydrochloric acid to 50–100 ml 2,2-dimethoxypropane. Cells fixed and rinsed in phosphate buffer should be rinsed in water before dehydration in dimethoxypropane, to avoid formation of a precipitate. The objects are then directly exposed to the dehydrating solution for a few minutes (we commonly use 5 min). The samples are either immediately embedded in the resin or passed through one additional step of propylene oxide. This step permits the use of any one of the described epoxy resins.

One other dehydration procedure employs a series of 3 successive changes for 6–12 h of absolute Methyl Cellosolve (2-ethoxyethanol), propanol, and propylene oxide at 4°C (Floyd et al. 1972).

E. Embedding

1. Infiltration. To assure adequate infiltration and even polymerization, the resins are gradually introduced with acetone or propylene oxide at room temperature:

1 part resin plus 3 parts solvent, 10 min to 5 h,
1 part resin plus 1 part solvent, 10 min to 5 h,
3 parts resin plus 1 part solvent (and occasionally 19 parts resin plus 1 part solvent), 10 min to 5 h,
pure resin, 10 min to 20 h,
resin plus catalyst, 20 min to 20 h.

Penetration times vary widely from a total time of 60 min for small cells, to several days for large cells and tissue. For the short penetration times, the catalyst can be included in all steps; but during the 20-h penetration time at room temperature, any resin will undergo a certain amount of polymerization; therefore, the initial penetration steps should be done with a resin formula lacking the catalyst.

2. Curing of blocks. The curing time for the resin depends upon the resin used; the instructions of the supplier should be followed. For Epon 812 formulas it is known that lower temperatures foster the formation of side branches and that higher temperatures stimulate the formation of long chains. The recommended time and temperature for final hardening is usually 60°C for 24–48 h. With increasing age, the blocks become increasingly harder. With very soft blocks, additional exposures to temperatures up to 80°C can sometimes result in additional hardening, thus saving an interesting specimen.

The choice of a polymerization mold depends on the specimen. One can use polyethylene molds (BEEM capsules, EMS brand capsules) of various sizes (preferably size 00 with a tip shape), or the less expensive gelatine capsules (size 0 or 00), which require predrying and storage in a desiccator. Several types of flat molds are also commercially available. Various specimens require special approaches to the precuring preparations.

a. Microscopic objects such as individual cells that are not encased in agar, may have to be concentrated in the tip of the mold by centrifugation. In this case, the capsules are filled two-thirds full with the final resin formula. One large drop containing the cell material (previously concentrated in the last resin step) is transferred to the top of the resin in the capsule with a Pasteur pipette. The capsule is then completely filled, and its contents are stirred carefully to avoid air-bubble formation. The filled capsule is then centrifuged for 15–30 min in a clinical centrifuge.

b. Larger objects (tissue parts or agar cubes) are transferred on the pointed end of a wooden stick to capsules or flat molds, positioned, and the molds are then filled with the final resin formula.

c. Orientation procedures. Precise orientation of a specimen may become necessary when the morphological arrangement of an object into specific planes of symmetry dictates the plane of the section. This may be the orientation of a diatom cell along one of the apical, transapical, or perivalvar planes, or it may be a cut perpendicular to the surface of a seaweed. Orientation can be aided, prior to embedding, by gluing the specimen (with a drop of hot agar) in the desired orientation at the end of a small strip of filter paper. Such a mount allows dehydration and resin infiltration as described above, and permits embedding in a plastic mold using the paper strip as a handle for orientation.

Thick sections (1-μm) can also be cut with glass knives, transferred on the tip of a fine needle to a drop of water on a microscope slide, dried on a hot plate (using moderate heat), and examined either unstained (with a phase contrast microscope) or stained (with the polychromate dye Paragon and a regular light microscope). Such an inspection can aid in reorientation.

Microscopically small objects can be embedded in flat aluminum dishes (weighing boats) in a layer of resin ca. 3 mm thick. The aluminum dish is removed after polymerization of the resin. This permits direct investigation of the objects in transmitted light either with a dissecting microscope or with a compound microscope. Objects of interest are marked and cut out with a jeweler's saw. If these cut-outs are large enough, they can be grasped directly in specimen holders (jaw types) designed to hold flat-molded specimens. They are further trimmed on a holder that permits the use of transmitted light (Reimann 1963). If, during trimming, rough surfaces obscure a clear view of the cells, the holder is submerged in a glycerol–water mixture. The high refractive index of this mixture matches that of the plastic and obscures the rough spots. The liquid level should just exceed the top of the plastic, so that it stays wet. In this fashion, light micrographs of the objects to be ultrathin-sectioned can be made with considerable image resolution. If the cut-out specimens are too small they are re-embedded or glued on the end of a glass or plastic rod with a special adhesive (e.g., Eastman 910).

F. Sectioning

The procedures for ultramicrotomy have to follow the characteristics inherent in the design of the instrument available. Every manufacturer provides detailed instructions for this task. In selecting such an instrument for the purpose of sectioning algae, the main emphasis should be placed upon the versatility of the specimen orientation. Blocks are trimmed to a cutting face of 0.2–0.3 mm^2 in either commercially available holders or holders prepared by the investigator (Reimann 1963).

The use of diamond knives is preferred over glass knives because glass knives rapidly dull when used in cutting epoxy resins. The sectioning of hard substances containing silicic acid or strontium compounds is no real problem as long as diamond knives are used. The problem more often encountered is loss of contact between a hard substance (less hydrated) and the embedding resin, which changes its character if water is taken up. In this respect, Maraglas has demonstrated the most suitable qualities.

Sections of material embedded in epoxy resins can be picked up on unsupported sections (that is, the sections are mounted without a special supporting layer such as Formvar). This decreases the chance of developing unwanted contamination. To aid adhesion of sections, grids can be exposed to a "glow discharge" procedure in a vacuum evaporator. Carbon reinforced Formvar or collodion films should, however, be used on single-hole specimen grids, which are necessary for studying serial sections. Small objects can be easily seen using 400-mesh grids. Survey pictures of large objects require specimen grids of 200-mesh size.

G. Section staining

Even material postfixed with osmium usually requires additional contrast. Therefore, additional staining is in general done on the ultrathin sections. A few of the most important conventionally used formulas are mentioned here.

1. Uranyl acetate [$UO_2(CH_3COO)_2 \cdot 2H_2O$]. This compound has been used widely. It can be used in absolute methanol (Stempak and Ward 1964) or as a 5% (w/v) aqueous solution (Watson 1958). A few precautions have to be observed in this procedure. The solubility of uranium acetate in water at room temperature is relatively low. (ca. 7.7 g in 100 ml H_2O) and is usually incomplete due to the presence of the basic salt. The solubility can be increased by acidification. The solution is prepared with stirring and consists of 2.5 g uranium acetate dissolved in 50 ml distilled water to which 0.5 ml glacial acetic acid has been added. The solution is kept in a dark bottle away from bright light, because uranium acetate is light-sensitive. Usually, drops of uranyl acetate are placed on a clean surface of dental wax or Parafilm, and the grids, with the sections facing the liquid surface, are placed on the drops for 15–60 min. This is carried out inside a petri dish to prevent loss by evaporation. The grids are removed from the drop and dipped vertically with a tweezer. If the stain was dissolved in methanol, the grids are rinsed in a dish with absolute methanol, and then in two dishes with 50% ethanol in each. If it was an aqueous solution, the rinses are done in distilled water. Between each of the rinses, the tweezer is replaced by a dry one, to avoid carry-over of liquid trapped between the jaws of the tweezer by capillary action. Finally, the grid is rinsed in a stream of 50% ethanol from a squeeze bottle and blotted on a piece of filter paper. The grid is allowed to dry completely before further staining.

2. *Lead compounds.* Lead components provide high electron contrast (Watson 1958; Karnovsky 1961) but often may form insoluble precipitates with the carbon dioxide of the air, causing contamination of the sections. Reynolds (1963) proposed the application of acetate as a chelating agent, which reduces the formation of lead carbonate.

The lead citrate reagent can either be prepared directly from a commercially available EM-grade substance or by conversion of lead nitrate with sodium citrate into lead citrate during the preparation of the solution.

a. Lead citrate, 0.2 g, is combined with 50 ml distilled water in a 50-ml volumetric flask. Sodium hydroxide, 0.5 ml (10 *N*), is added, and with the inversion of the flask a few times, causes the disappearance of initial white cloudiness; the solution should not be vigorously shaken.

b. In a 50-ml volumetric flask, 30 ml distilled water is combined with a 1.33 g lead nitrate [$Pb(NO_3)_2$] and 1.76 g sodium citrate [$Na_2(C_6H_5O_7) \cdot 2H_2O$]. This cloudy white suspension is intermittently shaken for the next 30 min, and 8 ml of 1 *N* sodium hydroxide is added plus distilled water to bring it to a total volume of 50 ml. The contents are mixed by inverting the flask; this causes clearing of the solution. If the solution is not totally cleared, more 1 *N* sodium hydroxide must be added.

With either lead mixture, the staining time ranges from 1 to 30 min. The procedure should be carried out quickly, and all liquids and substances to be used in the procedure should be carbonate free. The method with lead citrate is carried out in the same way as for uranium acetate, that is, by floating the specimen grids face down on a drop of staining solution. The petri dish in which the staining is carried out should contain some pellets of sodium hydroxide to reduce carbonate contamination. The first rinse is a squirt along the tweezer jaws with 0.02 *N* sodium hydroxide. Additional rinses are made with water by dipping the vertically held grid rapidly through the liquid surface a couple of times. The grid is eventually dried on a piece of filter paper (section face up).

3. Miscellaneous stains. The following three contrasting agents are of particular interest in the staining of algal cell walls. With the exception of ruthenium red, they can be used to provide general contrast for the sections, particularly in combination with lead citrate.

a. Permanganate. The solution used for the pernamganate contrast is a 1% aqueous unbuffered potassium permanganate solution, at least 1 day old. Sections are stained for 15–20 min, followed by distilled water rinses as in III.G.2.b. Precipitates can be removed by dipping the grids in a 0.25% solution of citric acid.

b. Phosphotungstic acid or phosphotungstates. Phosphotungstic acid is prepared as a 5% (w/v) solution either in water or acetone. The staining time of the aqueous solution is 10 min, and that of the acetone solution is 5 min. The aqueous solution, if desired, can also be neutralized with potassium hydroxide.

c. Ruthenium red. The application of ruthenium red as a contrasting agent on sections is limited. One of the prerequisites for usefulness of the stain is the presence of substances that have a strong affinity to it, such as polyglucuronic acid; but the contrasting intensity is never very high. The second prerequisite is that ruthenium red is the only contrasting agent (besides osmium tetroxide) used. A third factor is the purity of this reagent. It is imperative to obtain a properly prepared substance of fine, powderlike consistency and of a tannish color, which dissolves without residue. Dark red hygroscopic-appearing substances are not suitable. The concentration of the aqueous solution is not crucial, and solutions of 0.1% to 0.01% are often used with a staining time of a few minutes.

IV. References

Bennett, H. S., and Luft, J. H. 1959. s-Collidine as a basis for buffering fixatives. *J. Biophys. Biochem. Cytol.* 6, 113–14.

Bowles, F. A. 1968. Microstructure of sediments: investigation with ultrathin sections. *Science* 159, 1236–7.

Bowles, F. A., Bryant, W. R., and Wallin, C. 1969. Microstructure of unconsolidated and consolidated marine sediments. *J. Sedimen. Petrol.* 39, 1546–51.

Dalton, A. J. 1955. A chrome–osmium fixative for electron microscopy. *Anat. Record* 121, 281.

Erlandson, R. A. 1964. A new Maraglas, D.E.R. 732 embedment for electron microscopy. *J. Cell Biol.* 22, 704.

Floyd, G. L., Stewart, K. D., and Mattox, K. R. 1972. Comparative cytology of *Ulothrix* and *Stigeoclonium*. *J. Phycol.* 8, 68–81.

Franke, W. W., Krien, S., and Brown, R. M., Jr. 1969. Simultaneous glutaraldehyde–osmium tetroxide fixation with postosmication. An improved fixation procedure for electron microscopy of plant and animal cells. *Histochemie* 19, 162–4.

Freeman, J. A., and Spurlock, B. O. 1962a. A new epoxy embedment for electron microscopy. *J. Cell Biol.* 13, 437.

Freeman, J. A., and Spurlock, B. O. 1962b. *Maraglas Epoxy Embedding Media,* (P-11). *Fifth International Congress for Electron Microscopy.* Academic Press, New York.

Glauert, A. M., and Glauert, R. H. 1958. Araldite as an embedding medium for electron microscopy. *J. Biophys. Biochem. Cytol.* 4, 191–4.

Glauert, A. M., Robers, G. E., and Glauert, R. H. 1956. A new embedding medium for electron microscopy. *Nature* 178, 803.

Hayat, M. A. 1970. *Principles and Techniques of Electron Microscopy,* vol. 1. Van Nostrand-Reinhold, New York. 412 pp.

Hayat, M. A. 1972. *Basic Electron Micrsocopy Techniques.* Van Nostrand-Reinhold, New York.

Holt, S. J., and Hicks, R. M. 1961. Use of veronal buffers in formalin fixation. *Nature* 191, 832.

Karnovsky, M. J. 1961. Simple methods for "staining with lead" and high pH in electron microscopy. *J. Biophys. Biochem. Cytol.* 11, 729–32.

Karnovsky, M. J. 1965. A formaldehyde–glutaraldehyde fixative of high osmolality for use in electron microscopy. *J. Cell Biol.* 27, 137A–138A.

Kay, D. (ed.). 1965. *Techniques for Electron Microscopy,* 2nd ed. F. A. Davis, Philadelphia. 560 pp.

Kellenberger, E., Ryter, A., and Séchaud, J. 1958. Electron microscope study of DNA-containing plasms. II. Vegetative and mature phase DNA as compared with normal bacterial nucleoids in different physiological states. *J. Biophys. Biochem. Cytol.* 4, 671–8.

Luft, J. H. 1959. The use of acrolein as a fixative for light and electron microscopy. *Anat. Record* 133, 305.

Luft, J. H. 1961. Improvements in epoxy resin embedding methods. *J. Biophys. Biochem. Cytol.* 9, 409–14.

McLachlan, J. 1973. *Growth media–marine.* In Stein, J. R. (ed.), *Handbook of Phycological Methods: Culture Methods and Growth Measurements,* pp. 25–51. Cambridge University Press, Cambridge.

Meek, G. A. 1976. *Practical Electron Microscopy for Biologists,* 2nd ed. Wiley, New York.

Mollenhauer, H. H. 1964. Plastic embedding mixtures for use in electron microscopy. *Stain Technol.* 39, 111–15.

Muller, L. L., and Jacks, T. J. 1975. Rapid chemical dehydration of samples for electron microscopic examinations. *J. Histochem. Cytochem.* 23, 107–110.

Nichols, H. W. 1973. *Growth media–freshwater.* In Stein, J. R. (ed.), *Handbook of Phycological Methods: Culture Methods and Growth Measurements,* pp. 7–24. Cambridge University Press, Cambridge.

Pankratz, H. S., and Bowen, C. C. 1963. Cytology of blue-green algae. The cells of *Symploca muscorum. Amer. J. Bot.* 50, 387–99.

Pease, D. 1964. *Histological Techniques for Electron Microscopy,* 2nd ed. Academic Press, New York. 381 pp.

Reimann, B. E. F. 1963. A simple holder for trimming blocks for ultrathin sectioning. *Mikroskopie* 18, 162–3.

Reimann, B. E. F., and Volcani, B. E. 1968. Studies on the biochemistry and fine structure of silica shell formation in diatoms. III. The structure of the cell wall of *Phaeodactylum tricornutum* Bohlin. *J. Ultrastruct. Res.* 21, 182–93.

Reynolds, E. S. 1963. The use of lead citrate at high pH as an electron-opaque stain in electron microscopy. *J. Cell Biol.* 17, 208–12.

Sabatini, D. D., Bensch, K., and Barrnett, R. J. 1963. Cytochemistry and electron microscopy. The preservation of cellular ultrastructure and enzymatic activity by aldehyde fixation. *J. Cell Biol.* 17, 19–58.

Schnepf, E. 1960. Zur Feinstrucktur der Drüsen von *Drosophyllum lusitanicum. Planta* (Berlin) 54, 641–74.

Sjöstrand, F. S. 1967. *Electron Microscopy of Cells and Tissues,* Vol. 1. Instrumentation and Techniques. Academic Press, New York. 462 pp.

Spurr, A. R. 1969. A low-viscosity epoxy resin embedding medium for electron microscopy. *J. Ultrastruct. Res.* 26, 31–43.

Stempak, J. G., and Ward, R. T. 1964. An improved staining method for electron microscopy. *J. Cell Biol.* 22, 697–701.

Strugger, S. 1956. Die Uranylacetate-Kontrastierung für die elektronenmikroskopische Untersuchung von Pflanzenzellen. *Naturwiss.* 43, 357–8.

Watson, M. L. 1958. Staining of tissue sections for electron microscopy with heavy metals. II. Applications of solutions containing lead and barium. *J. Biophys. Biochem. Cytol.* 4, 727–9.

Wohlfahrth-Bottermann, K. E. 1957. Die Kontrastierung tierischer Zellen und Gewebe im Rahmen ihrer elektronenmikroskopischen Untersuchung an ultradünnen Schnitten. *Naturwiss.* 44, 287–8.

24: Cytochemical localization

RICHARD N. TRELEASE

Department of Botany and Microbiology, Arizona State University, Tempe, Arizona 85281

CONTENTS

I. Introduction

This chapter describes specific cytochemical methods that have been successfully applied to some algae for localizing certain enzymes and macromolecules at the ultrastructural level. Many different enzymes and macromolecules have been localized in higher plant and animal cells by cytochemical techniques, but comparatively few of these methods have been applied to algae. Because the techniques have not been concentrated in any taxon, both a general list of organisms for which the procedures have been used successfully and pertinent references to the procedures are given.

For enzyme localizations, cytochemical procedures are limited to those enzymatic reactions that produce an electron-dense product at the site of the enzyme. For example, hydrolases and oxidoreductases represent the typical classes of enzymes that are amenable to cytochemistry. In algae, the only hydrolase that has been reliably demonstrated in several taxa is acid phosphatase. Relatively few of the oxidoreductases have been localized in algal cells, but success can be achieved with the four described herein. It is interesting that three of the enzymes utilize 3,3′-diaminobenzidine (DAB) during the cytochemical reaction, because this compound readily penetrates gultaraldehyde-fixed membranes.

Procedures for detection and localization of macromolecules by specific positive staining procedures or enzyme digestions for the identification of proteins, mucopolysaccharides (glycoproteins), and DNA are also included. Although the procedures presented in this chapter are designed for electron-microscopic studies, many can be modified for light-microscopic studies and can be used in addition to those presented in Chap. 22.

II. Methods

A. General fixation and embedding procedures

Methods for general preparation of algal materials for electron microscopy are discussed in Chap. 23, which should be consulted for specific details and problems.

For techniques described herein, algal cells or small tissue blocks (1–3 mm^3) are usually fixed prior to treatment in the cytochemical reaction mixture in ca. 2.5% glutaraldehyde buffered with 0.05 *M* sodium cacodylate (pH 7.2) and containing 0.25 *M* sucrose. The sucrose may not be necessary but is often preferred for maintaining iso-osmatic conditions during fixation. Fixation, at room temperature or at 0–5°C for 30–120 min, is variable and depends on the results desired. Excess glutaraldehyde is removed by washing the organisms in several changes of the fixation buffer (including sucrose if used) over a 30–60 min period. Any additional treatments required before incubation in the reaction medium will be given with each procedure. Encasement in agar should not hinder cytochemical reactions or penetration of the reaction medium into the cells during incubation.

B. Enzyme localization

1. Acid phosphatase. This represents a group of nonspecific acid phosphatases that hydrolyze a variety of organic esters with the liberation of phosphate ions. The standard method for localizing acid phosphatase is the Gomori procedure (Gomori 1952). This involves glutaraldehyde-fixed cells in a medium containing β-glycerophosphate as substrate with lead nitrate as the trapping agent to precipitate the enzymatically released phosphate ions at the site of the enzyme. The electron-dense product is lead phosphate.

a. Reaction medium

sodium-β-glycerophosphate	30 mg
0.05 *M* sodium acetate buffer, pH 5.0	11 ml
0.36 *M* lead nitrate	0.1 ml

The components should be added in the order indicated. To curtail formation of lead carbonate, several precautions are observed. Lead nitrate is dissolved in distilled water (previously boiled), and this solution is added dropwise with gentle stirring to the buffered substrate. Preincubation of the medium for 1 h (37°C) allows precipitation of any foreign chemicals prior to tissue incubation. The reaction medium is filtered (Whatman No. 1) just before use.

b. Procedure

i. Fixed cells are first preincubated in 0.05 *M* sodium acetate buffer (pH 5.0) for 15–30 min at room temperature.

ii. They are then incubated (40–60 min) in the reaction mixture in a corked vial at 37°C. Intermittently, the mixture is gently agitated by hand; care must be taken not to mix in excess CO_2.

iii. Cells are rinsed in distilled water for several minutes, immersed in a 1% (v/v) acetic acid solution for 2–5 min, and rinsed again in distilled water. [The acid solution helps release nonspecific protein-bound lead in the cells (Gomori 1952)].

iv. Appropriate controls include omitting the substrate from the incubation medium or adding 0.01 *M* sodium fluoride to the complete medium just prior to adding the lead nitrate solution. Moreover, cytidine 5′-monophosphate and 0.01 *M* *p*-nitrophenylphosphate are typically used as substitute substrates; the former has not been tried with algae, but the latter was used by Sommer and Blum (1965) with essentially the same results as with β-glycerophosphate. One modification is to preincubate the cells in 0.1 *M* citrate buffer (pH 5.0) before incubation in the reaction mixture.

v. The cells are finally postfixed in buffered 2% osmium tetroxide, dehydrated, and embedded in either Epon-type or Spurr's plastic according to standard procedures. Sections may be viewed both with and without poststaining in uranium and lead solutions.

c. Results. The enzyme (indicated by a lead precipitate) is characteristically localized in Golgi cisternae and vesicles, and if such are present, in lysosomelike digestive vacuoles. Acid phosphatase reactivity has been described in several species of *Euglena* (Rougier 1972; Palisano and Walne 1972; Gomez et al. 1974), *Ulva* (Bråten 1975; Micalef 1975), and *Dunaliella* (Eyden 1975).

One must keep in mind that glutaraldehyde inhibits acid phosphatase activity. The degree of inhibition will vary with the particular species and conditions of fixation; it may also be related to the purity of the glutaraldehyde. However, sufficient acid phosphatase activity persists for cytochemical localization. This reactivity generally represents a small fraction of the original activity; it is assumed faithfully to reflect the distribution of enzyme present prior to fixation.

d. Application to light micrroscopy. The lead phosphate can be converted to lead sulphide by treating cells for 1–20 min with 0.5–2% aqueous ammonium sulphide after the acetic acid and water rinses. Postosmication is omitted. Section thickness should be from 0.5 to 1.0 μm for optimum viewing.

2. Catalase. This enzyme is capable of reducing hydrogen peroxide to water by two distinct mechanisms: (a) catalatic reactions whereby two molecules of H_2O_2 serve as electron donors; (b) peroxidatic reactions whereby H_2O_2 and an alternate electron donor are required. This is important to the cytochemistry of catalase because one routinely detects the peroxidatic reactivity of catalase rather than its catalatic activity. The standard procedure for localization was developed by Novikoff and Goldfischer (1969). Cells well fixed in glutaraldehyde, which inhibits the catalytic sites on the enzyme, are incubated in the substrates H_2O_2 and 3,3′-diaminobenzidine (DAB) at an alkaline pH. The high pH and alternate electron donor, DAB, serve to elicit the optimum peroxidatic activity of catalase.

a. Reaction medium

i. 0.05 *M* 2-amino-2-methyl-1,3-propandiol buffer (Sigma), pH 9.4.

ii. DAB, 20 mg, is dissolved in 10 ml buffer, and 0.2 ml of 3% (v/v) Superoxol (Merck 30% H_2O_2) is added and the solution is adjusted to pH 9.4. The solution should be made up immediately before use and should be light reddish-brown; it will darken with time and exposure to light.

b. Procedure

i. Cells fixed 1 h or more in cacodylate-buffered glutaraldehyde are rinsed 3 times (15 min each) in 0.05 *M* cacodylate buffer (pH 6.8–7.1), then once or twice (15–30 min) in propandiol buffer.

ii. Cells are incubated (40–60 min, 37°C) in tightly stoppered vials, nearly filled with the incubation medium. They are periodically agitated by inversion during the incubation.

iii. The cells are then rinsed several times (over 30 min) in 0.05 *M* cacodylate buffer (pH 7.2).

iv. Appropriate controls generally include: (a) omitting the substrate H_2O_2, (b) adding the specific catalase inhibitor aminotriazole (3-amino-1,2,4-triazole, Sigma) (AT), and/or (c) adding the mitochondrial oxidase inhibitor KCN. For AT controls, cells are preincubated for 20 min (37°C) in 0.02 *M* AT (buffered in 0.05 *M* propandiol, pH 9.4), then incubated in 10 ml fresh AT solution containing 20 mg DAB and 0.2 ml of 3% H_2O_2 adjusted to pH 9.4. The same procedure is used for KCN addition: 0.01 *M* KCN in propandiol buffers for preincubation and dissolution of the DAB and H_2O_2.

v. Cells are postfixed and embedded as for acid phosphatase.

c. Results. When DAB is oxidized by the enzyme reaction, it first polymerizes, then undergoes oxidative cyclization at the site of the enzyme. This molecular complex is then converted to osmium black after the cells are incubated in osmium tetroxide. This electron-dense product is indicative of catalase reactivity. The reactivity is generally found in microbodies (glyoxysomes, peroxisomes). Its cytochemistry has been examined in 37 taxa of chlorophytes (Silverberg 1975), *Euglena* (Brody and White 1973), *Acetabularia* (Menzel 1976), and *Ectocarpus* (Oliveira and Bisalputra 1976).

Very few problems accompany this technique. The localizations are usually clean, that is, background deposits are usually not present. Depending on the species, some reactivity is observed in the mitochondrial cristae. This is generally attributed to persistent cytochrome oxidase reactivity. The best control is to add KCN, which should abolish deposition in the mitochondria.

For a positive reaction, the content of DAB can be varied from 0.1 to 0.4%, and H_2O_2 can be varied from 0.003% to 0.5%. The above medium consisting of 0.2% DAB and 0.06% H_2O_2 is well within this range. Lowering the pH significantly below 9.4 typically abolishes the DAB peroxidatic activity of catalase.

3. Peroxidase. Peroxidase acts to decompose H_2O_2 to water by acting on H_2O_2 and an alternate electron donor. It is generally active near neutral pH, but inactive at the alkaline pH suitable for peroxidatic activity of catalase.

a. Reaction medium. This is the same as for catalase, except pH 7.6 is used in all cases.

b. Procedure. This is also as for catalase.

Controls involve omission of substrates, or varying the pH. Below pH 7.6, propandiol will not buffer adequately, so either phosphate or cacodylate should be used within their buffering range.

c. Results. The enzyme has been found in the cell walls, perinuclear space, endoplasmic reticulum, dictyosomes, and paramural space of the brown alga *Ectocarpus* (Oliveira and Bisalputra 1976). In their study, they used 2 ml of 3% H_2O_2 (final concentration 0.5%) rather than that prescribed for catalase (final concentration 0.06%).

4. Cytochrome oxidase. It serves as the terminal oxidase in the electron transport chain of mitochondria. The enzyme can be localized cytochemically with DAB (Seligman et al. 1968). The reactivity typically occurs on the inner mitochondrial membranes owing to cytochrome $a-a_3$ oxidation couples with cytochrome c_1 oxidation of DAB on the inner surface of the intracristate space (Seligman et al. 1968).

a. Reaction medium. DAB, 20 mg, is dissolved in 10 ml of 0.05 *M* sodium acetate buffer (pH 6.0) or in 0.05 *M* potassium phosphate (pH 7.4) buffer. If difficulty is encountered in dissolving the DAB at the above pH values, the pH can be temporarily increased until the DAB dissolves and then readjusts to either pH 6.0 or 7.4. The medium should be prepared just before use.

b. Procedure.

i. Incubation, washing, postosmication, dehydration, and embedding are the same as for catalase.

ii. Controls include adding 0.01 *M* KCN or 0.1 *M* sodium azide in both the preincubation and the incubation media.

c. Results. The enzyme has been localized in *Euglena* (Gomez et al. 1974); *Prototheca* and *Chlorella* (Pellegrini and deVecchi 1976); *Chlorogonium* and *Polytomella* (Gerhardt and Berger 1971); and *Chlamydomonas* (Silverberg and Sawa 1974). Few problems are encountered because of DAB's excellent penetration properites.

d. Application to light microscopy. The reaction product (osmium black) used for electron microscopy, is generally visible in 0.5- to 1.0-μm sections with the light microscope.

5. Glycolate dehydrogenase. Glycolate oxidation via this enzyme is characteristic of the photorespiratory system in many green algae. The method recently applied to *Chlamydomonas* (Breezley et al. 1976) involves a modified ferricyanide-reduction procedure commonly used for other oxidoreductases and is carried out with unfixed cells. The final electron-dense product, copper ferrocyanide, is formed by the enzymatic reduction of ferricyanide in the presence of copper ions.

a. Reaction mixture

0.20 *M* potassium phosphate (pH 7.2)	2.5 ml
0.025 *M* sodium potassium tartrate	2.0 ml
0.025 *M* copper sulfate	2.0 ml
0.05 *M* potassium ferricyanide	0.4 ml
0.50 *M* sodium glycolate	0.4 ml
distilled water	2.7 ml

Components are added in the order given immediately prior to use.

b. Procedure

i. Living cells harvested from the log phase of growth are rinsed twice with 0.05 *M* potassium phosphate buffer (pH 7.2) and resuspended in buffer (ca. 1 ml packed cell plus 24 ml buffer).

ii. The suspended cells are partially disrupted by sonication (5 sec at a setting of 3 with a Heat Systems sonifier W185).

iii. The partially disrupted cells are incubated in the reaction mixture in the dark for 20 min at room temperature.

iv. D(−)-Lactate (0.5 *M*, 0.4 ml) may be substituted for glycolate and give essentially the same results. L(+)-Lactate yields only a slightly increased deposition as compared to glycolate and D-lactate. Controls include omission of the substrates or addition of the specific inhibitor sodium oxamate. In this case, unfixed cells are preincubated for 30 min in 0.1 *M* sodium oxamate in 0.05 *M* phosphate buffer; then, 2.7 ml aqueous 0.37 *M* oxamate is added in place of water in the standard medium. Cyanide, a known inhibitor of the enzyme, cannot be used because it causes formation of premature precipitates in the ferricyanide medium.

v. The preparation is then rinsed with 0.05 *M* potassium phosphate buffer (pH 7.2), encased in 2% agar, and rinsed again.

vi. This is followed by the normal fixation (1 h) in 2–3% in glutaraldehyde, buffer washes, postfixation (1 h) in 2% osmium tetroxide, dehydration, embedding, and section staining.

c. Results. The reactivity is formed in the outer compartment of mitochondria, not in microbodies. Cytochemistry employing a ferricyanide reaction mixture typically results in considerably more background staining than with DAB mixtures. Small nonspecific deposits of the product, copper ferrocyanide (determined by control experiments), will be apparent throughout many parts of the cells. Penetration of the reactants is also a problem with this medium; one must use the partial disruption procedure to have reactivity in a large portion of the cells. Care must also be exercised to examine only those cells showing cellular alterations that are minimal compared to non-disrupted control cells.

d. Application to light microscopy. The reaction product for electron microscopy (Hatchett's brown) is visible in 0.5 to 1.0 μm sections under the light microscope.

C. *Mucopolysaccharide (glycoprotein) staining*

1. Ruthenium red (RR). This synthetically prepared crystalline compound (ammoniated ruthenium oxychloride) has been introduced to electron microscopy for demonstrating the sites of large numbers of polyanions typically found on extracellular surfaces and generally identified as mucopolysaccharides. Staining is prevented by prior treatment of cells with either hyaluronidase or quarternary ammonium compounds (which bind to mucopolysaccharides). The stain can bind the acid polysaccharides alone, but it is generally mixed with osmium tetroxide solutions to obtain higher contrast and to facilitate *en bloc* application (see also Chap. 23.III.C.1. and G.3.c). Osmium tetroxide and RR apparently interact at the site of binding to produce suitable electron images. Further details of the staining procedures and mechanisms for RR are given by Hayat (1975).

a. Reaction media

Solution A:	
aqueous glutaraldehyde (5%)	5.0 ml
0.2 *M* sodium cacodylate, pH 7.2	2.5 ml
RR (400 mg in 100 ml H_2O)	2.5 ml
Solution B:	
aqueous osmium tetroxide (4%)	5.0 ml
0.2 *M* sodium cacodylate, pH 7.2	2.5 ml
RR (400 mg in 100 ml H_2O)	2.5 ml

RR should be mixed with the fixatives just before use and discarded thereafter.

b. Procedure

i. Cells are fixed in solution A for 1 h at room temperature and rinsed (over 1 h) in several changes of 0.05 *M* cacodylate buffer (pH 7.2). Other buffers can also be used.

ii. This is followed by postfixation in solution B for 3–6 h at room temperature, rinsing in water, and embedding by standard procedures. Sections can be observed with or without staining.

c. Results. RR has been applied to a number of different algal species, primarily with the intent of looking at the mucilage components of sheaths, holdfasts, cell walls, and plasma membranes (Rougier 1972; Gomez et al. 1974; Bråten 1975; Dynesius and Walne 1975; Menge 1976; McCracken and Barcellona 1976). Long periods are generally required for staining; RR penetrates cells very slowly. Generally, one cannot rely on RR *en bloc* staining to depict mucopolysaccharides *within* algal cells; its applicability is more suited to extracellular mucilages. Intracellular sugars are more appropriately and reliably detected by staining sections by methods such as those described for silver stains (see below).

2. *Silver staining.* Several techniques have been developed to localize polysaccharides with silver as the electron-dense product (Hayat 1975). Two of the most common procedures have been applied to algal cells, and are described here.

a. Periodic acid–silver methenamine (PA–SM). The staining mechanism involves a reaction wherein metallic silver is precipitated from a methenamine complex by the reductive action of aldehyde radicals. To produce aldehydic groups, fixed cells or Epon-embedded sections are treated with periodic acid that oxidizes 1,2-glycol groups and alpha-amino groups. The affinity for the stain is strong for glycoproteins since they are rich in the two reactive groups.

i. Solutions for *en bloc* staining: (McCracken and Barcellona 1976):

5% aqueous silver nitrate	2 ml
3% aqueous hexamethylenetetramine	18 ml
2% borax (sodium borax in H_2O), pH 9.2	2 ml

The solutions are added with stirring in the order given (turbidity will clear after addition of the first two solutions). The mixture is filtered twice (Whatman No. 42 paper) and stored in dark bottles.

ii. Procedure. Cells fixed in glutaraldehyde and cacodylate buffer (pH 7.2) are oxidized in 1% periodic acid for 15 min at room temperature, and rinsed three times in distilled water. Cells are then stained in the above solution for 60 min at 60°C in the dark. After several rinses in distilled water, the cells are placed in a 3% sodium thiosulfate solution (5 min) at room temperature to stop the silver from reacting further. Further rinsing in water is followed by dehydration and embedding in plastic. Omission of periodic acid oxidization can serve as a control.

iii. Results and modifications. Cell walls, slime, starch, and vesicles stain. In two species of *Micrasterias,* with a modified silver methanamine procedure applied to embedded sections (Tutumi and Ueda 1975; Menge 1976), the same results have been obtained. *Note:* Spurr-embedded material cannot be used for silver staining, and sections should be mounted on gold or nickel grids because coarse precipitates result when copper grids are used.

b. Periodic acid–thiosemicarbazide–silver protein (PA–TSC–SP). This method is generally more specific and yields better staining than the above procedure. Periodic acid (PA) oxidation produces aldehyde groups in a similar fashion, but the thiosemicarbazide (TSC) reacts with the aldehyde groups rather than with silver in the silver methenamine complex. After binding, the TSC still retains its capacity to react with the silver salts (silver protein), producing metallic silver deposits. In essence, TSC acts as a ligand molecule between the exposed aldehydes and metallic silver. The TSC procedure gives reduced artificial background staining and yields generally more intense product than that from the silver methenamine reactions alone.

i. Procedure for staining sections. Sections of glutaraldehyde fixed cells, embedded in Epon or Epon–Araldite, are mounted on gold or nickel grids. The grids are floated (sections down) on a drop of 1% PA in a petri dish for 20 min at room temperature. Sections are then rinsed in three changes of distilled water and treated with 1% TSC in 10% acetic acid. Times of this treatment can be varied from 2–72 h. The length of time in TSC depends on the degree of polysaccharide complexing with other substances. For example, starch requires about 1 h, glycoproteins usually require 24 h or more. Excess TSC is removed by rinsing in an acetic acid series: 10 min in 10%, 5 min in 5%, 5 min in 1%, and 15 min in several changes of distilled water. Under a safelight (Kodak Wratten OC filter) sections are floated on a drop of 1% silver proteinate (Silver Protein Mild, Mallinckrodt) for 30 min at room temperature. Sections are then thoroughly rinsed under a fine jet-stream of water. All solutions should be freshly prepared before use. Silver protein solutions should be kept dark at all times. Controls for nonspecific staining can be made by omission of PA oxidation, omission of TSC treatment, or omission of both.

c. Results. An electron-dense product (visualized as small rosettes) indicates a positive reaction. The procedure has been applied to a diverse group of algae (Rougier 1972; Callow and Evans 1974; Chardard 1974; Tripodi and de Masi 1975).

3. Phosphotungistic acid–chromic acid(PTA–CrO_3). This mixture selectively stains plasma membranes (presumably the glycoproteins) of cells embedded in Epon and sectioned for electron microscopy. It has been applied to higher plants and, recently, to sections of whole cells and to plasma membrane fragments isolated from many algal groups (Sundberg and Lembi 1976). The technique's usefulness is not in localizing glycoprotein sites in the membranes, but in identifying plasma membranes, particularly among other cytomembranes in cell fractions. This is crucial because biochemical markers are not known for plant plasma membranes.

a. Staining solution. The solution consists of 1% phosphotungstic acid in 10% chromic acid (chromic acid is used to maintain the solution at pH 3.0 or lower, the pH necessary for staining glycoprotein moieties).

b. Procedure

i. Cells or membrane fractions are fixed (1 h) in glutaraldehyde (4%) in cacodylate buffer (pH 7.0) and 0.25 *M* sucrose. The material, encased in agar, is rinsed before postfixation in osmium tetroxide (1%) and is finally embedded in Epon.

ii. Sections are collected on a plastic loop or mounted on gold grids and bleached in 1% PA as described for the PA–TSC–SP procedure (II.C.2.b above).

iii. After rinsing, the sections are floated on the staining solution for 5–20 min, rinsed with water, dried, and examined directly without poststaining.

c. Results. Plasma membranes have been identified by their greater electron density after staining in PTA–CrO_3 in selected members of Euglenophyceae, Xanthophyceae, Bacillariophyceae, Chrysophyceae, Chlorophyceae, and Rhodophyceae. It is not applicable to Cyanophyceae (Sundberg and Lembi 1976). Caution is required in the analysis because some cell-wall and pellicle components may also show positive staining. Identification of the plasma membrane should be made independently on control section.

D. Enzymatic extraction

These methods are used *en bloc* and on sections. Their general use is as one of several controls for positive staining procedures. In other cases, the digestion techniques are applied to help elucidate the chemical composition of structures observed in normal sections. The primary problem encountered is one's ignorance of the different types of macromolecules bound and mixed together and forming a complex structure. Even though a structure may be composed primarily of one molecule, other bound molecules may mask the digestion of the applied enzyme. As a consequence, only general procedures are given below. Incubation times and the order of enzyme applications must be determined for each case.

1. DNA digestion. DNase-sensitive material has been digested within chloroplasts (*en bloc*) and within mitochondria (on sections). For *en-bloc* digestions, cells are fixed in glutaraldehyde, thoroughly washed, then incubated for about 6 h at 40°C in a 1.0–1.5 mg/ml solution of

DNase (in 0.2 *M* sodium-acetate buffer, pH 5.5). RNase-free DNase from Sigma Chemical Co. is generally satisfactory. For section staining, Epon–Araldite- embedded sections on copper or gold grids are floated on the same enzyme solution. If the cells are postfixed in OsO_4, the sections should first be floated on or submerged in a 10% H_2O_2 solution for 5–20 min. Sections are washed with water and acetate buffer before application of the enzyme.

2. *Protein digestions.* Proteolytic enzymes of various types have been applied to several different algal species both *en bloc* and on sections. The same procedures are used for handling the cells and sections as described for DNase digestion.

Proteolytic enzyme treatments are carried out at 37°C:

a. Pepsin (Sigma powder) 0.5% in 0.1 *N* HCl, for 15 min to 6 h.

b. Protease (Sigma, type VI) (or Pronase) 0.5% in 0.05 *M* potassium phosphate buffer (pH 7.4), for 15 min to 2 h.

c. Protease (Sigma, Subtilisin type VII) 0.5% in 0.05 *M* TRIS–HCl buffer (pH 7.8) for 15 min to 2 h.

d. Trypsin (Sigma, type I) 0.07% in 0.05 *M* TRIS–HCl buffer (pH 7.9) containing 0.001 *M* calcium chloride, for 1 h to 10 h.

3. *Polysaccharide digestions.* Only two enzymes have been used repeatedly on algae: hyaluronidase (for digesting mucopolysaccharides) and α-amylase (for starch).

a. Hyaluronidase (Sigma type I) 0.1% to 1% in 0.05 *M* sodium acetate (pH 5.5), for 2–20 h.

b. Alpha-amylase (*B. subtilis* type, Nutritional Biochemical Corp.) 0.1%, in a medium containing 12 ml acetic acid, 16.4 g sodium acetate and 100 ml distilled water (final pH 4.7) for 2–20 h.

To increase the contrast of the enzyme-treated sections, they can be exposed to osmium tetroxide fumes (in a closed petri dish) for 1 h at 60°C and then stained with lead citrate. **Caution:** Osmium tetroxide should be worked with only in a fume hood.

III. Acknowledgments

My sincere gratitude is extended to Dr. Milton R. Sommerfeld for his many helpful suggestions and critical review of the manuscript. Special thanks are also due Ms. H. P. Lin for her help in perusing the literature.

IV. References

Bråten, T. 1975. Observations on mechanisms of attachment in the green alga *Ulva mutabilis* Føyn. An ultrastructural and light microscopical study of zygotes and rhizoids. *Protoplasma* 84, 161–73.

Breezley, B. B., Gruber, P. J., and Frederick, S. E. 1976. Cytochemical localization of glycolate dehydrogenase in mitochondria of *Chlamydomonas. Plant Physiol.* 58, 315–19.

Brody, M., and White, J. E. 1973. Environmental regulation of enzymes in the microbodies and mitochondria of dark-grown, greening, and light-grown *Euglena gracilis. Develop. Biol.* 31, 348–61.

Callow, M. E., and Evans, L. V. 1974. Studies on the ship-fouling alga *Enteromorpha.* III. Cytochemistry and autoradiography of adhesive production. *Protoplasma* 80, 15–27.

Chardard, M. R. 1974. The cell wall of *Cosmarium lundelli* Delp. (Chlorophyceae, Desmidiales). Ultrastructure and attempt at cytochemical localization of constituents. *C. R. Acad. Sci. Ser. D.* 278, 609–1612.

Dynesius, R. A., and Walne, P. L. 1975. Ultrastructure of the reservoir and flagella in *Phacus pleuronectes* (Euglenophyceae). *J. Phycol.* 11, 125–30.

Eyden, B. P. 1975. Light and electron microscope study of *Dunaliella primolecta* Butcher (Volvocida). *J. Protozool.* 22, 336–44.

Gerhardt, B., and Berger, Ch. 1971. Microbodies und Diaminobenzidin-Reaktion in den Acetat-Flagellaten *Polytomella caeca* und *Chlorogonium elongatum. Planta* (Berl.) 100, 155–66.

Gomez, M. P., Harris, J. B., and Walne, P. L. 1974. Studies of *Euglena gracilis* in aging cultures. II. Ultrastructure. *Br. Phycol. J.* 9, 175–93.

Gomori, G. 1952. *Microscopic Histochemistry.* University of Chicago Press, Chicago. 273 pp.

Hayat, M. A. 1975. *Positive Staining for Electron Microscopy.* Van Nostrand-Reinhold, New York. 361 pp.

McCracken, M. D., and Barcellona, W. J. 1976. Electron histochemistry and ultrastructural localization of carbohydrate-containing substances in the sheath of *Volvox. J. Histochem. Cytochem.* 24, 668–73.

Menge, U. 1976. Ultracytochemische Untersuchungen an *Micrasterias denticulata* Bréb. *Protoplasma* 88, 287–303.

Menzel, D. 1976. Cytochemischer Nachweis von Katalase in Microbodies bei *Acetabularia mediterranea. Planta* (Berl.) 130, 181–4.

Micalef, H. 1975. Additional data on morphological and cytochemical features of the Golgi apparatus in vegetative cells of the chlorophycean alga *Ulva lactuca* L. (Ulvales). *C. R. Acad. Sci. Ser. D* 281, 775–8.

Novikoff, A. B., and Goldfischer, S. 1969. Visualization of peroxisomes (microbodies) and mitochondria with diaminobenzidine. *J. Histochem. Cytochem.* 17, 675–80.

Oliveira L., and Bisalputra, T. 1976. Studies in the brown alga *Ectocarpus* in culture: ultrastructural localization of enzymic activities. *Can. J. Bot.* 54, 913–22.

Palisano, J. R., and Walne, P. L. 1972. Acid phosphatase activity and ultrastructure of *Euglena granulata* from aged cultures. *J. Phycol.* 8, 81–8.

Pellegrini, S., and de Vecchi, L. 1976. Ultrastructural demonstration of cytochrome oxidase with 3,3′-diaminobenzidine reactions in *Prototheca moriformis, Chlorella vulgaris,* and in a yellow mutant of *Chlorella. J. Submicr. Cytol.* 8, 353–61.

Rougier, M. 1972. Etude cytochimique des squamules d'*Elodea canadensis.*

Mise en évidence de leur sécrétion polysaccharidique et de leur activité phosphatasique acide. *Protoplasma* 74, 113–31.

Seligman, A. M., Karnovsky, M. J., Wasserkrug, H. L., and Hanker, J. S. 1968. Nondroplet ultrastructural demonstration of cytochrome oxidase activity with a polymerizing osmiophilic reagent, diaminobenzidine (DAB). *J. Cell. Biol.* 38, 1–14.

Silverberg, B. A. 1975. An ultrastructural and cytochemical characterization of microbodies in the green algae. *Protoplasma* 83, 269–95.

Silverberg, B. A., and Sawa, T. 1974. A cytochemical and ultrastructural study of the echinate cytoplasmic inclusion in *Nitella flexilis* (Characeae). *Canad. J. Bot.* 52, 159–65.

Sommer, J. R., and Blum, J. J. 1965. Cytochemical localization of acid phosphatases in *Euglena gracilis. J. Cell. Biol.* 24, 235–51.

Sundberg, I., and Lembi, C. A. 1976. Phosphotungstic acid-chromic acid: A selective stain for algal plasma membranes. *J. Phycol.* 12, 48–54.

Tripodi, G., and de Masi, F. 1975. Cytological localization of polysaccharidic molecules in some red algae. *J. Submicr. Cytol.* 7, 197–209.

Tutumi, T., and Ueda, K. 1975. A cytochemical study of polysaccharides in cells of *Micrasterias americana. Cytologia* 40, 113–18.

25: Quantitative cytochemical measurement of DNA in Eudorina (Chlorophyceae)

C. LINDLEY KEMP AND K. K. NAIR

Department of Biological Sciences, Simon Fraser University, Burnaby, Canada, V5A 1S6

CONTENTS

I. Introduction

Cytophotometry and cytofluorometry are cytological tools for which detailed theoretical and practical foundations have been developed over the past 30–40 years. Commercially available instruments are capable of providing measurements from appropriately prepared material within reproducible accuracies of 3–5%. As a result, estimates of relative amounts of bound dye for comparative studies may now be done with relative ease. However, these useful techniques have only rarely been applied to problems of interest to phycologists (Hurdelbrink and Schwantes 1972; Koop 1975; Lee and Kemp 1975; Hopkins and McBride 1976). The scarcity of quantitative cytochemical data for the algae seems to originate from difficulties encountered in staining and visualizing small algal nuclei by the Feulgen procedure (Tautvydas 1976). Because the amount of DNA per nucleus in many of the algae is one tenth (or less) the amount found in vascular plants and vertebrates, the conditions for preparation and analysis must be optimal.

This chapter gives preparative protocols and analytical procedures for the Feulgen method as applied to quantitative measurements of DNA in two colonial green algae. Preliminary observations on a fluorescence technique, now being developed for quantitative DNA measurement in these same organisms will also be given.

For a detailed treatment of the theoretical and practical aspects of quantitative cytochemistry the volumes edited by Wied (1966) and Wied and Bahr (1970) are recommended.

II. Equipment

Instruments used for quantitative cytochemistry are relatively complex and expensive when compared with analagous equipment used for macro systems. They range from multipurpose systems to more modest systems of limited flexibility. For most of the studies, we have used a Carl Zeiss scanning microscope photometer (SMP) equipped with a precision scanning stage (Zimmer 1970) and linked to a PDP

12A digital computer (Digital Equipment Corp.) programmed by APAMOS (automatic analysis of microscope objects by scanning), a program similar to TICAS (taxanomic intracellular analytic system) (Wied et al. 1968).

III. Test organisms

Eudorina elegans (Ehrenberg) UTEX 1193 and *Eudorina californica* (Shaw) nov. comb. (Goldstein) obtained as *Pleodorina indica* UTEX 1990 were used in these studies. Both were maintained and grown under conditions previously described for *E. elegans* (Kemp and Wentworth 1971).

IV. Methods

Chemically clean glassware, distilled, deionized water, and reagent grade chemicals were used for all solutions. Hydrolysis, staining, and rinsing were all done in screw-cap staining dishes with the caps firmly in place.

A. Stain recipes

We have used standard recipes; they are given here for ease of reference.

1. Schiff's reagent (see Deitch 1966)

a. Basic fuchsin (Harleco), 0.05 g, is dissolved in 100 ml of boiling water.

b. After cooling to 50°C, it is filtered using fluted filter paper.

c. Potassium metabisulphite ($K_2S_2O_5$), 2.0 g, and 10 ml 1 *N* HCl are added with shaking. The solution is stoppered tightly and stored for 24 h at room temperature in the dark.

d. Activated charcoal, 0.25 g, is added to the straw-colored liquid with further shaking for 1 min, and the solution is rapidly filtered using a Buchner funnel.

e. The clear, colorless filtrate is stored in a tightly stoppered bottle at 4°C.

2. BAO Fluorochrome. This is an abbreviation for the fluorochrome 2,5-bis(4′-aminophenyl-1′)-1,3,4-oxidizole (Tridom Chemical) (Ruch 1970).

a. BAO, 10 mg, are added to 100 ml water with stirring for 5 min.

b. To this is added 10 ml 1 *N* HCl and 5 ml 10% $NaHSO_3$ with stirring (5–10 min) and filtering.

c. This is to be used immediately.

3. Sulfurous acid bleach. To 10 ml 1 *N* HCl are added 10 ml 10% anhydrous $K_2S_2O_5$, plus 180 ml water, for immediate use.

4. Subbing solution

a. An aqueous solution of 0.1% gelatin and 0.01% chromium potassium sulfate (chrome alum) is prepared and filtered.

b. Clean slides are dipped in the solution and dried in a vertical position in a dust-free area.

B. *Specimen preparation*

1. Standard procedure for Eudorina. The ultimate goal is to have cells flattened and not overlapping. The material being studied will determine the exact procedure to be used. The two procedures well suited for *Eudorina* are the Feulgen and BAO procedures.

a. The algae are concentrated by centrifugation, and 0.02 ml of concentrated algae are used for each slide.

b. The cells are placed on a "subbed" microscope slide, and covered with a cover glass applying gentle pressure to rupture the coenobia and to slightly flatten the cells. Although some cell rupture will occur, it should be kept to a minimum.

c. Appropriate preparations are frozen on dry ice and the cover glass is removed with a razor blade.

d. The adhering cells are fixed in 3 : 1 fixative (95% ethanol/glacial acetic acid) for 30 min at room temperature. If too much tissue loss is experienced at this step, the preparation can be air-dried for 10 min before fixation.

e. It is twice rinsed in 70% ethanol then hydrated through an ethanol series (5 min each step) to water, or stored in 70% ethanol at −20°C then hydrated when convenient.

f. After equilibration to 37°C in distilled water, the cells are hydrolyzed in 3.5 *N* HCl at 37°C for 20 min.

g. The hydrolysis is stopped by rinsing the material in 1 *N* HCl at 4°C.

h. It is then stained in Schiff's reagent at room temperature for 1 to 2 h in the dark, or in BAO for 2 h in the dark.

i. Stained material is washed in sulfurous acid bleach, 3 times for 5 min each, then rinsed in distilled water before mounting.

j. BAO-stained material is mounted in glycerine. The Feulgen-stained material is dehydrated in a graded series of ethanol, ethanol/xylene (1 : 1), and two changes of xylene; the sample is then mounted in oil of matching refractive index (n_D = 1.56) (Cargille Laboratories). For permanent preparations D.P.X. medium (ESBE Laboratory Supplies) is used.

2. *Alternate method.* For bulk fixation, the algae are concentrated by centrifugation and fixed as in B.1.d. above. For large algae, small pieces are cut, with material from different regions being kept separate to facilitate subsequent analysis. The fixed material is washed twice in 70% ethanol and stored either at −20°C in 70% ethanol, or hydrated, hydrolysized, and stained as above. Slides may be prepared (a) immediately following fixation, (b) following transfer to 70% ethanol, or (c) following bulk staining and sulfurous acid rinses. In each case, the slides must be subbed with the subbing solution (A.4.a.) prior to the fixed material's being squashed to prevent loss of material during subsequent steps. Squashing is best accomplished in 45% acetic acid; however, appropriate spreading and flattening of the material is more difficult once the algae have been fixed. The cover glass is removed by the freezing and the frozen preparation transferred to 100% ethanol. Subsequent steps will depend upon the point in the procedure when the slides were prepared and further handled as in B.1.

3. *Special notes*

a. Hydrolysis times must be determined for each type of material examined. It is very important that the optimal hydrolysis time be used. It is also advantageous if the hydrolysis curve has a fairly flat region around the optimal time, as in *E. elegans* (Lee and Kemp 1975). As noted by Fand (1970), optimal conditions of acid concentration and temperature for acid hydrolysis preceding the Feulgen staining procedure are at 3.5 *N* HCl and 37°C. In our experience, hydrolysis in 1 *N* HCl at 60°C has shown variable results. Less dye is bound, and the optimal hydrolysis time occurs as a peak rather than a plateau.

b. The refractive index of the mounting medium should be chosen so that the visibility of the cytoplasm is reduced to a minimum.

C. *Measurement*

Only a general statement will be given on measurements because the choice of equipment will determine the specific routine of appropriate measurement procedures. Initially, wavelengths are chosen to correspond to the maximum (λ_{max}) and half maximum absorptions ($\lambda_{(1/2)max}$) of the bound dye (560–580 nm and 500–510 nm respectively for Feulgen). This should be checked for each new system examined and for each dye lot used. For the BAO fluorescent technique, excitation is optimal at 360 nm with fluorescence emission at 450 nm.

1. *Scanning.* Once the object is located, positioned, focused, and the light source adjusted to give Köhler illumination (Chap. 20.III.A) at

A

				3						
	3	3				4				
							3			
3				3	4					
3		3	3	3	3					
							5	4		
	3					10	12	11	6	
		4	5	5	11	13	14	11	8	3
					10	15	15	12	7	
					9	14	14	11	6	
						4	7	7	4	
								3		

B

		5	4		
	10	12	11	6	
11	13	14	11	8	3
10	15	15	12	7	
9	14	14	11	6	
	4	7	7	4	
			3		

Fig. 25–1. A matrix of extinction values (×100) of an *E. californica* nucleus stained with Schiff's reagent: (A) nucleus and diffraction pattern surrounding a pair of pyrenoids; (B) same nucleus with the diffraction pattern erased.

λ_{max}, the instrument is then switched to automatic scan. The individual points, often more than 1,000, are processed by the digital computer and presented as X = points in the x axis, Y = lines in the y axis. A = area represented by points with a 5–95% transmission, TE = total extinction of this area, and ME = mean extinction of all the points in the area. Display of the points with a 5–95% transmission on a CRT display unit enables the operator to erase any extraneous absorbing material (e.g., other nuclei, diffraction artifacts) (Fig. 25–1) before filing the scan on linc tape for subsequent analysis.

2. *Two-wavelength procedure.* The nucleus to be measured is located and enclosed within the area of the measuring field diaphragm B in such a way as to exclude any nonnuclear absorbing material. The area of the measuring field diaphragm must be known and is most conveniently regulated by a series of fixed-aperture diaphragms. A pair of measurements of the nucleus are taken: one at the wavelength of maximum absorption ($\lambda_{max} = I_2$) and one at the wavelength of half maximum absorption ($\lambda_{(1/2)\ max} = I_1$). The measuring field diaphragm is located over an adjacent stain-free area, and another pair of readings is taken (I_2^0 and I_1^0). Transmission values $T_1 = I_1/I_1^0$, and

$T_2 = I_2/I_2^0$ are then used to calculate the relative amount of chromophore m (Patau 1952):

$$m = L_1CB/k_2 \ln 10 \qquad (1)$$

Because the extinction coefficient of the bound dye (k_2) is rarely determined in practice unless absolute values are sought, the term $1/k_2 \ln 10$ is taken to equal 1. Similarly, if all measurements to be compared use the same field-diaphragm setting the value of B is also taken as 1. The relative amount of chromophore is then

$$m = L_1C \qquad (2)$$

The value of L_1C is obtained directly from Mendelsohn's tables (Mendelsohn 1958) using the transmission values T_1 and T_2.

3. Cytofluorometry. Quantitative analysis of fluorochrome-stained algal nuclei has not been reported previously. We have recently stained nuclei of *E. californica* using BAO (Fig. 25–2) following the procedure outlined above. Cytofluorometry promises to become especially useful for measurement of relative DNA contents when small amounts of DNA are spread out. Its increased sensitivity and the development of a computer-controlled microspectrofluorometer, capable of providing corrected excitation and fluorescence spectra (Ploem et al. 1974) should make cytofluorometry a method of choice in future applications.

D. Treatment of data

1. When one is determining optimal hydrolysis times, one should choose a cell population in which the cell cycle stage is uniform. About 25–50 nuclei are measured for each hydrolysis time. The values within a specific set of measurements should give a normal distribution with a standard deviation s not greater than ±10% of the mean. Deviations greater than 10% usually indicate nonspecific absorption included in the measured area. For example, when the diffraction surrounding the pyrenoids in *E. californica* (Fig. 25–1A) is included in the calculations, the mean $\pm s$ is 2.85 ± 0.4 (N = 32). By removal of the pyrenoid diffraction (Fig. 25–1B) the value becomes 2.18 ± 0.19. The nonspecific absorption contributed by the pyrenoid is variable with a value of 0.67 ± 0.34.

2. Comparison of experimental populations should include about 50–100 nuclear measurements obtained from 3 or 4 slides prepared at the same time. Assuming measurement variability is demonstrably minimal, any statistically significant variability seen in the experimental populations should be attributable to differences in the DNA content of the nuclei measured. DNA levels in cell populations are

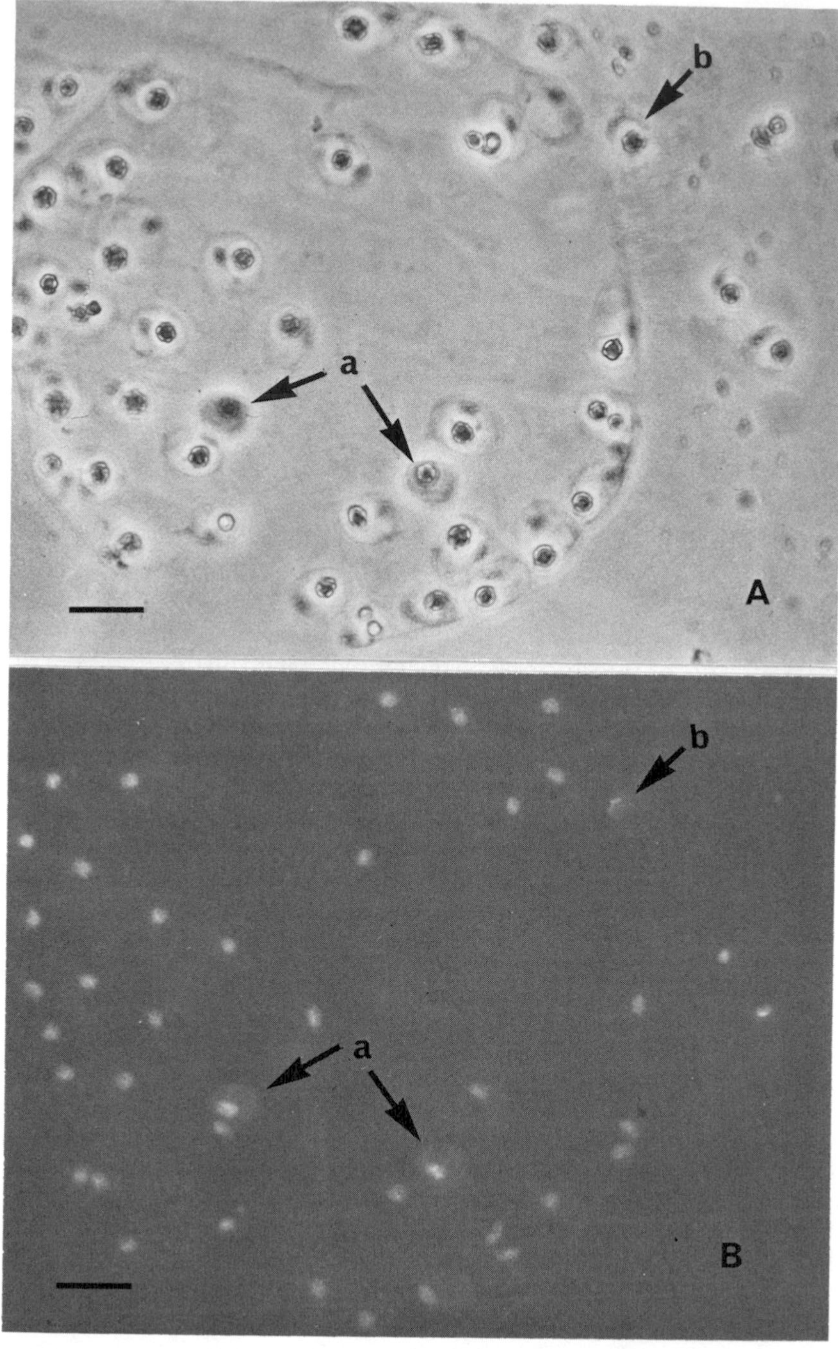

Fig. 25–2. BAO-stained *E. californica.* (A) Phase contrast. The refractile objects in each cell are pyrenoids. (B) Fluorescent pattern of the same coenobium (4-min exposure with Kodak Tri-X film). (a) Cells damaged during preparation; fluorescence is not confined to the nucleus. (b) Pyrenoid overlays most of the nucleus, masking much of its fluorescence. Scale = 10 μm.

usually expressed as class *C* values, with *1C* the base level, *2C* double the base value, *4C* quadruple the base value, and so on. Comparison of the *C* values is readily achieved by converting the relative chromophore units to $\log_2$. By plotting a frequency histogram of these values, one can visualize variations in DNA classes within the cell population. An increase in one unit in the $\log_2$ value represents a doubling of the real value. The extent of the variations will indicate whether additional measurements are necessary, as well as the nature of the statistical comparisons to be performed. Most problems involving the application of quantitative cytochemistry require comparisons of populations of cells. A clear understanding of the problems, both biological and technical, careful choice of material, and sufficient sample size, should enable analysis by standard statistical procedures.

V. Acknowledgment

We wish to thank Mr. G. Doyle for his excellent technical work in gathering the data for *E. californica.*

VI. References

Deitch, A. D. 1966. Cytophotometry of nucleic acids. In Wied, G. L. (ed.), *Introduction to Quantitative Cytochemistry,* pp. 327–54. Academic Press, New York.

Fand, S. B. 1970. Environmental conditions for optimal Feulgen hydrolysis. In Wied, G. L., and Bahr, G. F. (eds.), *Introduction to Quantitative Cytochemistry II,* pp. 209–21. Academic Press, New York.

Hopkins, A. W., and McBride, G. E. 1976. The life history of *Coleochaete scutata* (Chlorophyceae) studied by a Feulgen microspectrophotometric analysis of the DNA cycle. *J. Phycol.* 12, 29–35.

Hurdelbrink, L., and Schwantes, H. O. 1972. Sur le cycle de développement de *Batrachospermum. Soc. bot. Fran., Mémoires* 1972, 269–74.

Kemp, C. L., and Wentworth, J. W. 1971. Ultraviolet radiation studies on the colonial alga, *Eudorina elegans. Can. J. Microbiol.* 17, 1417–24.

Koop, H. -U. 1975. Über den Ort der Meiose bei *Acetabularia mediterranea. Protoplasma* 85, 109–14.

Lee, K. A., and Kemp, C. L. 1975. Microspectrophotometric analysis of DNA replication in *Eudorina elegans* (Volvocales, Chlorophyta). *Phycologia* 14, 247–52.

Mendelsohn, M. L. 1958. The two-wavelength method of microspectrophotometry II. A set of tables to facilitate the calculations. *J. Biophys. Biochem. Cytol.* 4, 415–24.

Patau, K. 1952. Absorption microphotometry of irregular-shaped objects. *Chromosoma* 5, 341–62.

Ploem, J. S., de Sterke, J. A., Bonnet, J., and Wasmund, H. 1974. A microspectrofluorometer with epi-illumination operated under computer control. *J. Histochem. Cytochem.* 22, 668–77.

Ruch, F. 1970. Principles and some applications of cytofluorometry. In Wied, G. L., and Bahr, G. F. (eds.), *Introduction to Quantitative Cytochemistry II,* pp. 431–50. Academic Press, New York.

Tautvydas, K. J. 1976. Evidence for chromosome endoreduplication in *Eudorina californica,* a colonial alga. *Differentiation* 5, 35–42.

Wied, G. L. (ed.). 1966. *Introduction to Quantitative Cytochemistry.* Academic Press, New York. 623 pp.

Wied, G. L., and Bahr, G. F. (eds.). 1970. *Introduction to Quantitative Cytochemistry II.* Academic Press, New York. 551 pp.

Wied, G. L., Bartels, P. H., Bahr, G. F., and Oldfield, D. G. 1968. Taxonomic intra-cellular analytic system (TICAS) for cell identification. *Acta Cytol.* 12, 180–204.

Zimmer, H. -G. 1970. Automatic analysis of microscopic images. *Zeiss Information Bulletin* 74, pp. 126–31.

26: Autoradiography for light and electron microscopy

RUTH E. SCHMITTER

Biology Department, University of Massachusetts at Boston, Boston, Massachusetts 02125

CONTENTS

I. Introduction

Autoradiography is a demonstrably useful tool for studying developmental phenomena as well as biosynthetic and secretory pathways (Gibbs 1968; Callow and Evans 1974). Recently, it has been used for determining phytoplankton species productivity (Knoechel and Kalff 1976). Both light- and electron-microscope autoradiography employ a thin layer of photographic emulsion applied over a radioactively labelled specimen as a radiation detector. The number and location of silver grains in the specimen are determined after photographic development, and conclusions are drawn based upon statistical analyses of the data. For the theoretical bases of the techniques and the necessary statistical methods, references by Salpeter and Bachmann (1972) and Rogers (1973) should be consulted. In the application of the techniques described in this chapter, it has been assumed that the material is already labelled with compounds containing tritium [^{3}H] (Gibbs and Poole 1973), sulfate [$^{35}SO_4$] (Ramus and Groves 1972), or other suitable labels.

General caution: Precautions observed in handling radioactive material used in the labelling of the specimens should be continued in the following procedures. This is important for the safety of the investigators and will also serve to prevent nonspecific labelling caused by stray radioactive material.

II. Methods for light microscopy

A. Dipping technique

Emulsion-coating by dipping is one of the most widely used techniques and is for that reason described here. The pretreatments of the labelled material follow general methods described in more detail in other chapters.

1. Fixation and embedding. Standard methods of fixation and embedding can be used (see Chap. 22) providing the fixative precipitates radioactive compounds in such a way that they are not appreciably removed during dehydration and embedding. Fixatives containing mercury or lead are not suitable, because they affect the photographic emulsion adversely. Although glutaraldehyde and formaldehyde are commonly used, they may cause some desensitization of the photographic emulsions (Bogoroch 1972). Paraffin-embedded material has usually been used for large, multicellular organisms; however, plastic-embedded material (Araldite, Epon, methacrylate, Spurr's medium) can be used for both light and electron microscopy (Gibbs and Poole 1973).

2. Sectioning and mounting

a. Microscope slides, with a frosted end, are cleaned by soaking overnight in a chromate–sulfuric acid solution, rinsed well in distilled water, and dried in an oven.

b. These cleaned slides are then "subbed," to provide maximum adhesion of sections and emulsion. The subbing solution (5.0 g gelatin and 0.5 g chrome alum dissolved in 1 liter distilled water) is heated 3 h (40°C) in a water bath with stirring. After filtration and cooling of the solution to room temperature, the cleaned slides are dipped into it and air-dried.

c. Sections are cut (1–10 μm) on a microtome or ultramicrotome. For any one experiment they must be of uniform thickness (Chap. 22.III.D).

d. Sections are applied toward the outer third of the slide (opposite the frosted area), so that a shallower emulsion solution can be used in dipping.

e. Extra slides (ca. 10–15) of each sample are prepared to allow a sufficient number for empirical determination of the precise exposure time, and for controls (II.A.4.f).

3. Prestaining sections. Sections for DNA detection by the Feulgen procedure (Chap. 22.III.F.13) must be stained prior to emulsion-coating because the acid hydrolysis removes deposited silver grains. Generally, sections are overstained (with toluidine blue, methylene blue, and safranin, Chap. 22.III.E and F) because stain will be lost during the subsequent processing of the sections. Some stains are not compatible with autoradiography (Thurston and Joftes 1963; Bogoroch 1972).

4. Emulsion-coating specimens. Kodak Nuclear Track emulsions NTB2 and NTB3 are most commonly used. NTB3 has greater sensitivity to β-particles than does NTB2. Bulk emulsion can be stored refrigerated for at least 3 months.

Coating and handling of the emulsion is carried out under photographic darkroom conditions with a safelight (Wratten Series 2 filter, 15 W bulb) or in darkness.

The background of each new emulsion batch is checked by emulsion coating, developing, and viewing a sample slide *before* applying emulsion to the experimental preparations.

a. Paraffin is removed from the sections with xylene, and slides are brought to water through an alcohol dilution series.

b. The emulsion is prepared in a Coplin jar or a 50-ml beaker at 43°C in a water bath. With a clean spatula, emulsion gel is added to give a volume sufficient to cover half to two thirds of a slide.

c. About 30–60 min are allowed for the emulsion to melt, then a blank slide is dipped into the emulsion to check for air bubbles. The bubbles must be dissipated before one proceeds.

d. A pair of experimental slides (back-to-back) is slowly dipped (1–2 sec) into the emulsion, with the slides held vertically by the frosted ends. They are then slowly and steadily withdrawn, being kept in the vertical position. (If a single slide is dipped, the emulsion is wiped from the back before drying).

e. The excess emulsion is allowed to drain on moist filter paper, and the slides are separated and allowed to dry vertically. This takes 30–60 min, depending on temperature and relative humidity. [Drying too rapidly can cause stress in emulsion, with subsequent clumping of grains (Bogoroch 1972)].

f. Control slides. Background control slides are dipped at the same time as the samples. These contain sections of unlabelled tissue, are stored for the same exposure times as the experimental slides, and are developed with them. As a control for fading of latent image, one experimental slide from each batch is exposed to light, then stored with other autoradiograms until development. Fading has occurred if grain density decreases with increased storage time.

If a small number of slides is to be coated, the emulsion is melted in a plastic slide mailer (Bogoroch 1972), or in a glass container as described by Gude (1968). Undiluted NTB3 results in emulsion layers of 3–4 μm. Dilution of 2 parts emultion with 1 part distilled water results in 1–2 μm layers, which are suitable for high-resolution studies. Rogers (1973) discusses selection and determination of emulsion thickness.

5. *Storing autoradiograms.* Autoradiograms are stored, when *dry,* in black slide boxes sealed with black tape, each box containing a small packet of silica gel. The coated slides are refrigerated until development, but not frozen. Test slides are developed at 7–10 day intervals to determine when the final autoradiograms should be developed. Exposure time is usually 2 weeks to 2 months, but longer exposures may be necessary.

6. *Processing autoradiograms.* Development is carried out at room tem-

perature under the same conditions as emulsion coating and is timed with a nonfluorescent timer. Many autoradiograms can be developed at once in staining dishes with movable slide racks.

a. Development in Kodak Dektol developer, diluted 1 part stock to 2 parts distilled water, is carried out for 2 min.

b. The slides are rinsed in distilled water.

c. Fixation is done in 24–30% sodium thiosulphate solution for 8 min. (Kodak Fixer can be used, but may remove prestain.)

d. Sections are then washed in running tap water for 15 min.

7. *Staining sections.* Preparations can be stained through the emulsion (Bogoroch 1972; Thurston and Joftes 1963), but staining should be attempted only after the fixer has been removed by thorough rinsing. Care should be taken that any stain used will color the sections, but not the gelatin, will not obscure silver grains, and will not remove developed silver grains. Aqueous 0.1% toluidine blue O is a useful poststain. After staining, sections can be dehydrated and made permanent by standard histological techniques (Chap. 22.III.E. and F.)

B. *Alternative methods*

Stripping film and nuclear-track techniques are both particularly suitable for use with high-energy isotopes (Gahan 1972). Nuclear-track techniques are also desirable if disintegration rates of labelled sources must be determined accurately (Rogers 1973). Special techniques involving freezing have been developed for handling readily diffusible labelled materials. For these Appleton (1972) and Roth and Stumpf (1969) should be consulted.

III. Methods for electron microscopy

A. *Flat substrate method (collodion film)*

The method detailed here is basically that of Salpeter and Bachmann (1972). Thin sections of radioactively labelled tissue are sandwiched between a collodion film on a glass slide and a coating of liquid emulsion. After development, plastic films are floated from the slides, and samples are picked up for viewing.

1. *Fixation and embedding.* Specimens are normally fixed in buffered glutaraldehyde (ca. 4%), washed extensively, and postfixed in 1–2% buffered OsO_4. After dehydration, they are embedded in an Epon–Araldite mixture or in Araldite alone (Chap. 23.II.F.2). Unlabelled tissue must also be prepared for use in controls. Details of fixation, embedding, and sectioning are extensively covered in Chapter 23.

2. Cleaning and collodion-coating slides

a. Slides are washed in dilute Alconox (Alconox, Inc.), well rinsed in distilled water, and hand-dried using facial tissue, *not* Kimwipes.

b. A 0.7% (v/v) collodion solution in amyl acetate is freshly prepared by dilution from a 2% (w/v) stock solution (which can be stored for several months).

c. Before each use, the 0.7% solution is filtered several times through the same filter in a fume hood.

d. The slides are held vertically and dipped into the 0.7% solution to cover about one-half of the slide and allowed to air-dry in the hood.

e. Samples from each batch of coated slides must be tested to check release of the film and proper thickness. This is done as follows. The edges of the slide are scored with a razor blade. Touching the surface of the water, the coated side is then gently lowered (ca. 30° angle with the water surface) into a large finger bowl filled with distilled water. The free-floating film should have a silver interference color. Other colors signify excessive thickness.

3. Sectioning and specimen mounting

a. Sections (1 μm) are cut on an ultramicrotome; special cleanliness should be observed.

b. Only ribbons of uniform section thickness are used (gray or silver sections).

c. The ribbons are transferred to a drop of distilled water or 10% acetone on a collodion coated slide. For this purpose, a platinum wire loop (ca. 3 mm in diameter) is useful. A glass rod or a sharpened applicator stick can also be used. Care must be taken not to touch the fragile collodion film.

d. The drop is drawn away from the ribbon with a sharpened stick and then removed with filter paper, thus allowing the ribbon to settle on the plastic substrate.

e. Location of the section is indicated by etching a circle (with a diamond marking pencil) on the back of the slide.

f. It is important to prepare sections of unlabelled tissue for controls in a similar manner.

4. Staining sections. Staining provides necessary additional contrast. It can be done either before application of the photographic emulsion or after photographic processing. In the former case, a thin carbon coating must be applied to prevent destaining during development (III.A.5). Autoradiograms of stained, unlabelled tissue and unstained, labelled tissue to serve as controls for stain-induced artifacts must also be prepared.

a. For staining before emulsion coating, a few drops of *clean,* lead stain (Millonig 1961; Reynolds 1963) are used to cover the sections for

10–20 min. They are then well rinsed in water, dried, and stored in a covered container.

b. Staining after photographic processing is carried out as above, but requires mounting of the specimen on an electron-microscope grid (III.A.9). Also, the collodion must be removed with amyl acetate. Salpeter and Bachmann (1972) discuss in detail some problems which may attend staining.

5. *Carbon-coating specimens.* Application of a thin (5–10 nm) layer of carbon over specimens is recommended before one applies the emulsion. It prevents destaining and provides a more uniform surface for formation of an even emulsion layer. Salpeter and Bachmann (1972) suggest use of SPK spectroscopic carbon, but satisfactory results have been obtained by the author without taking that precaution. Standard operating procedures for a vacuum evaporator are followed. A 5–10 nm layer of carbon, appearing light gray on a white test object, is deposited on the grids. Intense heat generated during evaporation may make it difficult to float specimens free later and should be avoided.

6. *Emulsion-coating specimens*

a. Under a Wratten OC safelight filter (15 W bulb) the Ilford L4 emulsion is diluted with distilled water at 45°C and maintained at that temperature until coating. Dilution varies with the desired emulsion thickness. Usually one starts with 1 g to 4 ml distilled water, adding water in one milliliter aliquots as needed. The emulsion is very gently mixed to prevent air bubble entrapment. After successive dilutions, test slides are coated, then observed under white light; the thickness is chosen by interference color. A dilution of 1 g/10 ml was found suitable by the author.

b. The emulsion is applied by dropping the emulsion onto slides with a medicine dropper. The dropper is held horizontally and emulsion dropped on slide until the film-coated area is flooded; one should be careful not to create air bubbles. After 5 sec, the excess is drained back into the diluted emulsion. (Alternatively, slides can also be dipped, drained, and dried.)

c. Each slide is drained on filter paper and dried vertically. Slides must be *dry* before storage.

d. At the same time that experimental specimens are coated, controls are prepared. As with light microscopy, control slides accompany each set of autoradiograms. Emulsion background should ordinarily be 1 grain/100 μm^2 or less.

7. *Exposing autoradiograms.* Dried specimens are stored (exposed) as described in section II.A.5.

8. Processing autoradiograms. Autoradiograms are developed at room temperature under the conditions used for emulsion-coating; a nonfluorescent timer is used. All solutions are prepared in distilled water.

a. Each slide is developed singly in undiluted *Microdol-X (Kodak)* for 3 min (for alternate developer see A.10, below).

b. The slide is rinsed in distilled water.

c. It is placed into 3% acetic acid (stop bath) for 15 sec, and again rinsed.

d. Fixation is for 1 min in *nonhardening* fixer: 20% sodium thiosulphate (w/v), or 20% sodium thiosulphate plus 2.5% potassium metabisulfite (w/v).

e. This is followed by repeated rinsing in distilled water. Specimen should *not* be dried before stripping from the slide.

9. Mounting autoradiograms

a. After developing, the autoradiogram (collodion + sections + carbon + developed emulsion) is floated off the glass slide as described in III.A.2.e.

b. Electron-microscope grids are gently placed (face down) over the sections in the autoradiogram. Sections are located by the interference colors reflected from a nearby desk lamp.

c. The autoradiograms on the grids are picked up on small pieces of Parafilm. From the top, the edge of the Parafilm is aligned with the floating film. Then with a slight tilt of the Parafilm, contact is made at one edge and continued until the entire film adheres to the Parafilm. The Parafilm is then picked up and placed grid side up to dry in a petri dish.

d. When grids are dry they are ready for examination in the electron microscope.

If difficulty is encountered in floating the autoradiograms free, 10% hydrofluoric acid (v/v) can be used to promote stripping (Salpeter and Bachmann 1972).

10. Alternate developing method for autoradiograms. Microdol-X developer (Kodak) is generally used because it gives the most reliable results and low background counts; but it produces filamentous type grains. For nonfilamentous grains of smaller size, phenidone developer may be used; however, it is less reliable in its background. A modification of the method by Lettré and Paweletz (1966) has been found useful by the author. Phenidone is prepared fresh by sequentially adding 1.5 g ascorbic acid, 0.25 g phenidone, 0.6 g potassium bromide, 1.3 g sodium carbonate, 20.0 g sodium sulfite, 6.0 g potassium thiocyanate and water to a final volume of 100 ml followed by filtration. Development is carried out for 1–2 min at 25°C.

Table 26–1. *Distribution of silver grains and random circles over sections of Porphyridium aerugineum*

Structure	Total grains (%)	Random circles (%)
Chloroplast	44	38
Vesicle	15	11
Starch grain	12	16
Golgi dictyosome	2	1
Cell surface[a]	21	23
Other	6	10

Note: Total grains = 205. All electron micrographs were analyzed at × 25,500. Random circles (250 nm diameter) around intersects of grid.
[a] Cell surface includes cell membrane, adjacent endoplasmic reticulum, and extracellular polysaccharide.

B. *Alternative methods*

The method of Caro and van Tubergen (1962; 1964) has been successfully used by Gibbs (1968) on *Ochromonas.* In this technique, the sections of radioactively labelled specimens are picked up on grids coated with collodion and carbon, affixed to a microscope slide, and the liquid emulsion is applied to the grids with a thin wire loop.

In the membrane method used by Budd and Pelc (1964), formvar films borne on perforated lucite slides are used. Sections are transferred to the formvar membranes bridging each hole, and liquid emulsion is applied to the membranes. After exposure and photographic development, electron microscope grids are applied over the sections, and the specimens are freed by piercing the film around each grid.

C. *Considerations in data analysis*

An example using the method of Williams (1969) on *Porphyridium aerugineum* is illustrated in Table 26–1. It is important to have a large number of *randomly* photographed relevant areas, that is, areas *not* selected for high label, low label, cleanliness, or other feature. Quantitative analysis of grain density in electron-microscope autoradiographs can be done (a) by dividing the number of developed grains overlying a structure by the total area occupied by the structure (grains/unit area) or (b) by using probability circles drawn around developed grains. A grain is at least partly associated with any structure falling within its probability circle. Area corrections are derived from use of random circles of the same size (Williams 1969; Salpeter and Bachmann 1972; Blackett and Parry 1973).

IV. Acknowledgments

The author applied the major techniques described here on *Porphyridium aerugineum* (UTEX 755), while a Brown Postdoctoral Fellow in Botany at Yale in the laboratory of Dr. Joseph S. Ramus of the Biology Department of Yale University. Dr. Lewis J. Feldman of the Botany Department of the University of California at Berkeley provided helpful discussion during the course of the work.

V. References

Appleton, T. C. 1972. Autoradiography of diffusible substances. In Gahan, P. B. (ed.), *Autoradiography for Biologists,* pp. 51–64. Academic Press, New York.

Blackett, N. M., and Parry, D. M. 1973. A new method for analyzing electron microscope autoradiographs using hypothetical grain distributions. *J. Cell Biol.* 57, 9–15.

Bogoroch, R. 1972. Liquid emulsion autoradiography. In Gahan, P. B. (ed.), *Autoradiography for Biologists,* pp. 65–94. Academic Press, New York.

Budd, G. C., and Pelc, S. R. 1964. The membrane method of electron microscope autoradiography. *Stain Technol.* 39, 295–302.

Callow, M. E., and Evans. L. V. 1974. Studies on the ship-fouling alga *Enteromorpha.* III. Cytochemistry and autoradiography of adhesive production. *Protoplasma* 80, 15–27.

Caro, L. G. 1964. High-resolution autoradiography. In Prescott, D. M. (ed.), *Methods in Cell Physiology,* vol. 1, pp. 327–63. Academic Press, New York.

Caro, L. G., and van Tubergen, R. P. 1962. High resolution autoradiography. I. Methods. *J. Cell Biol.* 15, 173–88.

Gahan, P. B. (ed.). 1972. *Autoradiography for Biologists.* Academic Press, New York, 124 pp.

Gibbs, S. P. 1968. Autoradiographic evidence for the in situ synthesis of chloroplast and mitochondrial RNA. *J. Cell Sci.* 3, 327–40.

Gibbs, S. P., and Poole, R. J. 1973. Autoradiographic evidence for many segregating DNA molecules in the chloroplast of *Ochromonas danica. J. Cell Biol.* 59, 318–28.

Gude, W. D. 1968. *Autoradiographic Techniques; Localization of Radioisotopes in Biological Material.* Prentice-Hall, Englewood Cliffs, N.J. 113 pp.

Knoechel, R. and Kalff, J. 1976. Track autoradiography: A method for the determination of phytoplankton species productivity. *Limnol. Oceanogr.* 21, 590–6.

Lettré, H., and Paweletz, N. 1966. Probleme der elektronenmikroskopischen Autoradiographie. Naturwiss. 53, 268–71.

Millonig, G. 1961. A modified procedure for lead-staining of thin sections. *J. Biophys. Biochem. Cytol.* 11, 736–9.

Ramus, J., and Groves, S. T. 1972. Incorporation of sulfate into the capsular polysaccharide of the red alga *Porphyridium*. *J. Cell Biol.* 54, 399–407.

Reynolds, E. S. 1963. The use of lead citrate at high pH as an electron-opaque stain in electron microscopy. *J. Cell Biol.* 17, 208–12.

Rogers, A. W. 1973. *Techniques of Autoradiography.* Elsevier, Amsterdam. 372 pp.

Roth, L. J., and Stumpf, W. E. (ed.). 1969. *Autoradiography of Diffusible Substances.* Academic Press, New York. 371 pp.

Salpeter, M. M., and Bachmann, L. 1972. Autoradiography. In Hayat, M. A. (ed.), *Principles and Techniques of Electron Microscopy, Biological Applications,* vol. 2, pp. 221–78. Van Nostrand Reinhold, New York.

Thurston, J. M., and Joftes, D. I. 1963. Stains compatible with dipping radioautography. *Stain Technol.* 38, 231–5.

Williams, M. A. 1969. The assessment of electron microscopic autoradiographs. In Barer, R. and Cosslett, V. E. (eds.), *Advances in Optical and Electron Microscopy,* vol. 3, pp. 219–72.Academic Press, New York.

27: Immunochemistry: labeled antibodies

ESTHER L. McCANDLESS AND
ELIZABETH GORDON-MILLS*

Department of Biology, McMaster University, Hamilton, Ontario, Canada

VALERIE VREELAND

Department of Botany, University of California, Berkeley, California 94720

CONTENTS

* Present address: c/o Department of Botany, University of Adelaide, Adelaide, South Australia 5001, Australia.

I. Introduction

Immunohistochemical techniques represent a valuable addition to tools now available to characterize the organic components of biological structures, but they have thus far received little application in phycology. Techniques of immunology have traditionally been associated with the field of medicine, rather than with botany. It is anticipated, however, that the use of labelled antibodies will increase as their advantages become better known.

Two particular advantages of the immunochemical method are obvious, their sensitivity and their specificity. Very little antigen (Ag) needs to be present in a biological structure for it to be detectable by its reaction with the appropriate specific antibody (Ab); its occurrence in extremely low concentrations or on specific cell organelles can be demonstrated by immunochemistry at the electron microscope level, if Ab of appropriate specificity can be prepared.

Antibodies produced against particular macromolecules are usually highly specific. Antibodies have been produced that differentiate between α- and β-anomeric configurations of constituent hexose units in a polysaccharide (Torii et al. 1964). Immunohistochemical methods extend the usefulness of this specificity to the tissue level and provide information on the composition of subcellular structures. The degree of specificity or nonspecificity can be easily demonstrated by immunodiffusion, a necessary step in characterizing antisera before use in histochemical exploration.

Information on these techniques and their applications can be found in general immunology texts and in selected references by Axelsen et al. (1973); Crowle (1973); Kabat and Mayer (1961); Sternberger (1974); Weir (1973); Williams and Chase (1967; 1968; 1971); and Wisse et al. (1974).

II. Method

A. Antiserum or antibody

1. Production

a. Preparation of Ag for injection. For specific antisera, purity of the material to be injected is of utmost importance. The Ab-producing animal does not necessarily distinguish between the Ag of interest to the investigator and a contaminant present in the preparation. If the material to be injected cannot be purified, then it is important to characterize its impurities.

Proteins are usually antigenic when administered in appropriate dosage schedules, but for acidic macromolecules or for small reactive groups, immunogenicity may be enhanced or conferred by complexing the material with a carrier such as methylated bovine serum albumin (mBSA), which can be purchased or prepared. It is prepared by dissolving 100 mg BSA in 10 ml absolute methanol and adding 84 μl 12 *N* HCl. The mixture is allowed to stand at room temperature in the dark for 3 days, with occasional mixing. The mBSA will precipitate from solution and can be removed by centrifugation. The precipitate is twice washed with methanol, suspended in water, and neutralized with NaOH. The mBSA precipitate is stored dry after lyophilization, or as a neutral solution after sterile filtration (Millipore). To complex with Ag, equal weights of mBSA (in solution) and Ag are mixed. To enhance the immunological reaction, the complexed Ag is emulsified with an equal volume of adjuvant (complete or incomplete Freund's) immediately before injection.

b. Immunization of animal with Ag. Animals traditionally used for the production of antisera are mice, guinea pigs, rabbits, goats, sheep, and horses. Selection of the species to be used depends on the facilities available and the responsiveness of the animal to the Ag under study, a factor which can be determined only by trial.

A preimmunization blood sample (5–20 ml) is obtained for control purposes. To prepare serum, the blood is allowed to clot for several hours at room temperature. The clot is then freed from its container and refrigerated overnight. The serum fraction (the supernatant fraction) is separated by centrifugation. It is then tested against the Ag to be used to check for other crossreacting material that might complicate conculsions after immunization (see II.A.2.a). Addition of 1% by volume of 10% sodium azide or of a 1% thimerosal (merthiolate) stock solution, assures preservation. The serum is then passed through a filter (Millipore, 0.45 μm) into sterile containers and stored at −20°C to −70°C or lyophilized and refrigerated (4°C).

The dosage schedule to be employed for immunization of the animal must be determined empirically. Doses that are too high or too low may produce tolerance to the antigen. In general, the immunogenic dose of protein (commonly 25–1,000 μg/kg/injection) is lower than that of other types of antigens; Vreeland found a total of 0.5 mg alginic acid complexed with mBSA to be antigenic in rabbits; Gordon-Mills and McCandless injected 25–30 mg carrageenan–mBSA into goats; both procedures used complete Freund's adjuvant and administered the material intramuscularly, or subcutaneously and intradermally.

A suggested schedule is as follows: One third of the total immunizing dose is injected intramuscularly (into the shoulder or hip muscles). This is followed with two similar injections (minus the adjuvant) at 2- or 3-week intervals. Blood is obtained in 4 weeks and subsequently at weekly intervals until specific Ab activity is demonstrated in the serum. When sufficiently high antibody levels have been reached, the animal is anesthetized and exsanguinated. The serum is prepared as above. Alternatively, the animal may be used as an Ab source for extended periods by giving "booster" injections of Ag when the Ab level drops. However, in some systems, specificity of Ab decreases with time. Neat antiserum may be used directly in the fluorescent Ab indirect method.

c. Isolation of γ-globulin from antiserum by precipitation. To isolate Ab from antiserum, finely ground $(NH_4)_2SO_4$ is gradually added (30–45 min) with stirring. IgG will precipitate as the $(NH_4)_2SO_4$ concentration reaches 33% saturation, that is after addition of 17.9 g/100 ml serum at room temperature. (To obtain other γ-globulins, continue to 50% saturation). The precipitate is collected by centrifugation (12,000*g*, 30 min, ca. 5°C) and is washed 2 or 3 times with 40% saturated $(NH_4)_2SO_4$ (21.4 g/100 ml). Following the final centrifugation, the precipitate (IgG) is dissolved in cold 0.01 *M* phosphate-buffered 0.15 *M* NaCl, pH 7.5. To obtain higher Ab titre, the IgG is dissolved in a lesser volume than that of the starting serum. The preparation is dialyzed overnight against 3 liters cold 0.01 *M* phosphate-buffered 0.15 *M* NaCl, pH 7.5, is sterilized by filtration, and stored frozen or lyophilized.

2. *Testing*

a. Immunodiffusion. One of the simplest assays that has been described for demonstrating the presence and specificity of antisera is the Ouchterlony immunodiffusion technique. A solution of 0.75–1% agarose in 0.01 *M* phosphate-buffered 0.15 *M* NaCl, pH 7.5, is heated to 100°C while being stirred. Thimerosal is added to a final concentration of 1 : 1,000, and the melted agarose is poured into small (5.5 cm) petri plates to a depth of about 5 mm and allowed to gel. Gels can also be prepared on microscope slides and stored in a moist chamber. Wells are cut into the gel with a gel cutter template or a cork cutter. The gel plates are allowed to "dry" for 30 min before being used. The antiserum or Ab to be tested is placed in one well, and homologous Ag and possible crossreacting substances at a concentration of 1–10 mg/ml are placed in adjacent wells. The materials are allowed to diffuse at room temperature for a suitable period of time (6–24 h is usual). A specific antiserum should give precipitin bands only with the homologous Ag and related compounds possessing a similar chemical structure.

Different buffers (TRIS, barbital), lower salt concentrations, or development of agarose plates at 4°C may yield better results with some antigens (Crowle 1973). Diffusion plates may also be prepared with 5% gelatin in TRIS-buffered saline rather than agarose. Diffusion in these plates is slower but resolution is good.

For a permanent record, and if precipitin bands are weak, the immunodiffusion plates are filled with water and photographed with Kodalith type 3, or Panatomic-X film, with transmitted light from a light box. Kodalith film is developed with undiluted D-19 developer, and Panatomic-X with Microdol-X (diluted 1:3). Alternatively, contact prints of the diffusion plate can be prepared (using it as a negative), and the contrast can be enhanced by using high contrast paper. Diffusion plates may also be stained with dyes and photographed (Axelsen et al. 1973).

b. Rocket immunoelectrophoresis. If a horizontal electrophoresis apparatus is available, this is a good way to follow development of specific Ab during immunization but is more complicated than immunodiffusion. It is, however, quantitative and highly sensitive. It may give good resolution of a mixture of antigenic components (Axelsen et al. 1973).

3. Purification. If immunodiffusion or immunoelectrophoretic techniques reveal the presence of antibodies to antigens other than the one under investigation, these may be removed by absorption of the antiserum with the unwanted Ag(s) with which it reacts. The time of absorption and the amount of absorbing Ag must be determined empirically. To a small volume of serum, an equal volume of Ag is added to a concentration range of 1 to 10 mg/ml. After mixing, it is first incubated at 37°C for 1 h and then refrigerated for 1–5 days. By low-speed centrifugation, the precipitate is removed, and the supernatant (absorbed) serum is then tested by immunodiffusion to determine that the absorption has removed the unwanted Ab and that it has not affected the Ab which is to be used.

4. Labeling procedure

a. Fluorescent-labeled antibody. Fluorescein isothiocyanate (FITC) or rhodamine B isothiocyanate (RBITC) can be conjugated to the immunoglobulins. For the direct method, the IgG used is that produced by injection of Ag. A 1% solution (w/v) of γ-globulin is chilled to 4°C. Cold 0.5 *M* Na_2CO_3 buffer, pH 9.5, is added with stirring to 9 parts of γ-globulin. The mixture is allowed to stand for 2 h at 4°C and then dry FITC or RBITC is slowly stirred into it, to a final concentration of 10–40 μg dye/mg protein. Stirring is continued at room temperature for 45 min. To remove the unreacted dye, 15 ml of conjugate is passed through a Sephadex G25 column, previously equilibrated with 0.01 *M* phosphate buffer, pH 7.0. The fluorescent-labeled antibody appears as a discrete band detectable with a UV lamp and is eluted just after the void volume. The Sephadex column can be reused after eluting the unreacted dye with phosphate buffer. The conjugated sample is stored frozen as 1 ml aliquots in containers protected from light.

For the indirect method, labeled IgG against γ-globulin of the animal species in which specific antiserum was raised can be purchased commercially.

b. Peroxidase-labeled Ab (Nakane 1975). Peroxidase-aldehyde is prepared first, before conjugation to the immunoglobulin. Five mg of horseradish peroxidase (R.Z. = 3.0) are dissolved in 1 ml freshly prepared 0.3 *M* $NaHCO_3$. To this is added 0.1 ml 1% 1-fluoro-2,4-dinitrobenzene (FDNB) in absolute ethanol; these are allowed to react for 1 h at room temperature, after which time 1 ml 0.04–0.08 *M* sodium metaperiodate is added. The reaction is stopped 30 min later by addition of 1 ml 0.16 *M* ethylene glycol and continuous stirring for 1 h. This is followed by exhaustive dialysis in 0.01 *M* Na_2CO_3 buffer, pH 9.5, prior to conjugation with the immunoglobulin. Five mg of immunoglobulin against IgG of the species in which specific Ab has been raised are mixed with the peroxidase-aldehyde prepared above and allowed to react for 3 h. Five mg $NaBH_4$ are added and the mixture is stored at 4°C for 4–24 h. This is followed by dialysis at 4°C for ca. 24 h against 0.01 *M* phosphate-buffered 0.15 *M* NaCl and chromatography on Sephadex (G-100 or G-200). Conjugated peroxidase-labeled Ab is eluted from the column first, and the superfluous peroxidase is retarded and separation results.

c. Unlabeled Ab-peroxidase. For this preparation, Sternberger (1974) should be consulted.

B. Tissue preparation

1. Fixation. A major problem in fixing plant tissues is their high content of polysaccharide, which may be extracted during fixation and embedding. Several different methods of fixation have been used in the preparation of algal specimens for fluorescent- and peroxidase-labeled antibody procedures. One of the most satisfactory fixations has been obtained with the Karnovsky reagent. In this case, small tissue segments are fixed for 2 h at room temperature in 5% glutaraldehyde–4% formaldehyde (Karnovsky 1965) in 0.025 *M* phosphate buffer. If the material is to be used for fluorescent Ab staining, it is then transferred to buffered 1% OsO_4 (**Caution:** Toxic!), and postfixed for 2 days at room temperature.

2. *Embedding and sectioning.* Several commonly used embedding media are unsatisfactory for fluorescent antibody procedures because they either quench specific fluorescence and thus obscure the results or they confer nonspecific fluorescence. Two of the most satisfactory are Araldite (Chap. 23.II.F.2.a) and glycolmethacrylate (Chap. 22.III.C.2.a). Cryostat sections, frequently used for animal tissue, were unsatisfactory for algal tissue (Vreeland 1972; Gordon-Mills and McCandless 1975). Sections are cut (dry, for glycolmethacrylate) 1-μm thick on a microtome or ultramicrotome and floated on water droplets. They are dried at 60°C onto glass microscope slides.

C. Staining procedures

1. Fluorescent antibody

a. Experimental material. Slight differences in procedure are necessary depending upon whether the fluorescent label (FITC or rhodamine) is attached to specific Ab (direct method), or whether it is attached to an Ab against serum proteins of the animal species in which specific Ab has been raised (indirect or sandwich method). The latter method requires a smaller amount of specific antiserum and is more sensitive.

For the indirect method, the tissue section is covered with a drop of specific Ab solution. It is incubated at room temperature for 30–60 min in a moist environment. The slides are gently rinsed with 0.01 *M* phosphate-buffered 0.15 *M* NaCl, pH 7.0, or with 0.05 *M* TRIS–saline, pH 7.0–7.6, to remove unreacted Ab or nonspecific background staining. The washing is repeated 3 times, after which the sections are covered with a drop of labelled (FITC or rhodamine) Ab, made against the IgG of the animal species in which the specific Ab has been raised. They are then incubated at room temperature for 30–60 min in a moist, dark environment. They are rinsed as above, drained, and mounted in glycerol.

For the direct method, one uses fluorescent Ab made against the Ag under study in the first step and omits the second incubation and wash.

b. Controls. The importance of appropriate controls for evaluation of nonspecific staining and autofluorescence cannot be overemphasized. Specific Ab in the first step of the procedure is omitted and one of the following is substituted: Ab against a different Ag (heterologous Ag); specific Ab which has been absorbed with specific Ag; pre-immune serum from the same species and, if possible, from the same animal as that used to produce specific Ab; serum from a different species of animal; phosphate-buffered saline. More than one type of control should be used. For the indirect procedure, preincubation with pre-immune serum from the species in which the labeled Ab is made eliminates nonspecific fluorescent staining (Pratt and Coleman 1971). The rest of the procedure is carried out as above.

2. Peroxidase-labeled Ab. To remove phenolic substances if present, the sections are pretreated for 12 h with chlorous acid (Rappay and van Duijn 1965). To remove activity of endogenous peroxidase, if present, Weir et al. (1974) pretreated sections with 0.074% HCl in ethanol for 5 min (See also Straus 1971).

a. Experimental material. For the indirect method, the tissue sections are incubated with specific Ab at room temperature for 30–60 min in a moist environment. The slides are carefully rinsed with 0.01 *M* phosphate-buffered 0.15 *M* NaCl, or 0.05 *M* TRIS–saline, pH 7.0–7.6, to remove unreacted antibody, and so forth. Washing is repeated three times, and the sections are covered with peroxidase-labeled Ab raised against serum proteins of IgG of the animal species in which specific Ab has been raised. Sections are incubated at room temperature for 30–60 min and rinsed as above. Staining for the peroxidase enzyme is as follows (Sternberger, 1974). The sections are incubated for 5–30 min at room temperature in freshly prepared 0.05% 3,3′-diaminobenzidine tetrahydrochloride in 0.05 *M* TRIS–HCl buffer, pH 7.6, containing 0.001–0.01% H_2O_2. (**Caution:** Diaminobenzidine may be carcinogenic.) The sections are washed in 3 changes of distilled water and postfixed in 1–2% OsO_4 containing 5% sucrose buffered to pH 7.2.

b. Controls. Control reactants are as in section C.1.b. Additional controls that should be included are omission of either antiserum, and of H_2O_2 or diaminobenzidine.

3. Unlabeled Ab-peroxidase. This method is described in detail by Sternberger (1974). It uses unlabeled Ab to peroxidase complexed with horseradish peroxidase (PAP soluble complex) and Ab against IgG of species in which antiperoxidase was raised. Specific Ab is raised in the same species as the antiperoxidase (the anti-IgG couples the two).

D. Specimen examination

1. Fluorescent antibody procedure. Stained sections are examined after preparation in a darkened room under dark-field illumination with a fluorescence microscope, using suitable filters (Chap. 20.C.). Non-fluorescing immersion oil should be applied between the condenser and the slide, and for high-power magnification between the objective and the slide as well, although the latter may give difficulty with a liquid mount.

Slides can be photographed on Ilford Pan F or Kodak Tri-X film using exposures of several minutes. Longer exposure reduces the fluorescence.

2. *Peroxidase-labeled antibody procedure.* Sections treated by this procedure are examined with a standard light microscope.

E. Extension of the method to the electron microscope level

Techniques of electron microscopy and immunochemical labeling using Ab directed against tissue components have been combined, using procedures similar to those described for light or fluorescence microscopy, but there are special problems.

The labels for antibodies used in electron immunohistochemistry must be electron opaque; usual labels are ferritin or reaction products of enzymes like peroxidase. Postfixation or poststaining with OsO_4 is a common additional step. For conjugation and other procedures, see Sternberger (1974) or Wisse et al. (1974).

The size of the labeled Ab molecule constitutes a barrier to penetration into tissues for studies of intracellular localization at the electron microscope level. To overcome this, Singer et al. (1974) used ultrathin frozen sections, and Nakane (1975) used labeled Ab fragments.

Adequate fixation for electron microscopy utilizes crosslinking fixatives, which adversely affect either cellular ultrastructure (e.g., paraformaldehyde) or antigenicity of cellular components of interest (e.g., glutaraldehyde). Most embedding media used for electron microscopy absorb the ferritin-labeled Ab conjugate extensively and nonspecifically. Nakane (1975) and McLean and Singer (1970) have suggested additional methods for fixation and embedding.

III. Sample data

Immunochemical techniques are most commonly used to demonstrate the presence (or absence) of a known tissue component in particular cellular structures, or in extracellular components (e.g., cuticles, matrix, etc.). At the present time, these techniques yield qualitative rather than quantitative information, the quality of which is limited by the purity of the antigen used in producing specific antiserum.

A. Localization of antigen

Application of FITC-labeled rabbit anti-goat γ-globulin to tissue sections of *Chondrus crispus* preincubated with specific goat antisera to κ- or λ-carrageenan demonstrated the localization of κ-carrageenan in cell walls and matrix of gametophytic plants and of λ-carrageenan in cell walls and matrix of sporophytes (Gordon-Mills and McCandless 1973; 1975; 1978). Similarily, FITC- or peroxidase-labeled goat anti-rabbit γ-globulin stained cell walls and some parts of the matrix of *Fucus distichus* sections pretreated with specific rabbit antiserum to alginic acid (Vreeland 1970; 1972). In later work Vreeland (1978), using FITC- and rhodamine-labeled antibodies and specific antisera against alginic acid and against a sulfated fucan, showed that the cell wall was two layered, usually with more alginic acid in the inner layer and more sulfated fucan in the outer layer. Differences in proportions of alginate and sulfated fucan could be detected by this method in young and mature parts of the thalli, in different cell types, and in different species. Alginate and fucan were demonstrated in walls of some cells of all brown algae examined, and alginate was demonstrated intracellularly in large amounts in certain cells.

Gantt and Lipschultz (1977) used antisera specific to B-phycoerythrin, R-phycocyanin, and allophycocyanin to probe the structure of phycobilisomes in *Porphyridium purpureum* by immuno-electron microscopy. Brown et al. (1976) investigated the flagellar rootlet system in *Polytomella agilis* with specific antitubulin Ab.

B. *Qualitative information*

An important aspect of the usefulness of the immunochemical structure of algal tissues is the fact that (with luck and hard work) different specific antibodies can be raised against macromolecules that are biochemically similar. This is particularly true if it is possible to separate the macromolecules involved and to use a very pure preparation for immunization. Alternatively, it may be possible to remove unwanted Ab components by absorption of the antiserum. With the latter method, Gordon-Mills and McCandless (1975) were able to show that the gametophyte and sporophyte generations of *C. crispus* produced carrageenans that differ only slightly in their primary chemical structure and that the change in synthetic capacity occurred in the first cells of each new generation.

IV. Problems of the method

A. *Nonspecificity of antibody*

At the present state of the art, it is impossible to specify what aspect(s) of a particular macromolecule will serve as the immunodominant group(s) when it serves as Ag in the animal being immunized. Macromolecules present in tissue possess common elements of chemical structure. The recognition of a minor contaminant in the Ag preparation may provide a family of antibodies that are specific for that contaminant but not for the Ag of interest. It is therefore important to characterize the Ab produced in terms of the crossreactions in which it participates and to recognize the complication this may introduce in the interpretation of results.

B. *Autofluorescence*

Autofluorescence in a tissue section may be derived from an element of the tissue itself, or it may result from the fixative or embedding material used. A red chloroplast (chlorophyll) autofluorescence constitutes a problem in all botanical studies but can be eliminated by postfixation in OsO_4. Chlorous acid removed a red autofluorescence believed to be due to phenolic compounds in *F. distichus*. The availability of both FITC- and RBITC-labeled antibodies makes it possible to choose for the Ab label a specific fluorescence of different color from the autofluorescence and to reduce nonspecific fluorescence with appropriate filters.

C. *Nonspecific staining*

Nonspecific fluorescence, present only when fluorescent Ab is applied, is distinguished from autofluorescence but constitutes a similar problem. It is the raison d'être for certain of the controls employed in the fluorescent Ab method. Sulfated polysaccharides present in many algae may react nonspecifically with protein, hence it is wise to use as a control Ab produced against a different Ag from that under investigation.

With the peroxidase method, the problem of reactive phenolic compounds in tissues yields to chlorous acid treatment. Nonspecific staining with ferritin-labeled Ab at the electron microscope level is minimized by embedding with a protein such as bovine serum albumin crosslinked with glutaraldehyde (Singer et al. 1974).

Nonspecific staining due to increased polarity and hydrophobicity of reagents caused by the labeling procedure is eliminated by the use of native proteins in the hybrid antibody and labeled antibody enzyme procedures (Sternberger 1974).

D. *Fluorescence quenching*

Quenching of fluorescence occurs with exposure to light, sets a limit to the useful lifetime of sections prepared for fluorescent antibody staining, and requires that the sections be photographed immediately after staining. Stored in the dark, however, sections retain some fluorescence for prolonged periods. In this regard, peroxidase-labeled antibodies are superior in that the results are permanent.

V. Acknowledgments

We should like to express appreciation to Dr. M. E. McCully for assistance in early stages of the *C. crispus* studies. We acknowledge research support from the National Research Council of Canada, U.S. Public Health Service, and the National Science Foundation.

VI. References

Axelsen, N. H., Kroll, J., Weeke, B. (eds.). 1973. A Manual of Quantitative Immunoelectrophoresis, Methods and Applications, *Scand. J. Immunol. Suppl.* 2, 230 pp.

Brown, D. L., Massalski, A., and Patenaude, R. 1976. Organization of the flagellar apparatus and associated cytoplasmic microtubules in the quadriflagellate alga *Polytomella agilis. J. Cell Bio.* 69, 106–25.

Crowle, A. J. 1973. *Immunodiffusion.* Academic Press, New York. 545 pp.

Gantt, E., and Lipschultz, C. A. 1977. Probing phycobilisome structure by immuno-electron microscopy. *J. Phycol.* 13, 185–92.

Gordon, E. M., and McCandless, E. L. 1973. Ultrastructure and histochemistry of *Chondrus crispus* Stackhouse. *Proc. Nova Scot. Inst. Sci.* 27, 111–33.

Gordon-Mills, E. M., and McCandless, E. L. 1975. Carrageenans in the cell walls of *Chondrus crispus* Stack. (Rhodophyceae, Gigartinales). I. Localization with fluorescent antibody. *Phycologia* 14, 275–81.

Gordon-Mills, E. M., and McCandless, E. L. 1978. Studies on carrageenans in the cell walls of *Chondrus crispus* Stack. In Fogg, G. E. (ed.), *Proceedings Internation Seaweed Symposium, Bangor,* 1974, vol. 8 (in press).

Kabat, E. A., and Mayer, M. M. 1961. *Experimental Immunochemistry,* 2nd ed. Thomas, Springfield. 905 pp.

Karnovsky, M. J. (1965). A formaldehyde–glutaraldehyde fixative of high osmolality for use in electron microscopy. *J. Cell Biol.* 27, 137A–138A.

McLean, J. S., and Singer S. J. 1970. A general method for the specific staining of intracellular antigens with ferritin-antibody conjugates. *Proc. Nat. Acad. Sci. U.S.* 65, 122–8.

Nakane, P. K. 1975. Recent progress in the peroxidase-labeled antibody method. *Ann. N.Y. Acad. Sci.* 254, 203–11.

Pratt, L. H., and Coleman, R. A. 1971. Immunocytochemical localization of phytochrome. *Proc. Nat. Acad. Sci. U.S.* 68, 2431–5.

Rappay, G., and van Duijn, P. 1965. Chlorous acid as an agent for blocking tissue aldehydes. *Stain Technol.* 40, 275–7.

Singer, S. J., Painter, R. J., and Tokuyasu, K. T. 1974. Ferritin antibody staining of ultrathin frozen sections. In Wisse, E., Daems, W. T., and Molenaar, I., and van Duijn, P. (eds.), *Electron Microscopy and Cytochemistry,* North Holland, Amsterdam.

Sternberger, L. A. 1974. *Immunocytochemistry.* Prentice-Hall, Englewood Cliffs, N.J. 246 pp.

Straus, W. 1971. Inhibition of peroxidase by methanol and by methanol-nitroferricyanide for use in immunoperoxidase procedures. *J. Histochem. Cytochem.* 19, 682–8.

Torii, M., Kabat, E. A., and Bezer, A. E. 1964. Separation of teichoic acid of *Staphlococcus aureus* into two immunologically distinct specific polysaccharides with α- and β-*N*-acetylglucosaminyl linkages respectively. Antigenicity of teichoic acid in man. *J. Exper. Med.* 120, 13–29.

Vreeland, V. 1970. Localization of a cell wall polysaccharide in a brown alga with labeled antibody. *J. Histochem. Cytochem.* 18, 371–3.

Vreeland, V. 1972. Immunocytochemical localization of the extracellular polysaccharide alginic acid in the brown seaweed *Fucus distichus. J. Histochem. Cytochem.* 20, 358–367.

Vreeland, V. 1978. Alginates and sulfated fucans in brown algal walls. In Fogg. G. E. (ed.), *Proceedings of the International Seaweed Symposium, Bangor, 1974,* vol. 8 (in press).

Weir, D. M. (ed.). 1973. *Handbook of Experimental Immunology.* 2nd ed. Blackwell, London. 1245 pp.

Weir, E. E., Pretlow, T. G., Pitts, A., and Williams, E. E. 1974. Destruction of edogenous peroxidase activity in order to locate cellular antigens by peroxidase-labeled antibodies. *J. Histochem. Cytochem.* 22, 51–4.

Williams, C. A., and Chase, M. W. (eds.). 1967, 1968, 1971. *Methods in Immunology and Immunochemistry,* vols. 1, 2, 3. Academic Press, New York, 479; 459; 515 pp.

Wisse, E., Daems, W. T., Molenaar, I., and van Duijn, P. (eds.). 1974. *Electron Microscopy and Cytochemistry.* North Holland, Amsterdam. 406 pp.

28: Freeze-fracture and freeze-etch techniques

L. ANDREW STAEHELIN

Department of Molecular, Cellular and Developmental Biology, University of Colorado, Boulder, Colorado 80309

CONTENTS

I. Introduction

Freeze-fracturing and freeze-etching techniques are based on the fact that biological materials cannot only be fixed and stabilized by chemical means but also by purely physical techniques, that is, by rapid freezing to low temperatures ($< -100°C$), which limits the movements of molecules. In recent years, this has become very important for studying the supramolecular architecture of biological membranes, because it allows the investigator to visualize the spatial distribution of specific components both within the plane and on both surfaces of a given membrane (Staehelin 1976). Although freeze-etching has been used for algae, the method has not been used to fullest advantage until recently when Staehelin and Pickett-Heaps (1975) and Brown and Montezinos (1976) applied it to the investigation of the structure and formation of algal cell walls.

This chapter describes and explains the reasoning behind the basic manipulations necessary to freeze-fracture and freeze-etch biological samples: pretreatment and freezing, fracturing, etching, replica formation, and cleaning. Although the description of the preparation steps applies most directly to the Balzers freeze-etch apparatus (Balzer High Vacuum Corp.), the general method, and certainly the theory, of freezing are applicable to other instruments. Additional information may be found in the reviews of Moor (1969), Koehler (1972), Bullivant (1973), and Southworth et al. (1975), and in the book on freeze-etching techniques edited by Benedetti and Favard (1973).

II. Pretreatment and freezing

A. Why pretreatment?

Pretreatment of the sample involves processing the material into a form and a state that allows optimal freezing to liquid nitrogen temperature to assure minimum disruption due to ice-crystal formation and, subsequently, to obtain useful and accurate images of the structures to be investigated. When an aqueous sample is cooled to liquid nitrogen temperatures, it passes through the freezing point (f.p.) and then through the recrystallization temperature (r.t.), the temperature below which large ice crystals cease to grow by a transfer of water molecules. Ice-crystal growth may occur between f.p. and r.t., the critical temperature interval. Ice-crystal size can be minimized by either reducing the critical temperature interval or by increasing the rate at which the sample passes through the interval (Moor 1964).

B. *Prefixation and glycerol infiltration*

Reduction of free water available for ice crystal formation can be accomplished by drying the specimen or by infiltrating it with a natural or an artificial water-binding cryoprotective agent. Glycerol was first employed in a freeze-etch experiment by Moor and Mühlethaler (1963) and has remained the most widely used cryoprotectant in freeze-etching because it seems to have the best overall cryoprotective properties of the many antifreeze agents so far tested. However, infiltration of cells with glycerol can produce clumping of intramembranous particles (McIntyre et al. 1974), disintegration of cell-wall fibril-synthesizing complexes (Brown and Montezinos 1976), rounding up of mitochondria and vesiculation of the endoplasmic reticulum (Staehelin, unpublished results), as well as many less obvious changes. For these reasons, samples are frequently chemically prefixed with glutaraldehyde (e.g., 1% for 15–60 min) before glycerol infiltration, despite the fact that glutaraldehyde can affect the cleaving mode of membrane components (Staehelin 1973). Prefixation with OsO_4 is less desirable because it reduces the number and size of the membrane fracture faces that are exposed (James and Branton 1971). Most cells are infiltrated with a 20–30% glycerol solution (glycerol–growth medium, buffer, or water; v/v); this may take anywhere from 15 min to 2 days depending on the organisms' permeability.

C. *Pretreatment for etching experiments*

With samples to be used for freeze-etching (deep-etching) experiments, in which true membrane or cell-wall surfaces are to be exposed by sublimation, the concentration of solute molecules in the sample medium must be reduced to less than 10 m*M*, and cryoprotectant agents cannot be used. Specimens are rinsed to remove solute molecules, which may interfere with the etching process. For rinsing membrane preparations, 2 m*M* $MgCl_2$ and 3–5 m*M* tricine buffer should be added to the rinse to avoid formation of artifacts that might occur with the use of distilled water alone (Staehelin 1976). Prefixation of cells and membranes with glutaraldehyde (1%) is often beneficial in reducing changes in the sample during the washing steps.

D. *Specimen supports*

The standard specimen support for the Balzers machine is a flat disc 3 mm in diameter, punched from approximately 0.1-mm thick copper foil. A grid pattern is scratched with a razor into the central two thirds of the disc to provide better adhesion between sample and disc. Before use, the supports are sonicated in acetone or cleaned in a solution of alcoholic NaOH (ca. 1% NaOH), rinsed thoroughly in distilled water, and dried. These flat specimen supports often have the disadvantage of becoming deformed during the scratching process. Subsequently, when such a support with a frozen sample is attached to the specimen stage by means of a screwcap, it may flatten, and the brittle sample break away. To overcome this problem we have modified the punch to produce hat-shaped specimen supports (diameter 3 mm; rim, ca. 0.5 mm wide; step height ca. 0.3 mm) that are held in an appropriately machined small depression in a brass block for scratching. Specimen supports punched from cardboard or matte-finish cellulose acetate can also be used, but they have poor thermal conductivity and are less desirable. If several types of specimens are to be stored together, the bottom sides of the discs can be color-coded with felt pens.

E. *Mounting of samples*

The sample, a droplet of suspension (about 1 μl) or a small piece of tissue (about 1 mm^3), is placed on the scratched area of the disc. The fluid should not flow over the clamping area of the support, otherwise the sample may break off during attachment to the specimen stage. Cell suspensions may be concentrated by centrifugation into a dense pellet before freezing. To prevent redilution of the sample following decantation of the supernatant, the walls of the centrifuge tube can be dried with a rolled piece of filter paper. Solutions or suspensions may be micropipetted onto the disc, or if the suspension is thick enough, applied with a dissecting needle. If tissues exhibit a tendency to break off the supports during the fracturing process, they can be "glued" to the supports with a paste of yeast cells suspended in the same solution as the sample. The samples should be mounted as quickly as possible and immediately frozen (see next section) to prevent dehydration artifacts from developing.

F. *Freezing and storage of specimens*

Freon-12 and Freon-22 (with boiling points of ca. −35°C and freezing points around −160°C) provide the best freezing rates and are the most widely used coolants. The freezing setup supplied by Balzers consists of a small Dewar flask and a freezing post: a 2-cm o.d.

chromed copper tube with a 2-cm-deep well in the top end. The freezing post is placed in the Dewar and surrounded with liquid nitrogen. Gaseous Freon is liquified by holding the tip of the Freon dispensing tube against the bottom of the cold well. Solidification of the Freon should occur within 5 min. Freezing of the sample is achieved by holding the support disc with tweezers and plunging it into a small pool of freezing liquid Freon generated by melting a portion of the frozen Freon with a metal rod [freezing rate achieved under these conditions −600° to −1000°C/sec, (Staehelin, unpublished results)]. The disc with the specimen is held in the pool for a few seconds and then quickly transferred to a second small, liquid-N_2-filled Dewar with a holding net. For long-term storage, containers can be placed in a storage can (usually a film can with a perforated lid) and the can hung in a nitrogen storage Dewar. The film-can storage method is suitable for most smaller freeze-etch laboratories. If larger numbers (>20) are to be stored, a specially designed storage container as described by Krah et al. (1973) should be considered.

G. *Special freezing techniques*

The spray freezing method developed by Bachmann and Schmitt (1971) allows vitrification of samples up to 30 μm in diameter. Plattner et al. (1973) have demonstrated that single cells, such as bacteria, algae, and some protozoa, as well as macromolecular solutions, can be successfully vitrified by this method without any antifreeze treatment (estimated freezing rate − 10,000°C/sec). The principle of the technique is as follows: The sample is sprayed into liquid propane at −190°C. The liquid propane vessel is then transferred to a cryostat −85°C) inside a glove box, and the propane is evaporated under reduced pressure. The remaining minute droplets (10–50 μm in diameter) of frozen sample are mixed with a drop of cold n-butylbenzene (−85°C), and little droplets of this mixture are transferred with a cold platinum wire onto precooled conventional specimen support discs. The discs are then dropped into liquid N_2 to solidify the butylbenzene. The specimens can be processed like regular samples. This freezing method appears to have great potential, but preliminary experiments in several laboratories have shown that the technical difficulties are substantial. The most serious problem appears to be the low yield of useful sample images per replica.

Since the critical temperature interval between f.p. and r.t. can also be reduced by depressing the f.p. with hydrostatic pressure, freezing under pressure has also been attempted (Riehle and Hoechli 1973). Although encouraging results have been obtained with this freezing method, the monetary investment in instruments still seems out of line with the derivable benefits.

III. Fracturing

The fracturing of a frozen sample can be produced by a cold "knife" or by tearing the sample apart as in double replica preparations. The liquid-nitrogen-cooled "knife", which can be any good quality single-edged injector razor blade, does not cut the specimen but chips it instead. The fracture planes thus initiated follow natural planes of weakness within the frozen specimen. In frozen cells, these planes are associated with structures stabilized at physiological temperatures by entropic interactions (hydrophobic bonds). When all the water surrounding such structures is frozen, the hydrophobic regions are weakened and provide less resistance to cleavage than do structures bonded by covalent and polar interactions. Since lipid bilayers and biological membranes are held together along their central plane by hydrophobic interactions, such structures are prime sources of natural fracture planes in frozen specimens. Consequently, biological membranes are split internally during freeze-fracturing (Branton 1966; 1969), and the resulting fracture-face images have to be supplemented with views of etched membranes if surface detail is also to be investigated (Staehelin 1976).

Before use, the razor blade has to be degreased with an organic solvent and then bolted onto the microtome arm after the clamp has been coated with a thin layer of Apiezon N (Apiezon Products) grease to ensure firm support and good thermal contact. After the specimen has been mounted on the precooled specimen stage and bell jar evacuation has been started the microtome arm can be cooled to − 196°C to speed up the pumping. Fracturing of the sample with the knife can start when the vacuum is approximately 2×10^{-6} torr. The "cutting" speed and the knife advance have to be experimentally determined for each type of specimen. As a rule, one should start with slow speed and small advance. High speed and/or a large advance can lead to shattering or breaking away of the whole sample. Specimens infiltrated with glycerol tend to chip more evenly than those frozen in water, which frequently crack and break apart. The surface of well fractured specimens exhibits a pearly shine when viewed through the magnifying glass in reflected light. When the knife becomes blunt, it scratches the sample more than it fractures. The resulting surface appears more mirrorlike than pearly. Scratching of the sample surface should be avoided insofar as possible, because the associated melting blurs all structural details in those areas. Double-replica preparations exhibit the same type of fracture faces as those produced by conventional techniques. A detailed discussion of double-replica methods can be found in the review of Koehler (1972).

If only no-etch freeze-fracture images of the sample are desired, the specimen stage is kept at $-110°C$, and the platinum–carbon shadowing is initiated immediately after the last pass of the knife. At $-110°C$ and lower, contamination of the specimen fracture faces by condensing water vapors may become a problem. Temperature-dependent contamination occurs when the partial pressure of a condensable gas in the vacuum system is greater than its equilibrium pressure above its liquid or solid phase. Practical hints of how to recognize and how to prevent contamination artifacts in freeze-etch replicas can be found in Staehelin and Bertaud (1971).

IV. Etching

Sublimation of ice can lead to the exposure of true membrane or cell-wall surfaces that are not revealed by the fracturing process. This generally occurs at the normal etching temperature of $-100°C$ at 2×10^{-6} torr. In the Balzers machine, optimal etching conditions are achieved by placing the liquid-nitrogen-cooled microtome arm, which acts as a cold trap (equilibrium vapor pressure of water at $-196°C$ is about 10^{-24} torr), immediately above the sample. Under these conditions, between 1 and 2 nm of water molecules are removed from a pure ice surface at $-100°C$/sec. Optimum etching times have to be experimentally determined for each type of specimen and may vary between 30 sec and 30 min.

V. Replica formation

Upon completion of etching, or immediately after fracturing, the exposed specimen surface is first shadowed at an angle with a mixture of platinum (Pt) and carbon (C) to produce the electron-scattering component of the replica, which is responsible for the contrast and three-dimensional relief seen in the electron microscope. This is followed by the evaporation of pure carbon from above to produce the low-contrast carbon film that holds the platinum–carbon "shadows" together. In normal replicas the Pt–C layer is about 2 nm thick, the carbon film 5–10 nm.

A. Resistance-heating evaporation

The conventional method for evaporating Pt–C is by means of resistance heating. Setting up the Pt–C shadowing unit and evaporating the Pt–C mixture is still as much an art as a science, due to the fact that a delicate balance in the evaporation rates of the two components has to be achieved. Several configurations of carbon-rod electrodes for Pt–C evaporation have been reported in the literature (Fig. 28–1).

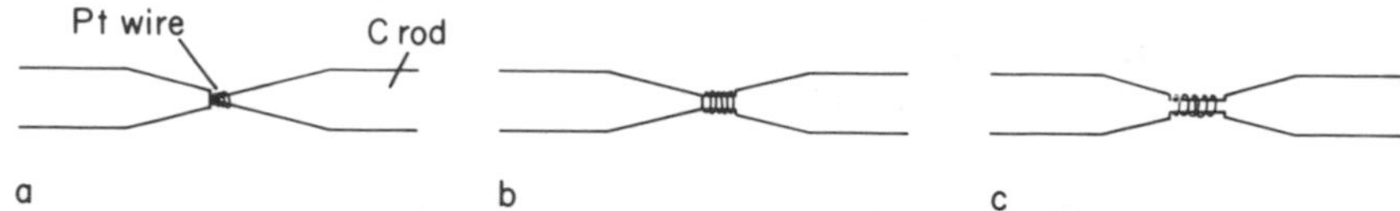

Fig. 28–1. Diagram showing different electrode configurations used for resistance heating evaporation of Pt–C.

The carbon rods (Specpure graphite rods, 6.5 mm in diameter and 30 cm long, grade II available from Johnson Matthey Chemicals Ltd., or Jarrell-Ash Division of Fisher Scientific) are sharpened by a pencil sharpener, blunted where needed with emery paper, and the nipple (0.9 mm o.d. × 2.0 mm long) formed by grinding a blunt-tipped rod with a Balzers BKS 610 grinding tool. A 0.1 mm × ca. 6.0 cm long (length can be varied to adjust for distance between evaporator and specimen) Pt wire (super pure 99.99%) is brought to glow in a gas flame to reduce its springiness, then wound around a mandrel (1.0 mm o.d.) and pushed onto the spring-loaded rods as shown in Fig. 28–1a. Alternatively, the Pt coil may be directly wound around the aligned carbon rods. Fig. 28–1c shows the configuration used by the author from 1966–1974 in conjunction with a specially designed shadowing unit that allowed easy and precise alignment of two nippled carbon rods and control of the advance distance of the rods (ca. 2.3 nm), which determined the amount of material that could be evaporated. The evaporation units supplied by Balzers lack the second feature, and the amount of evaporation has to be regulated by timing, with the observed rate of evaporation being taken into account (3–6 sec after the first sparks are seen). A piece of white filter paper (the "shadow paper") placed under the specimen can be watched during firing to evaluate the evaporation rate; it also provides a record of the shadow. The Pt–C electrode is fired by heating to white heat at low voltage (7.2–8.2 V) and high current (50–60 amp). Lower voltages generally provide more consistent, but also coarser and therefore lower quality, replicas. During the assembly of the Pt–C unit, one should: (1) Check the spring-pressure drive to ensure smooth advance of the rods; (2) align the abutting points very carefully; (3) flatten the contact points by grinding to ensure good contact; (4) tighten the Pt coil around carbon points by gently tugging at both ends with tweezers; (5) standardize the evaporation conditions (voltage, rate of heating, etc.) as far as possible.

Immediately after Pt–C evaporation the C–C electrode is fired. The setting up and evaporation of the C–C electrode is relatively simple. The fixed carbon rod is sharpened to a blunt tip, the moving rod to a point (Fig. 28–1a minus Pt wire). Heating of the electrode should be regulated with the Variac to produce evaporation of the carbon with minimal sparking. Evaporation times may vary from 10 to 60 sec, the shadow paper providing a monitor for the amount of carbon deposited.

B. Electron-gun shadowing (alternative evaporation technique)

Electron-gun shadowing involves heating the material to be evaporated by bombardment with electrons. Electron-gun shadowing units have the advantage over resistance heating units of producing more reproducible results and being able to evaporate the highest-melting-point metals. During two years of operation, the author has found the Balzers electron-gun shadowing unit, coupled with a quartz crystal thin-film monitor, to produce good quality replicas on a fairly regular basis. Conventional electron-gun shadowing devices emit intensive radiation, which warms up the fracture face and thereby obliterates fine structural details of the sample. The Balzers gun has been designed to reduce significantly or completely eliminate radiation from the three major sources: the heavy-metal source, the cathode filament of the gun, and backscattered and secondary electrons (Moor 1971).

The electron gun for the Pt–C evaporation can be set up in a manner appropriate to the unit. We have found it desirable to clean the unit and readjust the height of the anode after every second or third run. The evaporation of Pt–C involves setting the voltage to 1,800 V, heating the cathode filament to an emission current of 70 mA (voltage drops to about 1,450 V), and allowing evaporation to occur for 4–6 sec (= 150–180 Hz on monitor). To prevent radiation damage during the heating of the gun, we have added a shutter in front of the unit with a 4 sec time-delay switch.

VI. Replica cleaning

Upon completion of the replication process, the bell jar is vented and opened, and the specimens removed with forceps. As soon as each specimen has thawed, its replica can be floated off onto distilled water or onto other cleaning solutions, whose composition is determined by the nature of the material being investigated. Commercial bleach (undiluted) is the best cleaning solution for biological materials (30 min to 12 h). If it proves inadequate, the replica should be transferred with a Pt loop to distilled water for 1–2 h and then to either 70% H_2SO_4, fuming HNO_3, 40% CrO_3, or any other suitable cleaning solution for 1 to 12 h. Finally, the replica after several rinses in distilled water, is picked up from below on EM grids (uncoated or coated with Formvar, Chap. 30.II). If the replica sinks during the cleaning procedure, it should be left submerged until all cleaning steps are completed. It can then be gently teased to the fluid surface with a Pt loop and picked up as usual. If the submerged replica is curled up, it may be opened by replacing the water with alcohol or acetone and then transferring the replica onto a fresh water surface. The difference in surface tension between water and alcohol (or acetone) will uncurl or break the replica.

VII. Freeze-etch artifacts and interpretation of images

The most extensive compendium of freeze-etch artifacts can be found in the publication "Artifacts and Specimen Preparation Faults in Freeze-etch Technology" by S. Böhler, which may be purchased from Balzers (Balzers, Principality of Liechtenstein). Balzers Literature Service also publishes a yearly summary of references that includes all publications on freeze-etching that come to their attention. Beginners should try to evaluate their progress by comparing their micrographs with those contained in articles published in journals having very strict quality standards, such as the *Journal of Cell Biology*. Such articles usually present the most reliable and advanced interpretations of freeze-etch images.

VIII. Acknowledgments

The author thanks Dr. David Carter for his comments on the manuscript. This work was supported by the National Institutes of General Medical Sciences under Grant GM22912.

IX. References

Bachmann, L. and Schmitt, W. W. 1971. Improved cryofixation applicable to freeze-etching. *Proc. Nat. Acad. Sci. U.S.* 68, 2149–52.

Benedetti, E. L., and Favard, P. 1973. *Freeze-Etching Techniques and Applications.* Societé Française de Microscopie Electronique, Paris.

Branton, D. 1966. Fracture faces of frozen membranes. *Proc. Nat. Acad. Sci. U.S.* 55, 1048–56.

Branton, D. 1969. Membrane structure. *Ann. Rev. Plant Physiol.* 20, 209–38.

Brown, R. M., and Montezinos, D. 1976. Cellulose microfibrils: Visualization of biosynthetic and orienting complexes in association with the plasma membrane. *Proc. Nat. Acad. Sci. U.S.* 73, 143–7.

Bullivant, S. 1973. Freeze-etching and freeze-fracturing. In Koehler, J. (ed.), *Advanced Techniques in Biological Electron Microscopy,* pp. 67–112. Springer-Verlag, New York.

James, R., and Branton, D. 1971. The correlation between the saturation of membrane fatty acids and the presence of membrane fracture faces after fixation. *Biochim. Biophys. Acta* 288, 504–12.

Koehler, J. 1972. The freeze-etching technique. In Hyatt, M. A. (ed.), *Principles and Techniques of Electron Microscopy,* vol. 2., pp. 53–98. Van Nostrand Reinhold, New York.

Krah, S., Staehelin, L. A., and Nettesheim, G. 1973. A new type of storage container for freeze-etch specimens. *J. Microscopy* 99, 349–52.

McIntyre, J. A., Gilula, N. B., and Karnovsky, M. J. 1974. Cryoprotectant-induced redistribution of intramembranous particles in mouse lymphocytes. *J. Cell Biol.* 60, 192–203.

Moor, H. 1964. Die Gefrier-fixation lebender Zellen und ihre Anwendung in der Elektronenmikroskopie. *Z. Zellforsch.* 62, 546–80.

Moor, H. 1969. Freeze-etching. *Int. Rev. Cytol.* 25, 391–412.

Moor, H. 1971. Recent progress in the freeze-etching technique. *Phil. Trans. Roy. Soc. Lond. B* 261, 121–31.

Moor, H. and Mühlethaler, K. 1963. Fine structure of frozen-etched yeast cells, *J. Cell Biol.* 17, 609–28.

Plattner, H., Schmitt-Fumian, W. W., and Bachmann, L. 1973. Cryofixation of single cells by spray-freezing. In Benedetti, E. L. and Favard, P. (eds.), *Freeze-Etching Techniques and Applications,* pp. 81–100. Societé Française de Microscopie Electronique, Paris.

Riehle, U., and Hoechli, M. 1973. The theory and technique of high pressure freezing. In Benedetti, E. L., and Favard, P. (eds.), *Freeze-Etching Techniques and Applications,* pp. 32–61. Societé Française de Microscopie Electronique, Paris.

Southworth, D., Fisher, K., and Branton, D. 1975. Principles of freeze-fracturing and etching. In Glick, D., and Rosenbaum, R. (eds.), *Techniques of Biochemical and Biophysical Morphology,* vol. 2, pp. 247–82. Wiley, New York.

Staehelin, L. A. 1973. Analysis and critical evaluation of the information contained in freeze-etch micrographs. In Benedetti, E. L., and Favard, P. (eds.), *Freeze-Etching Techniques and Applicants,* pp. 113–34. Societé Française de Microscopie Electronique, Paris.

Staehelin, L. A. 1976. Reversible particle movements associated with unstacking and restacking of chloroplast membranes. *J. Cell Biol.* 71, 136–58.

Staehelin, L. A., and Bertaud, W. S. 1971. Temperature and contamination dependent freeze-etch images of frozen water and glycerol solutions. *J. Ultrastruct. Res.* 37, 146–68.

Staehelin, L. A., and Pickett-Heaps, J. D. 1975. The ultrastructure of *Scenedesmus* (Chlorophyceae). I. Species with the reticulate or "warty" type of ornamental layer. *J. Phycol.* 11, 163–85.

29: Preparation of algae for scanning electron microscopy

JEREMY D. PICKETT-HEAPS

Department of Molecular, Cellular and Developmental Biology, University of Colorado, Boulder, Colorado 80309

CONTENTS

I. Introduction

Recent developments in all aspects of scanning electron microscopy (SEM) have made this technique reliable and useful for the study of a wide variety of cells. It has been used primarily to study surface characteristics of cells and has thus become an invaluable tool in studying the taxonomy, morphology, and even the ecological relationships of many algae.

There have been many books published recently on techniques for all aspects of SEM (e.g., Hayat 1974; Wells 1974; Goldstein and Yakowitz 1975); I will not attempt to review either the techniques or the literature; I will concentrate on the preservation of surface structure of algae. A basic protocol is suggested for routine use; this is followed by a discussion of difficulties likely to be encountered. These comments serve only as guidelines; with each alga, problems are quite likely to arise requiring modification of the methods. In preparing samples, it is always necessary to work with filtered solutions; this decreases the chances of secondary surface depositions.

II. Methods

A. Example: selected procedure for desmids

Small cells present special problems in SEM and have therefore been selected for illustration. Pickett-Heaps (1973) and Marchant (1973) have developed methods that have been used successfully with many small algae (e.g., *Hydrodictyon, Pediastrum, Cosmarium, Staurastrum*).

1. The cells are collected on Solvinert Millipore (1.5 μm) filters by gentle suction via a regulated vacuum line (this is used for all subsequent rinsing and dehydration steps). If Solvinert filters are not available, other acetone-resistant types are satisfactory substitutes. Desmids, grown in culture, are cleaned with a preparation of Glusulase (Endo Laboratories, Inc.) (1 part enzyme to 50 parts water) at room temperature for 30–90 min. They are then washed several times with filtered water before fixation.

2. The cells are fixed at room temperature for 30–60 min in 1%

OsO_4 made up in the culture medium. Fixation is followed by several washings with distilled water. [Glutaraldehyde prefixation seldom has merit, and usually, I advise against its use in SEM, (see II.B)].

3. The cells, on the Solvinert filter pads, are dehydrated slowly at 0°C. Each solution of acetone (10%, 25%, 50%, 75%, 100%) is successively added dropwise over 5–10 min, until the solution over the sample approaches that being added. Some of the solution is then removed from the sample, and the next highest acetone solution is added in a similar fashion. Two to four changes in absolute acetone complete dehydration. Care should be taken at all steps (until CO_2 replacement) not to expose the specimens to air: evaporation of volatile solvents could distort the specimens.

4. The specimens are then ready for critical-point drying. The filters with the algae are cut into small, round circles (using, for instance, corkborers) in the last dehydration solution. The mounted specimens are held in a simple perforated chamber constructed from a polypropylene vial or a BEEM capsule (Marchant 1973). Critical-point drying (CPD) can be carried out on any of the commerically available machines (such as Sorvall, Denton, and others), using standard procedures. The acetone in the specimens is now replaced by liquid CO_2 under pressure; then the temperature is raised above the critical point of liquid CO_2 so that it then passes directly into the gaseous phase. Because the CO_2 is under pressure, great care is necessary when first admitting the CO_2 to the specimen chamber to prevent physical removal of the cells from their substrates.

5. After CPD, the filter circles, with the specimens on the upper surface, are attached to an aluminum stub with double-sided adhesive tape and omnidirectionally shadowed in a vacuum evaporator (according to instructions for the apparatus) with carbon (5 nm) and gold (15 nm). By ringing the filter with colloidal silver paint, good electrical contact can be assured between the stub and surface of the filter pad. The coated specimens on the filter are now ready for examination in a scanning electron microscope. They can be stored in an airtight container over a desiccant. Many workers prefer to store them under vacuum, but we do not find this important.

B. General comments

Because scanning microscopy has been almost entirely concerned with viewing the external morphology of cells and tissues, many problems encountered in conventional microscopy (in particular, the disruption or destruction of internal cell components) need not be considered. However, different types of problems may arise. For example, partial disruption of the plasma membrane of naked cells produces unsightly and serious defects in the SEM image; such dis-

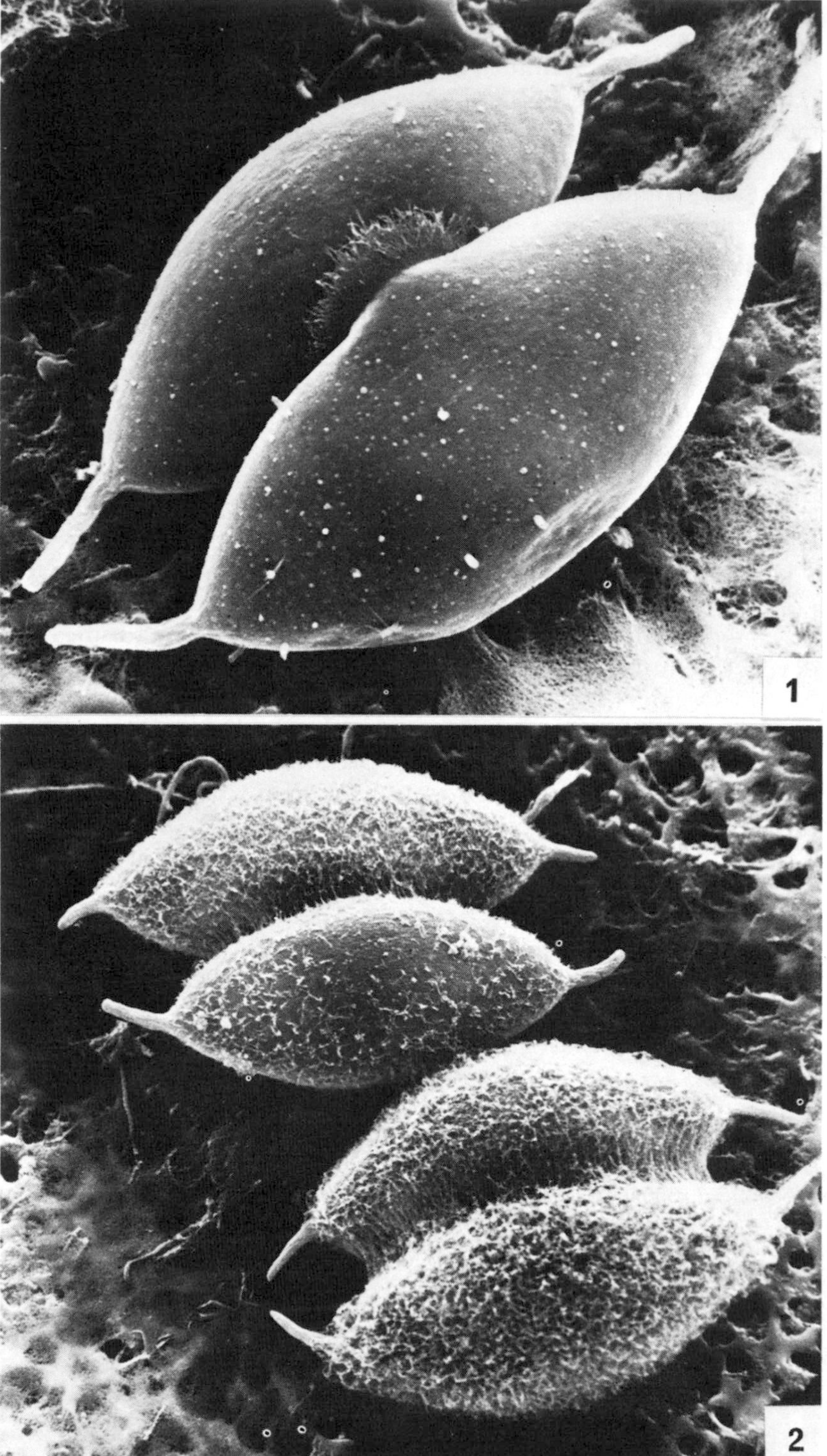
1
2

ruptions may not be particularly noticeable in thin-sectioned material. The presence of particulate and colloidal matter in culture solutions (washing and dehydrating agents, for example) can be serious in SEM, because such material may stick to the outside of the cell and become highly visible under the microscope; most particulate contamination is seldom of concern with thin-sectioned material. Thus, it pays to wash cells prepared for SEM in filtered water and also to filter the dehydrating agents if necessary. It is well to remember the difference in objectives between scanning and transmission microscopy and to modify processing techniques accordingly. For example, many microscopists fix cells for SEM as they do for transmission microscopy, which is not necessarily a good idea. In my opinion, other techniques that are sometimes widely used may also be suspect. For example, mechanical shakers are a fixture in some electron microscopic labs; not only are they entirely unnecessary in achieving dehydration and infiltration of the majority of specimens for transmission microscopy, but their use unnecessarily disrupts brittle and fragile specimens that have been fixed. Thoughtful consideration of the objectives of each step in the procedure and analysis of the problems encountered for different types of specimens usually yield a protocol that is successful even with the most difficult and delicate tissues.

III. Potential problems and possible solutions

A. Extracellular layers and contamination

All cells prepared for SEM should be washed in filtered water before processing. Most algal cells characteristically secrete various quantities of mucilaginous material. Such secretions can undoubtedly be considered an integral component of the cell, but they pose particular problems for the scanning microscopist because they may condense during dehydration into a fibrous or amorphous layer, partially or totally obscuring the surface of the cell or its true cell wall (Figs. 29–1 and 29–2). Various measures can be attempted to rid the cell of this material, either before or after fixation, and considerable experimentation may be necessary to achieve satisfactory results. Because fixa-

Figs. 29–1 and 29–2. The desmid *Arthrodesmus* sp. prepared for SEM. These cells, for cleaning, had been prepared by treatment in warm, dilute sodium hydroxide for about 10 minutes after fixation. In Fig. 29–1, the investment of mucilage had been completely removed, revealing the surface of the cell wall; such cells were very uncommon in this particular sample. Fig. 29–2 shows the more usual image obtained, with varying amounts of the mucilaginous sheath still left about the cells; this sheath, upon dehydration in acetone, and drying, is typically seen as an irregular fibrous coat that obscures the surface of the wall.

tion may affect such layers (probably rendering them less easy to deal with) they should be removed prior to fixation.

Some mucilaginous layers are soluble in acetone–water or alcohol–water mixtures and may therefore be extracted in the course of dehydration. Warm water (30°C), with or without chelating agents (e.g., 0.02 *M* EDTA) will solubilize pectic fractions of higher plant cells and are sometimes successful in removing surface layers from algae (unpublished data). Although hot (50°C) or cold (0°C) solutions of sodium hydroxide (0.1 *N*) are more strongly extractive, their use presents an increasing likelihood of damaging the cell wall proper. Polysaccharides are often susceptible to oxidation and solubilization in alkaline conditions, and only trial and error can indicate what the useful limits of such treatments are with given specimens. This is also true when one is using certain wall-degrading enzyme preparations (e.g., snail gut juice, pectinases) to remove mucilage layers. Acid solutions may also prove useful, but we tend to avoid them, except specifically for organisms with siliceous walls or scales (e.g., Chrysophytes, diatoms). Silica resists acid, but it is slowly dissolved in alkali; thus, fine detail of such wall components may be affected by sodium hydroxide treatment. A wide variety of methods have been devised to clean diatoms and are described in the literature (e.g., Cupp 1943); robust cells can be simply treated with strong oxidizing acids (e.g., concentrated nitric acid) for a few minutes, after which the acid is diluted (carefully) and washed away. Hydrogen peroxide (20–30%) also works well. Smaller diatoms may be more physically fragile and will probably need gentler treatment if the walls are not to be broken up into segments.

Some cells strenuously resist attempts to remove these interfering layers. We have never suceeded, for example, in getting good micrographs of filamentous desmids. Considerable variability may also be encountered between cells in a population; for example, a few cells of our strain of *Arthrodesmus* could be cleaned in dilute alkali, but a majority remained completely enshrouded in their mucilage (Figs. 29–1 and 2).

B. *Collection and mounting of cells*

Collection and mounting of cells for SEM may constitute one of the most difficult problems for the microscopist, particularly when he is dealing with delicate cells that are not numerous in culture or with collections from nature. Some cells (e.g., diatoms) can be dried down directly onto a convenient mount (e.g., small, circular coverslips), a procedure that destroys most other cell types. One of the most successful methods involves use of Millipore filters of an appropriate pore size (see II.A).

Adhesion of cells to a filter disc is highly variable with the species whether cells are cleaned before or after filtration. For improving adhesion of the cells to the filter disc, polylysine (Mazia et al. 1974) has been advantageous. Filter discs or coverslips can simple be moistened with an aqueous polylysine solution (1 mg/ml) and dried before cells are applied.

If abundant cells are available, we prefer to make up several lightly covered pads and process them together. We expect removal of many of the cells and can tolerate considerable loss of specimens – for example, even a 90% loss is not particularly critical if one commences with 10^5 cells per stub. Larger filamentous algae (e.g., *Bulbochaete*) can be fixed and collected in this fashion, and later, clumps of material can be pressed onto the pad with fine forceps as necessary. This procedure, though breaking up many cells, usually ensures that the remainder will not be removed during subsequent processing.

C. *Fixation*

Using glutaraldehyde followed by osmium postfixation for SEM specimens is not usually necessary and may be deleterious. Glutaraldehyde usually offers only improved fixation for observing intracellular detail. Membranes are probably not well stabilized by glutaraldehyde and may remain osmotically active. Thus, the cells may swell (particularly when naked) and are likely to suffer damage to their surface membranes. For most algae we use osmium fixation alone for SEM, because it offers improved images of delicate, naked cells. Many thick-walled cells, however, appear equally good following glutaraldehyde fixation, but this extra step seldom offers any additional advantages.

D. *Dehydration*

Acetone is a convenient dehydrating agent. For delicate specimens, it is safest to increase the concentration of the dehydrating agent as smoothly as possible, preferably at 0°C. Standard procedures involving the transfer of specimens from water to increasing concentrations of acetone in water (e.g. 25%, 50%, 75%) should be avoided. Such precautions are usually unnecessary with larger, thick-walled cells.

Air-drying of washed specimens is usually unsuccessful with algae, because few cells can withstand drastic disruption if they are dried down directly from water onto a mount (e.g., a coverslip); diatoms, with their tough siliceous walls, are one of the exceptions. The CPD procedure is by far the safest and best technique to use when removing the solvent from specimens fixed for SEM. This method was introduced by Anderson (1951) and its principles are set out in any book of electron microscopical technique (see also Horridge and Tamm 1969; Boyde and Wood 1969).

E. Coating the specimens

Dried specimens are usually coated in a vacuum evaporator. Large unicells (e.g., desmids) and bushy filamentous algae may have little area of contact with the specimen stub, and thus the cells may tend to "charge up" under the electron beam. To remedy this problem, a heavy coating with metal and/or carbon is necessary, and occasionally, we have found it necessary to coat specimens up to three times before suitable conductance was achieved. Such coatings may obscure surface detail.

F. Stereo scanning microscopy

SEM also offers the possibility of obtaining stereo images of specimens. This is achieved by photographing the cell at two slightly different angles but at precisely the same magnification. The resultant micrographs are then carefully mounted side by side and observed in a stereo viewer. The eyes can easily be trained to see such images without the viewer: a simple technique is set out by Pickett-Heaps (1975, Appendix I). Some modern scanning microscopes now offer the possibility of direct viewing in three dimensions.

IV. Conclusion

The techniques for SEM and the scanning electron microscope itself are continually being improved. Considerable advances have been made in achieving high resolution (e.g., 25 Å or better) with commercially available machines. It now appears possible to examine unfixed and uncoated plant surfaces with the scanning microscope; for example, Ledbetter's (1976) images of such objects are as good as, if not better than, the best of those obtained after fixation and coating. SEM remains, however, as much an art as a science, and with difficult subjects, the phycologist must be prepared to invent new ways of handling cells or organisms if standard procedures are not adequate.

V. References

Anderson, T. F. 1951. Techniques for the preservation of three-dimensional structures in preparing specimens for the electron microscope. *Trans. N.Y. Acad. Sci., Ser. II.* 13, 130–4.

Boyde, A., and Wood, C. 1969. Preparation of animal tissues for surface-scanning electron microscopy. *J. Microscopy* 90, 221–49.

Cupp, E. E. 1943. *Marine Plankton Diatoms of the West Coast of North America.* University of California Press, Berkeley.

Goldstein, J., and Yakowitz, H. (eds.). 1975. *Practical Scanning Electron Microscopy: Electron and Ion Microprobe Analysis.* Plenum, New York. 582 pp.

Hayat, M. (ed.). 1974. *Principles and Techniques of Scanning Electron Microscopy; Biological Application.* Van Nostrand Reinhold, New York. 412 pp.

Horridge, G. A., and Tamm, S. L. 1969. Critical point drying for scanning electron microscopic study of ciliary motion. *Science* 163, 817–18.

Ledbetter, M. C. 1976. Practical problems in observation of unfixed, uncoated plant surfaces by SEM. In Johari, O. (ed.), *Scanning Electron Microscopy/1976/II.* Chicago Press Corporation, Chicago. 453 pp.

Marchant, H. J. 1973. Processing small, delicate biological specimens for scanning electron microscopy. *J. Microscopy* 97, 369–71.

Mazia, D., Sale, W. S., and Schatten, G. 1974. Polylysine as an adhesive for electron microscopy. (Abstract). *J. Cell Biol.* 63, 212

Pickett-Heaps, J. D. 1973. Stereo scanning electron microscopy of desmids. *J. Microscopy* 99, 109–16.

Pickett-Heaps, J. D. 1974. Scanning electron microscopy of some cultured desmids. *Trans. Am. Microscop. Soc.* 93, 1–23.

Pickett-Heaps, J. D. 1975. Green Algae. Sinauer Assoc., Sunderland, Mass. 606 pp.

Wells, O. C. 1974. Scanning Electron Microscopy. McGraw-Hill, New York. 421 pp.

30: Replica production and negative staining

ELISABETH GANTT

Radiation Biology Laboratory, Smithsonian Institution,
Rockville, Maryland 20852

CONTENTS

I. Introduction

Replica production and negative staining are two rather simple and inexpensive techniques for elucidating the surface fine structure of algal cells. They have been under-utilized, partly because of the popularity of scanning electron microscopy (Chap. 29) and the freeze-fracture technique (Chap. 28). The techniques presented here can complement the freeze-fracturing and scanning and, in many cases, can serve as alternatives.

Production of replicas involves the in vacuo deposition of electron-dense metal in such a way as to improve the contrast of the material being studied and to produce a three-dimensional impression. It involves the same basic steps as shadowcasting (Chap. 31), except that the organic (or inorganic) material is dissolved away as in freeze-fracture replicas. Replicas can be produced from small cells, from portions of large cells, and even from crystalline material, providing the structures being replicated can be somehow dissolved in water, chromic acid, or sodium hypochlorite. This technique has the advantage over scanning electron microscopy in yielding considerably greater details. For fine-structural studies of algal surfaces (Gantt 1971; Faust 1974), material can be prepared directly in a vacuum evaporator, which is already available in most electron microscopy laboratories, and does not require a costly freeze-fracture apparatus.

In negative staining, solutions of heavy metals salts (electron dense) are applied; these aid in outlining small structures. It is particularly useful for viewing fresh material, often without fixation (flagellar and attached hairs, surface fibrils, scales, trichocysts, and of course, internal structures such as membrane particles and ribosomes). It can provide greater resolution than any other "staining" or metal-shadowing technique.

II. Replica production

Replicas can be produced in several ways, but only the single-stage technique will be described here. For detailed treatments of additional replica techniques and variations, there are several excellent chapters in standard electron microscopy technique books (Goodhew 1972; Henderson and Griffiths 1972).

A. Preparation of support grids

A standard method for preparing support film on grids as detailed by Hayat (1970), has been routinely used in our laboratory.

1. Microscope slides, cleaned by rinsing in water and wiping with lint-free tissue, are dipped into a 0.5% (w/v) solution of Formvar in ethylene dichloride (in a well ventilated area).
2. After drying (10–20 min), the edges of the slide are scraped with a razor blade, and the Formvar film from the top of the slide is floated off on dust-free distilled water at an angle of ca. 20°. Breathing on the slide aids in the release of the film.
3. Electron microscope grids (200–400 mesh, copper) are then gently placed on the floating film (silver interference color), with the shiny side toward the Formvar.
4. The grids, with the film, are then picked up with the clean side of a piece of Parafilm or with another coated slide, and dried in a petri dish.
5. Carbon is deposited on the Formvar-coated side of the grids by evaporation in a vacuum evaporator (operational procedure given with most instruments). The thickness is determined experimentally but is considered sufficient when a shiny brownish metallic look appears on the margin of the slide. Grids thus prepared can be stored for months in a covered petri dish.
6. Alternatively, films can be cast as a carbon film on glass (Towe 1965) or on mica (Pease 1960). Such films are more desirable for high-resolution work of negatively stained material, because they are thinner due to the absence of Formvar.

B. Specimen preparation and mounting

Cells with rigid surfaces need not be fixed and can be mounted directly; however, fixation is desirable for cells with thin walls or periplasts (Gantt 1971; Faust 1974).

1. Fixation

a. Cell fixation can be accomplished by directly immersing the cells for 1–2 h in phosphate-buffered (pH 7.0) 2–4% glutaraldehyde. Glutaraldehyde can also be added dropwise to the cells in the culture medium over a 15-min period, for a final concentration of 1% glutaraldehyde and total fixation time of 1 h. This is useful for organisms such as some Cryptophyceae, which tend to destructively release their ejectosomes if they are pelleted before fixation.

b. Fixed cells are rinsed three times in distilled water and are ready for mounting.

2. Mounting on mica

a. Pieces (ca. 3 cm^2) of high-quality mica (Ladd Research Industries) are freshly cleaved by splitting with a razor blade from one corner. The freshly cleaved surfaces are kept clean by storage (underlaid by filter paper) in covered petri plates.

b. Small drops of the fixed cell suspension are placed on the freshly cleaved mica surface; care must be taken not to wet the margins – this would cause run-off on the filter paper.

c. The water is allowed to evaporate at room temperature or in an oven (ca. 50°C), and the material is ready for shadowing.

C. Shadowing and cleaning of replicas

1. Shadowing

a. The specimens, on the mica, are placed in the vacuum evaporator on a slide or on filter paper. The angle of shadowing is chosen emperically and is dependent on the size of the structure being studied. [Good general guidelines are found in Pease (1960).]

b. For resolution of finer details, platinum–carbon pellets (Ladd Research Industries) or platinum wire (see Chap. 28.V.A) are used. A platinum–carbon pellet (2.0–3.0 mm long) is mounted between two carbon rods having flattened surfaces. The pellet is first "degassed" to prevent excess sputtering by a brief preliminary heating. After a few minutes the vacuum reaches ca. 1.5×10^{-5} torr and the platinum–carbon pellet reaches white heat, when it will evaporate.

c. This is immediately followed by depositing carbon from an alternate set of electrodes in the evaporator (such as manufactured by Denton Vacuum). One carbon rod should be presharpened to a fine point, and the point should abut the flattened surface of another carbon rod. The rods should be pushed against one another either by tension loading or by gravity feeding (when the rod assembly is vertically mounted with the pointed rod on top). Evaporation is carried out as with the platinum–carbon pellets, except that it is done vertically (directly overhead) to avoid additional shadows and to give the replica even overall strength.

2. Cleaning of replicas

a. The mica sheets are cut with scissors into smaller strips (ca. 5–10 mm wide), and the replicas are floated off onto the cleaning solution. The cleaning solution varies with the material to be dissolved. For algae such as *Chroomonas, Porphyridium,* or filamentous blue-greens, a 30–60 min treatment in commercially available acid dichromate solution (1 : 1 dilution with water) or full-strength bleach (ca. 5%

sodium hypochlorite) have been found satisfactory. Long exposure (overnight) to bleach may cause weakening of the carbon film.

b. The replicas are rinsed by transfer to distilled water with a clean platinum loop or with a fine wire mesh (previously tested against corrosion in the appropriate solution).

c. After several additional rinses in distilled water, the replicas are ready to be picked up on uncoated or Formvar–carbon-coated grids (prepared as in II.A). The grids are aligned (shiny side up) on a wire mesh, which is resting on the bottom of the rinsing dish. With forceps, the mesh is gradually brought up underneath the replicas until contact is made between the replicas and the grids, which are then lifted out and placed on filter paper. Upon drying, they are ready for examination or for storage in a covered petri dish. (For additional discussion of replica cleaning, see Chap. 28.II.)

D. Possible improvements

One of the main problems with this technique is cell shrinkage and consequent distortion of surface detail. Fixation can sometimes relieve this problem, but cannot prevent it. Preparation of material by critical-point drying would be desirable for cells that do not have a rigid surface. If very large cells are to be used, they should be cut into smaller sections before mounting.

For some cells, it may be necessary to prepare double replicas by making a plastic cast (double-stage replicas) as detailed by Henderson and Griffiths (1972), Bradley (1965), and Goodhew (1972). In these, however, the resolution is usually less than that obtained with the above procedure.

III. Negative staining

This technique can give a great deal of additional information on the fine structure of flagella and on surface fibrils characteristic of small algae.

A. Stains

The two most generally useful negative stains are phosphotungstic acid and uranyl acetate. Others, such as ammonium molybdate (0.5–3.0% aqueous solution), and tungstate salts 0.5–2.0% aqueous solution (sodium tungstate, lithium tungstate, sodium silicotungstate) may also be used but are generally not superior in the author's experience.

1. Phosphotungstic acid (0.5–2.0% w/v) is dissolved in distilled water, the pH is adjusted to neutrality with 1 *N* potassium or sodium

hydroxide, and the solution is stored. Before use, the stain is filtered (Whatman No. 1 paper) to remove any precipitates or microorganisms that may have accumulated during storage.

2. Uranyl acetate (0.5–1.0% w/v), when dissolved in distilled water, is acidic (ca. pH 4.0). This may restrict its use, because proteins tend to precipitate at such a low pH. If this is a problem, a uranyl oxalate solution (pH 6.5–6.8) can be prepared by the technique of Mellema et al. (1967).

B. *Staining procedure*

1. Grids to be used are prepared as in II.A. To enhance spreading of the material, the grids are exposed (2–3 min) to glow-discharge ("ion cleaning") in the vacuum evaporator. If the evaporator is not equipped with a glow-discharge transformer, a high-frequency induction coil (Tesla type) can be attached to an external lead on the evaporator (Towe 1965).

2. For staining, each grid is held at the edge by a pair of fine watchmaker's forceps. (These are held closed by pushing the back of the forceps through a paper clip. This also serves to balance the forceps on the table or in a large petri dish.)

3. A small drop of unfixed or aldehyde-fixed material is applied to the grid in its suspending medium. After 1–5 min, the excess liquid is drawn off by touching the flat side of a piece of filter paper (Whatman No. 1 or No. 4) against the edge of the grid. Rinsing with medium or dilute buffer (0.01 *M* phosphate or 0.05 *M* ammonium acetate) is done in the same manner; in rinsing, one must be careful not to allow the surface to dry completely.

4. A first drop of negative stain is applied and withdrawn, and then a second drop is applied and left for 1–15 min. After removal of the stain with filter paper, as above, and drying at room temperature (ca. 10 min) the grids can be examined.

C. *Alternate procedures*

The stain can also be applied by spraying it on the grid after the specimen has been mounted to it. This procedure may, however, cause additional drying artifacts. Sometimes it may also be advantageous to float the mounted material on the staining solution. (Additional discussion of alternative techniques can be found in Horne (1965) and Haschemeyer and Myers (1972).

IV. References

Bradley, D. E. 1965. Replica and shadowing techniques. In Kay, D. H. (ed.), *Techniques for Electron Microscopy,* 2nd ed., pp. 96–165. Davis, Philadelphia.

Faust, M. A. 1974. Structure of the periplast of *Cryptomonas ovata* var. palustris. *J. Phycol.* 10, 121–4.

Gantt, E. 1971. Micromorphology of the periplast of *Chroomonas* sp. (Cryptophyceae). *J. Phycol.* 7, 177–84.

Goodhew, P. J. 1972. Replica techniques. In Glauert, A. M. (ed.), *Practical Methods in Electron Microscopy,* vol. 1, pp. 137–58. North Holland. Amsterdam.

Haschemeyer, R., and Myers, R. 1972. Negative staining. In Hayat, M. A. (ed.), *Principles and Techniques of Electron Microscopy: Biological Applications,* vol. 2, pp. 99–147. Van Nostrand Reinhold, New York.

Hayat, M. A. 1970. *Principles and Techniques of Electron Microscopy: Biological Applications,* vol. 1. Van Nostrand Reinhold, New York. 412 pp.

Henderson, W. J., and Griffiths, K. 1972. Shadow casting and replication. In Hayat, M. A., *Principles and Techniques of Electron Microscopy: Biological Microscopy,* vol. 2, pp. 149–93. Van Nostrand Reinhold, New York.

Horne, R. W. 1965. Negative staining methods. In Kay, D. H., (ed.) *Techniques for Electron Microscopy,* 2nd ed., pp. 328–55. Davis, Philadelphia.

Mellema, J. E., Van Bruggen, E. F. J., and Grubber, M. 1967. Uranyl oxalate as a negative stain for electron microscopy of proteins. *Biochim. Biophys. Acta* 140, 180–2.

Pease, D. C. 1960. *Histological Techniques for Electron Microscopy.* Academic Press, New York, 274 pp.

Towe, K. M. 1965. Carbon films for electron microscopy: a reliable method for stripping from glass surfaces. *Rev. Sci. Instr.* 36, 1247.

31: Preparation of shadow-cast whole mounts

ØJVIND MOESTRUP AND HELGE A. THOMSEN

Institut for Sporeplanter, Ø. Farimagsgade 2D,
1353 Copenhagen K., Denmark

CONTENTS

I. Objectives

The objectives of the shadow-casting technique are to enhance the contrast of material to be studied in the transmission electron microscope and to produce a three-dimensional impression. The technique was one of the first to be used successfully in electron microscopy, and since 1950 it has played a major role in ellucidating the 9 + 2 structure of cilia and flagella in algae and other plants. (For a review of these early studies, see Manton 1956a.)

Shadow-cast whole mounts have been used mainly to study cellular structures that are associated with the external surfaces and too small to be observed properly, if at all, by light microscopy. A large number of publications have appeared on flagellar appendages (hairs, spines, etc.) and flagellar lengths and on the many different and taxonomically important scale types (e.g., coccoliths) that cover the cell in several groups of flagellates. More recently the technique has become important in studies of choanoflagellates (Norris 1965; Leadbeater 1972). Although internal cell structures are not so readily studied by this technique, some internal structures, such as the eyespot of *Fucus* spermatozoids [Fig. 12 in Manton (1956b)] and the tiny mitochondria of *Dichotomosiphon* spermatozoids [Fig. 13 in Moestrup and Hoffman (1975)] have been observed in whole mounts.

A very important feature of the shadow-casting technique is the short preparation time; material may be ready for electron microscopy in about 1 hour. It remains very stable in the electron beam, and a carefully prepared grid may therefore by valuable for teaching purposes. If stored away from moisture, material prepared by this method will remain unchanged for many years.

II. Equipment and test organisms

The major items of equipment are a vacuum evaporator and a transmission microscope. Organisms that often give excellent results with this technique and that are useful for practice and teaching are small scale-covered flagellates, or choanoflagellates if these can be obtained (members of the Acanthoecaceae with their beautiful loricae are particularly useful). From the Cambridge Culture Collection can be suggested: *Paraphysomonas butcheri* Pennick and Clarke (LB 935/1); *Pyramimonas grossii* Park (LB 67/10 or LB 67/11); and *Mallomonas papillosa* Harris and Bradley (LB 929/2). Larger cells are much more difficult to work with, especially if they produce mucilage in large amounts. *Gonyostomum semen* is particularly ill suited.

III. Methods

A. Specimen preparation

1. Transfer of material to the grids. To facilitate handling, the grids should be attached to the edge of the tape (see details in Fig. 31–1A, B, C), support film upwards. The pair of slides, with the attached grids, are then transferred to a petri dish, and a small amount of liquid containing the material is pipetted onto each grid. To avoid any unwanted effect on the material (for example temperature rise when working with fragile flagellates) the material should be treated as rapidly as possible.

2. Killing and drying of the material. If the osmium solution (2% aqueous) in a bottle is inverted several times, a small amount of fixative will remain on the stopper. The inside of the petri dish cover is then touched in 2–3 places with the stopper, and this usually leaves the required amount of fixative on the cover (Fig. 31–1D). The material is fixed in the osmium vapour for 30 sec (see also III.B). After removal, the cover is placed in a fume hood to allow osmium evaporation). Most of the liquid on the grids is then slowly and carefully removed with absorbant dental points, preferably leaving just enough liquid to cover the film. During this procedure material is lost, and if the drop contains only little material (e.g., a few cells only) more liquid may have to be left on the grid (compare with III.B). Normal filter paper cannot replace the absorbant dental points: it usually absorbs all liquid instantaneously. To avoid dust particles the grids should never be left uncovered for any length of time. The grids (on the tape) can then be placed into an oven (ca. 45°C) to dry (ca. 10 min is usually sufficient for the liquid to evaporate).

3. Washing. The grids are removed from the tape with the fine-tipped forceps, care being taken not to bend the grids. They are then washed, although this step may be omitted for fresh-water material. As standard procedure, we always wash the grids by leaving them for 10 min, material upwards, on the wire mesh in distilled water in a petri dish. Before being taken from the water, each grid is further

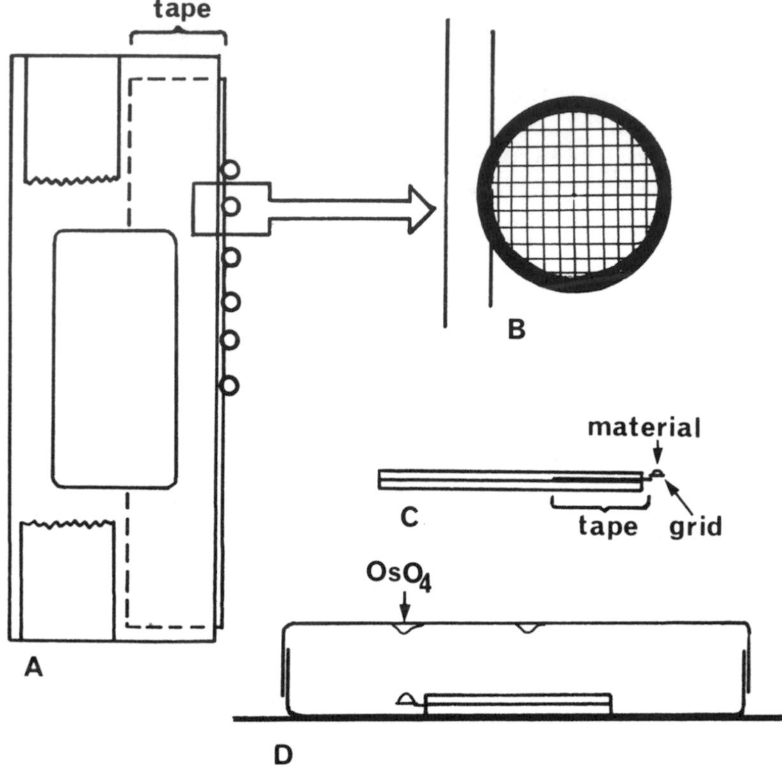

Fig. 31–1. Grids are shown attached to tape on the right side of the slide (A). They are seen in greater detail from the top (B) and side (C). The mounted grids are enclosed within a petri dish as for fixation with osmium tetroxide vapors (D).

washed in water (10–20 times) by moving it with forceps to and fro in the plane of the support film. The grids are then returned to the oven to dry again.

4. Shadow-casting. A gold–palladium wire (60/40) (ca. 5 mm long) is bent into a loop around a V-shaped stem wire (0.5 mm diameter). The grids are usually shadowed at an angle of about 20° by adjusting the distance between metals and grids. If the angle is not crucial, the grids are left, for example, in an oblique row; on one end they are 1 cm closer to the gold–palladium wire than the other. The grids nearest the wire will be more heavily shadowed. It is often useful to have a range of shadow thicknesses in order to be able to select the best for each purpose. For shadowing, the grids are placed on glass slides with a piece of filter paper underneath. The shadow, which will be formed on the paper behind the glass slide, indicates the thickness of the shadowing layer being deposited.

During the actual shadow-casting (in a standard evaporator, e.g., Edwards, Kinney, Denton), the voltage should be increased slowly until the gold–palladium wire has melted onto the tungsten. The voltage can then be increased more quickly to produce a bright white light for a few seconds. When the voltage has been turned down, the shadow is visible behind the glass slide and, if insufficient, can be increased by repeating the shadowing.

For chromium shadow-casting, a few pellets are placed in a tungsten wire basket. Otherwise the same procedure is used. For greater resolution platinum, or platinum–carbon pellets are used (Chap. 30).

5. Photography. Specimens prepared by this method are often very contrasty; this may cause problems during printing. If possible, one should switch to a higher voltage on the microscope to reduce contrast (e.g., 80 kV) and reduce exposure time. In this way, the need for a special film developer may be avoided. During printing, reversed prints may sometimes render certain details better.

B. Sample data, problems, and troubleshooting

For examples of the successful application of the whole-mount technique to the test organisms suggested above, see Pennick and Clarke (1972) (*Paraphysomonas*), Manton (1969) (*Pyramimonas*), and Belcher (1969) (*Mallomonas*). For choanoflagellates see Hibberd (1975) and Leadbeater (1972). With this technique, the results are not always as excellent as those given in these articles. Precautions to be observed and some solutions to problems encountered are listed below.

1. It is essential that everything be as clean as possible. Poor results are caused by dirt on the film and grids, by using dirty water for rinsing, or by impurities that might be contained in the tungsten or shadow-casting metals. If too much osmium tetroxide is used for fixation, black spots may arise. Insufficient washing of marine material will result in salt crystals' remaining behind.

2. Marine material on copper grids must never be stored for any length of time before being washed, because the seawater will corrode the copper and this will result in deposits being laid over the structures of interest.

3. Cells rupturing or the dropping off of certain appendages, such as hairs on the flagella, may be due to old fixative or other problems. The problems can usually be eliminated by changing the fixation time (50 or 60 rather than 30 sec), by drying the fixed material more rapidly, or by removing almost all liquid from the grid immediately after fixation. This may leave very little, but better-fixed, material.

4. Wrinkling around even relatively small particles may be caused by the support film's being too thin. If larger particles occur in the

preparation, the film will often break when the material is dried or when it is exposed to the electron beam.

5. Other fixatives may sometimes preserve certain details better. Standard EM-fixation techniques can be tried, and the material can be shadowed after it has been postosmicated and rinsed in buffer or distilled water.

C. Whole mounts without shadow-casting

Material that is in itself very contrasty may be viewed in the electron microscope without shadow-casting. This is often done with structures containing a large amount of silica or calcium, for example, diatom frustules or large scales, and it may be particularly advantageous when the metal grains in the shadow-casting are large and therefore give poor resolution.

IV. Acknowledgments

It is a privilege to thank the members of the Botany Department at Leeds University, and in particular Mr. K. Oates (now at the University of Lancaster), the benefit of whose technical expertise one of us, Ø. M., was able to enjoy during tenure of a British Council Scholarship at Leeds from September 1968 to July 1969.

V. References

Belcher, J. H. 1969. Some remarks upon *Mallomonas papillosa* Harris and Bradley and *M. calceolus* Bradley. *Nova Hedwigia* 18, 257–70.

Hibberd, D. J. 1975. Observations on the ultrastructure of the choanoflagellate *Codosiga botrytis* (Ehr.) Saville-Kent with special reference to the flagellar apparatus. *J. Cell Sci.* 17, 191–219.

Leadbeater, B. S. C. 1972. Fine-structural observations on some marine choanoflagellates from the coast of Norway. *J. Mar. Biol. Ass. U.K.* 52, 67–79.

Manton, I. 1956a. Plant cilia and associated organelles. In Rudnick, D. (ed.), *Cellular Mechanisms in Differentiation and Growth,* pp. 61–71. Princeton University Press, Princeton.

Manton, I. 1956b. Observations with the electron microscope on the internal structure of the spermatozoid of *Fucus. J. Exp. Bot.* 7, 416–32.

Manton, I. 1969. Tubular trichocysts in a species of *Pyramimonas* (*P. grossii Parke*). *Öst. Bot. Z.* 116, 378–92.

Moestrup, Ø., and Hoffman, L. R. 1975. A study of the spermatozoids of *Dichotomosiphon tuberosus* (Chlorophyceae). *J. Phycol.* 11, 225–35.

Norris, R. E. 1965. Neustonic marine Craspedomonadales (Choanoflagellates) from Washington and California. *J. Protozool.* 12, 589–602.

Pennick, N. C., and Clarke, K. J. 1972. *Paraphysomonas butcheri* sp. nov. a marine, colourless, scale-bearing member of the Chrysophyceae. *Br. Phycol. J.* 7, 45–8.

32: Stereology: quantitative electron microscopic analysis

WAYNE R. FAGERBERG

Department of Biology, The University of Texas at Arlington, Arlington, Texas 76019

CONTENTS

I. Introduction

Electron microscope studies usually deal with two-dimensional representations (micrographs) of three-dimensional objects (cells). Typically, the relationships that exist between the representation and the object are not treated in a quantitative manner. The use of stereology (quantitative electron microscopy) provides a method through which two-dimensional micrographs can be used to describe the three-dimensional objects (cells). *Stereology* is usually defined as a study of three-dimensional structure based on two-dimensional sections through the material (Underwood 1970). Stereological techniques are not new; geologists and mineralogists have used and expanded these methods since 1847 and have established the validity of the mathematical relationships through such use.

Application of stereological techniques can provide the means of statistically analyzing data allowing quantitatively meaningful comparisons and can establish the validity of small cytological changes observed in tissues independently from the investigator's subjectivity. This chapter is designed to give the investigator the guidelines for estimating the overall trend of morphological relationships. For more precise morphological quantitation, corrections are required which are beyond the scope of this chapter and for which specific references at the end of the chapter should be consulted.

II. Materials

The equipment used to perform stereology varies with the laboratory, but essentially all that is needed is a set of grid points and lines displayed on transparent plastic sheets, a hand-held counter, and a calculator.

A. Transparent grids

Transparent plastic sheets upon which a number of points are arranged in an orderly pattern (Fig. 32–1A), or with a series of parallel or randomly oriented lines enscribed (Fig. 32–1B) can be made by

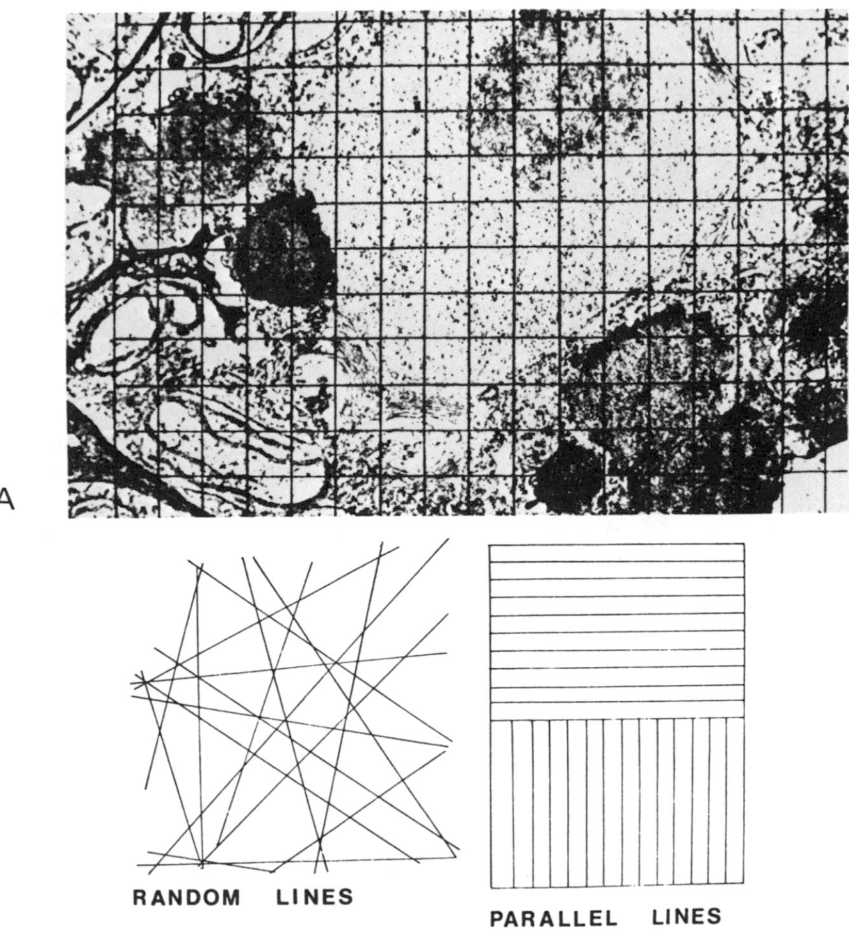

Fig. 32–1. (A) Regularly spaced line intersects points on a transparent sheet placed over a micrograph by intersections occurring in the organelles (P_{test}), plus the cytoplasm (P_{tot}). (B) Random and parallel line grids on transparent plastic sheets used in determining surface to volume ratios (S_v).

drawing line intersects or points on a sheet of white paper with India ink. This drawing is then used to make a Thermo-Fax or Xerox process transparency.

B. Calculator

The requirements in a calculator include a memory function, memory summation, square root, log, and *Y*-to-the-*x*-power-type function keys. The more sophisticated the calculator used, the less tedious the data manipulation will be.

III. Methods

A. Fixation, dehydration, and sectioning

The effect of the fixative on the tissue being studied and any variation in procedure between samples will alter results of the quantitative investigation by affecting the volumetric relationships of organelles to one another. When possible, the osmolarity of the fixative and that of the cells should be as close as possible. An approximation can be obtained by observing cells with the light microscope, adding solutions of known osmolarity, and noting the range in which plasmolysis does not occur. After appropriate fixation of the sample (Chap. 23), care should be taken to avoid swelling and shrinking during dehydration. Dehydration is best started with 70% ethanol or 70% acetone (Weibel and Bolender 1973), avoiding lower concentrations. Once the study has begun, the tissue-preparation procedures should not be changed, because this could introduce substantial variation between samples. Sections should always be of uniform thickness, and compression effects should be avoided.

B. Random sampling

Tissue must be selected and prepared for electron microscopy with as little investigator bias as possible. Usually, randomnization of samples can be achieved by preparing extra samples and avoiding preselection in trimming certain blocks or certain areas. Of equal, or perhaps even greater, importance is randomnization of areas photographed in the electron microscope. The same area of the grid or section should be consistently photographed (e.g., center, upper left corner, etc.), thus avoiding biased selection of "interesting" areas (Weibel and Bolender 1973).

In our laboratory, electron micrographs are enlarged to equal magnification and printed on 8 × 10 in. photographic paper. With large cells or tissue, the entire print area should be covered with the image. However, for measuring small cells, the boundary of each cell should be contained within the print area.

C. Analysis of electron micrographs

This involves placing a grid with N number of points, or lines, over a micrograph and counting the number of points, or line intersections, that fall within the boundary of the cytological structure of interest. The grid applied to the micrograph is held in place by tape at two diagonally opposite corners.

1. Criteria used in sampling. Several factors have to be decided before final counts for analysis are made: (a) total boundary to be sampled, (b) point density per sample, and (c) sample size.

a. Total boundary. A total boundary area can consist of a small cell whose size is small enough so that it fits within one electron micrograph, or it can consist of the area of a cell covered by the transparent grid. Each total boundary is defined by the number of points (P_{tot}) that intersect an object such as an organelle (P_{test}), plus those points occurring over the cytoplasm (Fig. 32–1). This is further explained in Section III.D.1.

b. Determination of adequate point density per sample. The array of grid points should be such that approximately one point falls in each test object (Underwood 1970; Weibel and Bolender 1973). This can be approximated by formulas (1) and (2) (Weibel and Bolender 1973):

$$d^2 = A\bar{m} \tag{1}$$

or

$$d = \sqrt{A\bar{m}} \tag{2}$$

where d is the distance between points, and $A\bar{m}$ is the mean area of several measured test objects. In situations where test structures show a regular periodicity, a grid of random points should be used. The total number of test points (P_t) necessary to yield a desired degree of precision in measuring the volume of a specific component of a cell can be approximated mathematically by the relationship (Weibel and Bolender 1973):

$$P_t = 0.453\,\frac{(1 - V_v)}{V_v \cdot E^2} \tag{3}$$

The point-spacing on the grid must be determined as described in relationships (1) and (2). The volume of a specific component (V_v) is determined by counting a few samples (see III.C.2.a); E is the acceptable error in decimal percent that the investigator is willing to accept, and P_t is the total number of points that must be applied to the sample to estimate the volumetric fraction of the test component within the error interval. As an example, 68 total points would be needed to achieve a 10% error in sampling when the volumetric component of the test structure is 40% [68 = (1 − 0.4)/(0.4 × 0.01) × 0.453]. This also provides a method of defining a sample as the number of micrographs that need to be examined in order to apply the optimal number of points (as determined above) based on a specific point distribution. It should also be noted that the optimal number of points applied to each sample varies with the size of the test component.

In addition to the relationship described here, others have been reported by Carpenter and Lazarow (1962), Weibel (1969), Underwood (1970), and Mayhew and Cruz-Orive (1974).

c. Sample size. A representative sample should be such that all the components of the test structure are present in measurable quantities. The total number of representative samples necessary in a particular study can be based upon statistically derived confidence intervals.

D. Calculating relationships

1. Volumetric determinations. The relationship of two volumes (such as that of a chloroplasts and that of a cell) can be expressed as a ratio by the general formula according to Underwood (1970)

$$V_v = \frac{P_{\text{test}}}{P_{\text{tot}}} \tag{4}$$

where P_{test} is the total number of points that fall within the test object, and P_{tot} is the total number of points falling on the entire area being tested. V_v is equal to the percent volume occupied by the test object.

Two examples will be given for determining the chloroplast to cytoplasm volume relationship in small cells (a), and in sections of large cells (b) both derived by Mayhew and Cruz-Orive (1974).

a. In small cells, whose cell boundaries can be accommodated within an electron micrographic print, the following expression can be used:

$$\bar{V}_v = \frac{\Sigma P_{\text{test}/N}}{\Sigma P_{\text{tot}/N}} = \frac{\bar{P}_{\text{test}}}{\bar{P}_{\text{tot}}} \tag{5}$$

In this case only points are taken that fall within the cell area. Here $\bar{V}_v$ represents the mean volume percent of the chloroplasts within all cell sections (N) measured, $\bar{P}_{\text{test}}$ is the mean number of points falling in all chloroplasts, and $\bar{P}_{\text{tot}}$ is the mean number of points falling in all the cell sections.

b. When dealing with sections of large cells, it is necessary to decide what will be the boundary area to be tested. This can be done by taking the entire micrograph as test area, or by letting a specific number of points define the boundary of the test area. By the formula

$$\bar{V}_v = \frac{\Sigma V_{vt}}{N} \tag{6}$$

one can obtain the mean volume percent of the chloroplasts in the cytoplasm ($\bar{V}_v$) from the test areas counted (N), and the percent volume of the chloroplasts (V_{vt}) in each individual sample (micrograph).

2. Surface-to-volume parameters. These can be determined with the point-counting grid described by Smith and Page (1976) or by a system of parallel- or random-line grids of definable length (Weibel et al. 1966; Underwood 1970; Weibel and Bolender 1973). Preparation of electron micrographs for analysis is similar to that presented for the point-counting (above). These methods are applicable in studying the internal membranes cells (e.g., endoplasmic reticulum) or organelles (e.g., chloroplast lamellae), as well as boundary membrane surface. A grid (of parallel or randomly oriented lines) is placed over the test area (cell, or organelle) and the number of intersections of the internal membranes per number of units of line (mm, mμ, etc.) is counted. The surface-to-volume ratio is solved for by the equation (Underwood 1970):

$$S_v = 2P_1 \,\text{mm}^2/\text{mm}^3 \tag{7}$$

where S_v is the surface-to-volume ratio in units of line length (mm^2/mm^3, $\text{m}\mu^2/\text{m}\mu^3$, etc.) and P_1 is the number of intersections with the test line divided by the total length of the line in units, within the test area (cell or organelle). If complete randomness is not obtained between the line-test system and the intersecting membranes, several test planes oriented in different directions should be used. It should also be noted that because the relationship between the number of intersections and line length varies inversely with magnification of the final print, a magnification correction factor (mf) needs to be applied to the data according to some standard format (e.g., final print magnification ÷ 1000 = mf; mf × S_v value = S_v corrected in mm^2/mm^3).

3. Determination of line length in an area. A useful measurement very similar to the S_v ratio is the measurement of total line length per micrograph. This measurement can be used to describe the total length of endoplasmic reticulum in various tissues. The measurement method is similar to that described for the S_v ratio: A grid of parallel lines is placed over the micrograph and the number of intersections with those lines are counted. These values are then solved for by the relationship (Underwood 1970):

$$L_a = \left(\frac{\pi}{2}\right) P_1 \,\text{mm}/\text{mm}^2 \tag{8}$$

where L_a is the line length per unit area; the other symbols are similar to those described for S_v ratio. The same criteria for randomness of test grid and test structure described for S_v must be met here.

IV. Data analysis

Values generated by the methods described are usually expressed with their standard deviation, standard error, and coefficient of variation. When the sample sizes are large, and the data fits a normal distribution, comparisons between tissues can be made using standard parametric tests such as Student's *t* correlation, regression analysis, or the commonly used methods of analysis of variance (ANOVA).

Where volumetric values were determined using the relationship described for a variable-boundary test area or the data, for one reason or another, do not fit the basic assumption of parametric statistics, (that is, that the data are not normally distributed) nonparametric statistics should be used. In such instances nonparametric statistics such as Wilcox's sign-rank test and the Mann-Whitney rank correlation can be used to test for significant differences between sample means, and such tests as Kendall's coefficient of rank correlation (Tau), Spearman's Rho, and the R × C contingency tables can be used to measure the degree of association between sampled data (Snedecor and Cochran 1967; Sokal and Rohlf 1969).

V. Problem areas

Normally, random tissue sampling is advised (Underwood 1970; Weibel 1973); however, special situations may arise, where randomness may be set aside. Sometimes the test structure may exhibit features that have a regularity corresponding to the regularity of the test grid. In many cases, this can be overcome through orientation of samples during sectioning (Eisenberg et al. 1974) or applying test grids with randomly distributed test points.

Error can also arise when the size of the organelle being counted is close to or less than the section thickness. Such an effect ("Holmes effect") results in a decreased count because there is an overlap of two or more structures. This can be reduced by obtaining the thinnest possible sections, and one can apply mathematical corrections (Weibel 1969; Underwood 1970; Mayhew 1972; Weibel and Bolender 1973; Smith and Page 1976; Weibel 1976). The thinner the section, the greater the compression artifacts (especially important in S_v calculations); a trade-off therefore exists between the Holmes effect and compression distortion.

Finally, if one uses photographically enlarged prints, the effect of paper shrinkage during preparation should be measured by processing a piece of paper of known size and measuring the change in size that occurs after processing (R. P. Bolender, personal communication). In all calculations using total line length (e.g., S_v and L_a calculations) the line length should be increased by the percent of paper shrinkage.

Example: $S_v = 2P_L$ mm²/mm³; and in this example $P = 10$, $L = 100$ mm therefore, $S_v = 0.20$. If paper shrinkage were 20% then the correction would be

$$100 \text{ mm} \times 0.20 = 20$$

and corrected line length would be

$$20 + 100 \text{ mm} = 120 \text{ mm}$$

Therefore

$$S_v = \frac{10}{120} \times 2 = 0.16$$

VI. Acknowledgments

The author would like to express his thanks to Dr. Howard Arnott for providing critical comments and suggestions concerning this manuscript and for introducing the author to the concepts of quantitative ultrastructure. The author would also like to thank Dr. Archibald Hopkins for his helpful review of this manuscript.

VII. References

Carpenter, A. M., and Lazarow, A. 1962. Component quantitation of tissue sections. II. A study of the factors which influence the accuracy of the method. *J. Histochem. Cytochem.* 10, 329–40.

Eisenberg, B. R., Kuda, A. M., and Peter, J. B. 1974. Stereological analysis of mammalian skeletal muscle. I. Soleus muscle of the adult Guinea Pig. *J. Cell Biol.* 60, 732–54.

Mayhew, T. M. 1972. A comparison of several methods for stereological determination of the numbers of organelles per unit volume of cytoplasm. *J. Microsc.* 96, 37–44.

Mayhew, T. M., and Cruz-Orive, L. M. 1974. Caveat on the use of the Delesse principle of areal analysis for estimating component volume densities *J. Microsc.* 102, 195–207.

Smith, H. E., and Page, E. 1976. Morphometry of rat heart mitochondrial subcompartments and membranes: application to myocardial cell atrophy after hypophysectomy. *J. Ultra. Res.* 55, 31–41.

Snedecor, G. W., and Cochran, W. G. 1967. *Statistical Methods.* Iowa State University Press, Ames, Iowa. 593 pp.

Sokal, R. R., and Rohlf, F. J. 1973. *Introduction to biostatistics.* Freeman, San Francisco. 368 pp.

Underwood, E. E. 1970. *Quantitative Stereology.* Addison-Wesley. Reading, Mass. 274 pp.

Underwood, E. E. 1972. The stereology of projected images. *J. Microsc.* 95, 25–44.

Weibel, E. R. 1969. Stereological principles for morphometry in electron microscopic cytology. *Int. Rev. Cytol.* 26, 235–302.

Weibel, E. R. 1973. Selection of the best method in stereology. *J. Microsc.* 100, 261–9.

Weibel, E. R. 1976. Progress, success and problems in applying stereology in biological research. In Underwood, E. E. (ed.), *The Proceedings of the Fourth International Congress for Stereology.* National Bureau of Standards special publication No. 431. pp. 341–350. U.S. Gov. Printing Office, Washington D. C.

Weibel, E. R., and Bolender, R. P. 1973. Stereological techniques for electron microscopic morphometry. In Hayat, M. A. (ed.), *Principles and Techniques of Electron Microscopy: Biological Applications,* vol. 3, pp. 237–96. Van Nostrand Reinhold, New York.

Weibel, E. R., Kistler, G. S., and Scherle, W. F. 1966. Practical stereological methods for morphometric cytology. *J. Cell Biol.* 30, 23–38.

Section III

Appendixes

Culture collections

Czechoslovakia

1. Culture Collection of Algae
   Department of Botany
   Charles University of Prague
   Prague

Federal Republic of Germany

1. Sammlung von Algenkulturen
   Universität Göttingen
   18 Nikolausbergerweg
   Göttingen

Japan

1. Algal Culture Collection
   Institute of Applied Microbiology
   University of Tokyo
   Tokyo

United Kingdom

1. Culture Centre of Algae and Protozoa (Cambridge)
   Institute of Terrestrial Ecology
   36 Storeys Way
   Cambridge, CB3 0DT
2. Freshwater Biological Association
   Supply Department
   Windermere Laboratory
   The Ferry House
   Ambleside, Westmorland
3. Marine Biological Association of the UK (MBA)
   The Laboratory
   Citadel Hill
   Plymouth, PL1 2PB

United States

1. American Type Culture Collection (ATCC)
   12301 Parklawn Drive
   Rockville, Maryland 20852
2. Carolina Biological Supply Company
   Burlington, North Carolina 27215
   and
   Gladstone, Oregon 97027

3. University of Texas Culture Collection (UTEX)
   Department of Botany
   University of Texas
   Austin, Texas 78712
4. Supply Department
   Marine Biology Laboratory
   Woods Hole, Massachusetts 02543

Suppliers

Accurate Chemical and Scientific Corp., 28 Tec Street, Hicksville, New York 11801
Alconox, Inc., 215 Park Avenue So., New York, New York 10003
Allied Chemical Corp., Specialty Chemicals Div., P. O. Box 1087R, Morristown, New Jersey 07960
American Cyanamid Co., Berdan Avenue, Wayne, New Jersey 07470
American Instrument Co., 8030 Georgia Avenue, Silver Spring, Maryland 20910
American Optical Corp., Sci. Instrument Div., Sugar and Eggert Roads, Buffalo, New York 14215
Amicon Corporation, Scientific Systems Division, 21 Hartwell Avenue, Lexington, Massachusetts 02173
Apiezon Products Ltd., 4 York Road, London SE1, England (In U.S.A: J. G. Biddle Co., Plymouth Meeting, Pennsylvania 19462)
Aquarium Systems, 33208 Lakeland Blvd., Eastlake, Ohio 44094
Bailey Instruments, 515 Victor Street, Saddle Brook, New Jersey 07662
Baird Atomic, Inc., 125 Middlesex Tpke., Bedford, Massachusetts 01730
J. T. Baker Chemical Co., 222 Red School Lane, Phillipsburg, New Jersey 08865
Balzer High Vacuum Corp., P. O. Box 10816, Santa Ana, California 92711
Bausch and Lomb, SOPD Div., 1400 N. Goodman, Rochester, New York 14602
BEEM, P. O. Box 132, Jerome Avenue Station, Bronx, New York 10468
Bio-Rad Laboratories, 2200 Wright Avenue, Richmond, California 94804
Buchler Instruments, Div. of Searle Diagnostics, Inc., 1327 16th Street, Fort Lee, New Jersey 07024
Calbiochem, 10933 N. Torrey Pines Road, La Jolla, California 92037
N. L. Cappel Laboratories, Inc., Center Hall Road, Cochranville, Pennsylvania 19330
R. P. Cargille Labs, Inc., 55 Commerce Road, Cedar Grove, New Jersey 07009
Ciba-Geigy Inc., Saw Mill River Road, Ardsley, New York 10502
Corning Glass Works, Houghton Park, Corning, New York 14830
Denton Vacuum, Cherry Hill Industrial Center, Cherry Hill, New Jersey 08003
Difco Laboratories, P. O. Box 1058A, Detroit, Michigan 48232
Digital Equipment Corp., 200 Forest Street, Marlborough, Massachusetts 01752
Dupont Instruments, Sorvall, Wilmington, Delaware 19898

Eastman Kodak Co., Eastman Organic Chemicals, 343 State Street, Rochester, New York 14650

Ebtec Corporation, C. W. French Division, 5 Shawsheen Avenue, Bedford, Massachusetts 01730

Electron Microscopy Sciences, P. O. Box 251, Fort Washington, Pennsylvania 19034

Endo Laboratories, Inc., 1000 Stewart Avenue, Garden City, New York 11530

Eppley Laboratory, Inc., 12 Sheffield Avenue, Newport, Rhode Island 02840

ESBE Laboratory Supplies, 401 Alness Street, Downsview, Ontario, Canada

Falcon Division, Becton Dickinson and Co., 1950 Williams Drive, Oxnard, California 93030

Fisher Scientific Company, 711 Forbes Avenue, Pittsburgh, Pennsylvania 15219

Ernest F. Fullam, Inc., P. O. Box 444, Schenectady, New York 12301

Gaymar Industries, One Bank Street, Orchard Park, Buffalo, New York 14127

Gelber Pump Company, 5806 N. Lincoln Avenue, Chicago, Illinois 60659

Gelman Instruments Co., 600 S. Wagner Road, Ann Arbor, Michigan 48106

Grafar Corp., 7340 Fenkell, Detroit, Michigan 48238

G. T. Gurr, Searle Scientific Services, Coronation Road, Cressex Industrial Estate, High Wycombe, Buckinghamshire, England

H-B Instrument Co., 4303 N. American Street, Philadelphia, Pennsylvania 19140

Hamamatsu Corp., 120 Wood Avenue, Middlesex, New Jersey 08846 (In Japan: Hamamatsu TV Co., Ltd., 1126 Ichino-cho, Hamamatsu City, Japan)

Harleco, 480 Democrat Road, Gibbstown, New Jersey 08027

Hartung Associates, Box 1344, Camden, New Jersey 08106

Hayakawa Electric Co., Ltd., Abenoku, Osaka, Japan

Heat Systems-Ultrasonics, Inc., 53 E Mall, Plainview, New York 11803

Helena Laboratories, P. O. Box 752, Beaumont, Texas 77704

Hyland Diagnostics Div. of Travenol Labs., Inc., 3300 Hyland Avenue, Costa Mesa, California 92626

Ilford, Inc., Paramus, New Jersey 07652 (In Canada: Ingram and Bell, Ltd., 20 Bond Avenue, Don Mills, Ontario, Canada)

ISCO, P. O. Box 5347, Lincoln, Nebraska 68505

Jarrell-Ash Div. of Fisher Scientific, 590 Lincoln Street, Waltham, Massachusetts 02154

Jenoptik Jena G.m.b.H., P. O. Box 190, DDR-69 Jena, East Germany Democratic Republic (In U.S.A: Charvoz-Carsen Corp. International, Micro-Optics Div., 5 Daniel Road, Fairfield, New Jersey 07006)

Johnson Matthey and Co., Ltd., 78 Hatton Garden, London, ECIP 1AE, United Kingdom

Kettering Scientific Instruments, Yellow Springs, Ohio 45387 (In Holland: Kipp and Zonen, N. V. Instrumentfabriek En-Handel, Delft, Holland)

Lab-Line Instruments, Inc., Lab-Line Plaza, 15th and Bloomingdale Avenues, Melrose Park, Illinois 60160

Ladd Research Industries, Inc., P. O. Box 901, Burlington, Vermont 05401

E. Leitz, Inc., Link Drive, Rockleigh, New Jersey 07647

LKB Instruments, Inc., 12221 Parklawn Drive, Rockville, Maryland 20852

Matheson, Coleman and Bell Co., Inc., 1275 Valley Brook Avenue, Lyndhurst, New Jersey 07071

MEER Corporation, 9500 Railroad Avenue, North Bergen, New Jersey 07047

Merck and Co., Inc., 126 E. Lincoln Avenue, P. O. Box 2000, Rahway, New Jersey 07065

Microbiological Associates, Biggs Ford Road, Walkersville, Maryland 21793

Miles Laboratories, Inc., Miles Research Products, 1127 Myrtle Street, Elkhart, Indiana 46515

Millipore Corp., Ashby Road, Bedford, Massachusetts 01730

Modulation Optics, Inc., 100 Forest Drive at East Hills, Greenvale, New York 11548

New Brunswick Scientific Co., Inc., 44 Talmadge Road, Edison, New Jersey 08817

Nutritional Biochem. Corp., ICN Pharmaceuticals, Inc., 26201 Miles Road, Cleveland, Ohio 44128

Olympus Corporation of America, 2 Nevada Drive, New Hyde Park, New York 11040 (In Japan: Olympus Optical Co., Ltd., 43-2 Hatagaya 2-chome, Shibuya-ku, Tokyo, Japan)

Oriel Optical Co., Oriel Corporation of America, 15 Market Street, Stamford, Connecticut 06902

Pharmacia Fine Chemicals, 800 Centennial Avenue, Piscataway, New Jersey 08854

Polaroid Corp., 549 Technology Sq., Cambridge, Massachusetts 02139

Polaron, 1202 Bethlehem Pike, Line Lexington, Pennsylvania 18932

Polysciences, Inc., Paul Valley Industrial Park, Warrington, Pennsylvania 18976

Redlake Corp., Photo Instrument Div., 1711 Dell Avenue, Santa Clara, California 95051

C. Reichert, American Optical Corp., Eggert and Sugar Rds., Buffalo, New York 14215 (In Austria: C. Reichert A. G., Hernalser Hauptstrasse 219, 1170 Vienna, Austria)

Rohm and Haas Co., Independence Mall West, Philadelphia, Pennsylvania 19105

Schleicher and Schüll, Inc., 543 Washington Street, Keene, New Hampshire 03431 (In Germany: Schleicher and Schüll, 3354 Dassel, Federal Republic of Germany)

Schott Optical Glass, Inc., 400 York Avenue, Durea, Pennsylvania 18642

Scientific Chemical Co., Inc., 15564 Producer Lane, Huntington Beach Industrial Park, Huntington Beach, California 92649

Shandon Southern Instruments, 515 Broad Street, Sewickley, Pennsylvania 15143

Sigma Chemical Co., P. O. Box 14508, St. Louis, Missouri 63178

SPI Supplies, P. O. Box 342, Westchester, Pennsylvania 19380

TAAB Labs., 52 Kidmore End Road, Emmer Green, Reading, England (In U.S.A.: EXTECH International Corp., 177 State Street, Boston, Massachusetts 02109)

Ted Pella and Co. (PELCO), P. O. Box 510, Tustin, California 92680

Tobler, Ernst and Traber, Inc., 71 Murray Street, New York, New York 10007

Toshiba Kasei, Kogyo Co., Ltd., Kambara, Shizuoka, Japan

Tousimis Research Corp (Biodynamics), P. O. Box 2189, Rockville, Maryland 20852

Tridon Chemical Inc., 255 Oser Avenue, Happauge, New York 11787

Ultra-Violet Products, Inc., 5100 Walnut Grove Avenue, San Gabriel, California 91778

VWR Scientific Division, P. O. Box 3200, San Francisco, California 94119

Worthington Biochemical Co., Box 650, Freehold, New Jersey 07728

Carl Zeiss, Inc., 444 5th Avenue, New York, New York 10018 (In Germany: Carl Zeiss, Postfach 1369/1380, 7082 Oberkochen, Federal Republic of Germany)

Subject index

Author index

*(Italic numbers indicate author citation in full; *indicates chapter author.)*

Taxonomic index